Jeserich et al. (Eds.)

Molecular Signaling and Regulation in Glial Cells

Springer

Berlin
Heidelberg
New York
Barcelona
Budapest
Hong Kong
London
Milan
Paris
Santa Clara
Singapur
Tokyo

Jeserich et al. (Eds.)

Molecular Signaling and Regulation in Glial Cells

A Key to Remyelination and Functional Repair

With 120 Figures and 17 Tables

Springer

PROF. DR. GUNNAR JESERICH
Universität Osnabrück
FB Biologie/Chemie
Barbarastr. 11
D-49069 Osnabrück

PROF. DR. CHRISTIANE RICHTER-LANDSBERG
Universität Oldenburg
FB Biologie
Postfach 2503
D-26111 Oldenburg

PROF. DR. HANS H. ALTHAUS
Max-Planck-Institut für
Experimentelle Medizin
Hermann-Reinsstr. 3
D-37075 Göttingen

PROF. DR. ROLF HEUMANN
Universität Bochum
Molekulare Neurobiochemie
Universitätsstr. 150
D-44801 Bochum

ISBN-13: 978-3-642-64501-3 e-ISBN-13: 978-3-642-60669-4
DOI: 10.1007/978-3-642-60669-4

CIP-Data applied for
Die Deutsche Bibliothek - CIP-Einheitsaufnahme
Molecular signaling and regulation in glial cells: a key to remyelination and functional repair /ed by G. Jeserich ...- Berlin ; Heidelberg ; New York ; Barcelona ; Budapest ; Hong Kong ; London ; Milan ; Paris ; Santa Clara ; Singapore ; Tokyo ; Springer 1997

ISBN-13: 978-3-642-64501-3
NE: Jeserich, Gunnar [Hrsg.]

Softcover reprint of the hardcover 1st edition 1997

Cover Design: D & P Heidelberg
Camera ready by editors
SPIN 10526684 31/3137 5 4 3 2 1 0 - Printed on acid free paper -

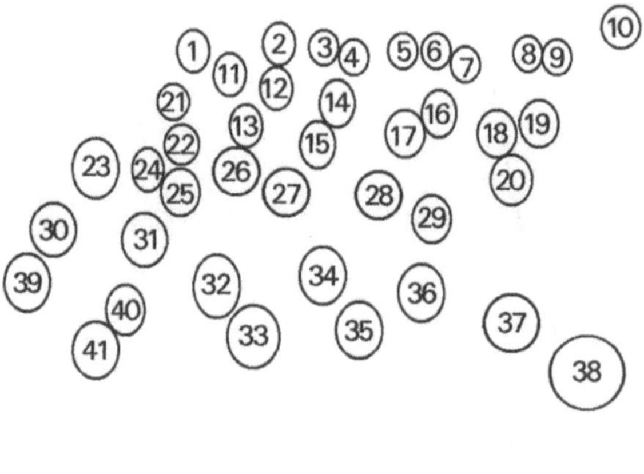
1 2 3 4 5 6 7 8 9 10
11 12 14 16 18 19
21 13 15 17
22 20
23 24 26 27 28 29
25
30 31
39 32 34 36
40 33 35 37
41 38

1 H.G. Luesse, 2 W. Brück, 3 H.W. Müller, 4 J.-M. Matthieu, 5 P. Rieckmann, 6 J. Strelau, 7 F. Aloisi, 8 M. Yoshida, 9 H.S. Keirstead, 10 N. Pringle, 11 K.A. Nave, 12 R. Ankerhold, 13 J.E. Merrill, 14 A.L. Gard, 15 P. Shrager, 16 G. Rohde, 17 S.J. Moorman, 18 R.M. Gould, 19 R. Matsas, 20 H. Rabe, 21 C. Stichel, 22 R. Heumann, 23 M. Heinrich, 24 P. Remde, 25 J. Trotter, 26 R. Franklin, 27 N. Scolding, 28 E. Yavin, 29 A. Brand, 30 R. Hempel, 31 J. de Vellis, 32 L. Hudson, 33 C. Richter-Landsberg, 34 S. Nona, 35 G. Almazan, 36 G. Jeserich, 37 A. Espinosa de los Monteros, 38 H.H. Althaus, 39 J. Gehrmann, 40 M. Dubois-Dalcq, 41 S.G. Amur-Umarjee.

Almazan, Guillermina
McGill University,
Dept. Pharmacology
3655 Drummond Street, Room 1321
Montreal, Quebec H3G1Y6, Canada
FAX:+ 1 - 514 - 398- 6690
Email: almazan@pharma1.pharma.mcgill.ca

Aloisi, Francesca
Istituto Superiore di Sanita
Viale Regina Elena 299
I-00161 Rome, Italy
FAX:+ 39 - 6- 4957821
Email: fos4@iss.it

Althaus, Hans H.
MPI Experimentelle Medizin
Hermann Reinstr. 3
D-37075 Göttingen
FAX:+ 49 - 551 - 3899 - 388
Email: althaus@emed1.dnet.gwdg.de

Ankerhold, Richard
Fakultät Biologie,
Universität Konstanz
Universitätsstr. 10
D-78464 Konstanz
FAX:+ 49 - 7531 88-38 94
Email: richard.ankerhold@uni-konstanz.de

Amur-Umarjee, Shashi G.
UCLA Los Angeles
Mental Retardation Center
School of Medicine, 760 Westwood Placa
Los Angeles, CA 90024-1759, USA
FAX:+ 1 - 310 - 206 - 5050
Email: samur@npimain.medsch.ucla.edu.

Brück, Wolfgang
Abteilung Neuropathologie
Neurologische Universitätsklinik
D-37075 Göttingen
FAX:+ 49 - 551 - 398572

de Vellis, Jean
UCLA Los Angeles
Mental Retardation Center
760 Westwood Placa
Los Angeles, CA 90024-1759, USA
FAX:+ 1 - 310 - 206 - 5061
Email: jedevellis@npimain.medsch.ucla.edu

Dermietzel, Rolf
Medizinische Fakultät
Universität Regensburg
Universitätsstr. 31
D-93053 Regensburg
FAX:+ 49 - 941-943-2868

Dubois-Dalcq, Monique
Unite de Neurovirologie et Regeneration
Institut Pasteur, 25 Rue du Dr. Roux
F-75724 Paris Cedex 15
FAX:+ 33 - 1 - 40613421
Email: mdalcq@pasteur.fr

Espinosa delos Monteros, Aracelie
UCLA Los Angeles
Mental Retardation Center
760 Westwood Placa
Los Angeles, CA 90024-1759, USA
FAX:+ 1 - 310 - 2065061

Finsen, Bente
Pharma Biotec Research Center
Department of Anatomy
Institute of Medical Biology
University of Odense
DK-5000 Odense
FAX:+ 45 - 65 -906321

Franklin, Robin
Dept. Clinical & Vet. Medicine
Univ. Cambridge
Cambridge CB2 2QQ, UK
FAX:+ 44 - 1223 - 337610
Email: rjf 1000@cam.ac.uk

Gard, Anthony L.
University of South Alabama
College of Medicine
Mobile, AL 36688, USA
FAX:+ 1 - 334 - 460 - 6771
Email: agard@jaguar1.usouthal.edu

Gehrmann, Jochen
Institut für Neuropathologie der
Abt. für Pathologie der Universität
CH-8091 Zürich
FAX:+ 41 - 12554402
Email: jge@pathol. unizh.ch

Gould, Robert M.
NY State Institute f. Basic Research
in Developmental Disabilities
150 Forest Hill Road
Staten Island, 10314 NY, USA
FAX:+ 1 - 718 - 698 - 7169

Heumann, Rolf
Lehrstuhl f. Molekulare
Neurobiochemie
Ruhr-Universität Bochum
Universitätsstr. 150
D-44801 Bochum
FAX:+ 49 - 234 - 7094105

Hudson, Lynn D.
NINDS, NIH
Buildg. 36, Rm 5D04
Bethesda, MD 20892, USA
FAX:+ 1 - 301 - 496 - 0899
Email: hudson@helix.nih.gov

Jeserich, Gunnar
AG Zoophysiologie
Universität Osnabrück, Barbarastr. 11
49069 Osnabrück
FAX:+ 49 - 541 - 9692870
Email:jeserich@sfbbio1.biologie.
uni-osnabrueck.de

Keirstead, Hans S.
MRC Cambridge Ctr f. Brain Repair
Dept. Clinical Veterinary Medicine
University of Cambridge
Cambridge CB2 2PY, UK
FAX:+ 44 - 1223 - 337610
Email: hsk20@cam.ac.uk

Matsas, Rebecca
Inst. Pasteur Hellenique
127, Vas Sofias Avenue
GR-11521 Athens. Greece
FAX:+ 30 - 16423498

Matthieu, Jean-Marie
Lab. de Neurochimie
Service de Pediatrie, CHUV
CH-1011 Lausanne
FAX:+ 41 - 21 - 3143477
Email: jmatthie@ulys.unil.ch

Merrill, Jean E.
UCSF San Francisco
Berlex Biosciences, POB 4099
5049 San Pablo Avenue
Richmond, CA 94804-0099, USA
FAX:+ 1- 510 - 669 - 4244

Moorman, Stephen J.
Dept. Anatomy and Cell Biology
University of North Texas
Health Sci Ctr at Fort Worth
Fort Worth, TX 76107, USA
FAX:+1 - 817 - 735 - 2610
Email: smoorman@molly.hsc.unt.edu

Müller, Hans Werner
Abt. Neurologie der
Heinrich-Heine Universität Düsseldorf
Moorenstr. 51
40225 Düsseldorf
FAX:+ 49 - 211 - 311 - 8485

Nave, Klaus Armin
Zentrum Molek. Biol. Heidelberg
Im Neuenheimer Feld
D-6900 Heidelberg
FAX:+ 49 - 6221 - 565894
Email:nave@sun∅.urz.uni-heidelberg.de

Nona, Sam
UMIST, Dept. Optomet.& Vision Sc.
Dev. Neurobiol. Lab., Box 88
Manchester M60 1QD, UK
FAX:+ 44 - 1611 - 236 - 0409
Email: sam.nona@umist.ac.uk

Pringle, Nigel
MRC- Lab Molec. Cell Biol.
Gowerstr.
London WCIE 6BT, UK
FAX:+ 44 - 171 - 380 - 7805
Email: dmcbnpp@ucl.ac.uk

Privat, Alain
U 336 INSERM
USTL Case Courrier 106
Place Eugene Batallion
F-34095 Montpellier Cedex 05
France
FAX:+ 33 - 67 - 143318

Richter-Landsberg, Christiane
Fachbereich Biologie
Abt. Molekulare Neurobiologie
Universiät Oldenburg, Postfach 2503
26111 Oldenburg
FAX:+ 49 - 441 - 798 - 3223
Email:crl@biologie.uni-oldenburg.de

Rieckmann, Paul
Neurologische Klinik der
Universität Göttingen
37075 Göttingen
FAX:+ 49 -551 - 398405

Rodriguez, Moses
Mayo Clinic/Foundation
Department of Immunology
200 First Sreet Souhtwest
Rochester, MN 55905, USA
FAX:+ 1 - 507 - 284 - 1637
Email:rodriguez@mayo.edu

Scolding, Neil
Neurology Unit , Univ. Cambridge
Addenbrookes Hospital, Hills Road
Cambridge CB2 2QQ, UK
FAX:+ 44 - 1223 - 336941

Shrager, Peter
Univ. Rochester, Medical Ctr.
Dept. Physiol. Box 642
601 Elmwood Avenue
Rochester, NY 14642, USA
FAX:+ 1 - 716 - 461 - 3259
Email: pshr@db1.cc.rochester.edu

Sontheimer, Harry W.
Neurobiol. Research Ctr.
University of Alabama
1719 6th Avenue S Circ
Birmingham, AL 35294, USA
FAX:+ 1 - 205 - 934 - 6571
Email:

Stichel, Christiane
Abt. Neurologie der
Heinrich-Heine Universität Düsseldorf
Moorenstr. 51
D-40225 Düsseldorf
FAX:+ 49 - 211 - 311 - 8485

Trotter, Jacqueline
Abt. Neurobiologie der
Universität Heidelberg
Im Neuenheimer Feld 364
69120 Heidelberg
FAX:+ 49 - 6221 56 3700
Email: jtrotter@suno.urz.uni-heidelberg.de

Unsicker, Klaus
Institut f. Anatomie u. Zellbiologie III
Universität Heidelberg
Im Neuenheimer Feld 307
D-69120 Heidelberg
FAX:+ 49 - 6221 - 565604

Yoshida, Mika
Mt. Sinai School of Medicine
Brookdale Center for Molec. Biology
Box 1126, 1 Gustave Levy Place
New York, NY 10021, USA
FAX:+ 1 - 212 - 423 - 0596
Email: yoshida@msvax.mssm.edu

Yavin, Ephraim
Dept. Neurobiology
The Weizmann Institute of Science
Rehovot 76100, Israel
FAX:+ 972 - 8 - 460225
Email: bnyavin@weizmann.ac.il.

Preface

Cellular and molecular interactions between nerve-and glial cells are of basic importance for a proper functioning of the nervous system. A disruption of this highly regulated interplay results in severe neurological and/or psychiatric deficits. The regenerative capacity of the mammalian CNS appears to be rather limited in contrast to the situation in the mammalian PNS and the CNS of lower vertebrates. However, recent results based on e.g. cell transplantation, cellular engeneering, and tissue culture techniques substantially expanded our knowledge about the complex interactions taking place after injury or immunological attacks.

A compilation of our present status of knowledge about the molecular and cellular mechanisms underlying degeneration and repair in the CNS was given by leading neuroscientists from Europe, Israel, USA and Canada at an international symposium held at Monastery Ohrbeck, near Osnabrück, August 28 - September 2, 1995. The symposium concentrated on two major aspects of regenerative attempts seen in the CNS: the responses after trauma and after demyelination. Since it has become increasingly clear that the glial environment of neurons plays an important part in the regeneration process, the emphasis was on the activities of glial cells and their interactions with their neuronal partners.

The program of the symposium included the following major topics: expression, secretion and effect of growth factors, cytokines and other signaling molecules for astrocytes and oligodendrocytes. In addition the molecular reactions of microglial cells after trauma or neuronal degeneration and regeneration were followed. Moreover, a quite different mode of cell-to-cell-signaling via ion channels and neurotransmitter receptors was addressed. Subsequently, the structure, regulation and function of genes and proteins of myelin were analysed in the normal brain as well as in diseased states. Thereby the identification of a novel myelin transcription factor was reported. Another topic of major interest was the immunocytochemical and functional characterization of human oligodendrocytes in cell culture and in the human MS brain. In this context current strategies to promote remyelination in the human brain were discussed. Finally the regenerative potential of oligodendrocytes transplanted into areas of demyelination was highlighted.

In general, the meeting was characterized by a most creative and stimulating atmosphere, which resulted in intense discussions during the scientific sessions and thereafter. The very intimate setting of the monastery and the limited number of participants consisting of leading experts in their respective field have greatly contributed to the success of the meeting.

We are particularly grateful to the Max-Planck-Society, the German Research Council (DFG) the Hertie-Foundation and the University of Osnabrück (SFB171, Graduate College) who gave support to this project. In addition the meeting was sponsored by the following companies: Biomol/UBI (Hamburg), Heraeus Instruments (Hanau), Jürgens Laborbedarf GmbH (Bremen), Merck AG

(Darmstadt), Nunc GmbH (Wiesbaden), Perkin Elmer/Applied Biosystems GmbH (Weiterstadt), Schering AG (Berlin), Schleicher & Schuell (Dassel).

Finally we would like to thank the staff of Monastery Ohrbeck for their kind hospitality. Last but not least the excellent help of our co-workers is gratefully acknowledged.

G. Jeserich   H.H. Althaus   C. Richter-Landsberg   R. Heumann

Contents

Part 1 Signaling for Glial Cells

Cytokines, Growth Factors and Their Intracellular Cascades

Ion Channels and Neurotransmitter Receptors

Part 2 Genes and Proteins of Myelin

Structure, Regulation and Function

Part 3 Neural Repair Mechanisms in the CNS

Cellular Responses to Injury

Remyelination of Lesions in Multiple Sclerosis

Part 1 Signaling for Glial Cells

Cytokines, Growth Factors and Their Intracellular Cascades

Origins and early development of oligodendrocyte precursor cells

Nigel Pringle[1], Wei-Ping Yu[2], Ellen J Collarini[1], Andy Calver[1], Anita Hall[1], Sarah Guthrie[3] and William D Richardson[1].

[1] MRC Laboratory for Molecular Cell Biology and Department of Biology, University College London, Gower Street, London WC1E 6BT, UK
[2]present address: Institute of Molecular and Cell Biology, National University of Singapore, 10 Kent Ridge Crescent, Singapore 0511.
[3]Division of Anatomy and Cell Biology, UMDS, Guy's Hospital Campus, St Thomas Street, London Bridge, London SE1 9RT

Keywords. Oligodendrocytes, O-2A progenitors, PDGF, PDGFRα, transgenic mice, migration, proliferation

Oligodendrocyte precursors: the O-2A progenitor cell

Oligodendrocytes, the myelinating cells of the CNS, develop from glial progenitor cells known as O-2A progenitors (for reviews see 1,2 and 3). O-2A progenitor cells are so-called because they can differentiate into either oligodendrocytes or type-2 astrocytes in vitro: in medium containing low ($\leq$ 0.5%) fetal calf serum (FCS) they differentiate into oligodendrocytes whereas in 10% FCS they differentiate into type-2 astrocytes (4). O-2A progenitors and their differentiated progeny can be distinguished in vitro by morphology and by their characteristic antigenic phenotypes. O-2A progenitors often have a bipolar morphology and label with monoclonal antibody A2B5 (5), which recognizes a specific set of gangliosides, and with antibodies to the NG2 chondroitin sulphate (6). As they mature, O-2A progenitors become multi-polar, their proliferative and migratory properties change (7 and 8) and they start to express sulphatide and other related antigens that are recognized by monoclonal antibody O4 (9). Differentiated oligodendrocytes have a complex, process-bearing morphology and label specifically with antibodies against galactocerebroside (GC) (10). Type-2 astrocytes are process-bearing cells in vitro that label with antibodies against the glial fibrillary acidic protein (GFAP). After they differentiate, oligodendrocytes lose the A2B5 antigen whereas type-2 astrocytes retain it. The oligodendrocyte differentiation pathway seems to be the default behaviour for O-2A progenitors because a single progenitor cell differentiates into an oligodendrocyte if it is cultured on its own in defined, low-serum medium in the absence of other cells (11). It is not known what the active ingredient in FCS is that can induce type-2 astrocyte differentiation in vitro, but the activity can be mimicked by pure ciliary

Molecular Signaling and Regulation in Glial Cells
A Key to Remyelination and Functional Repair

Figure 1. The life history of an oligodendrocyte

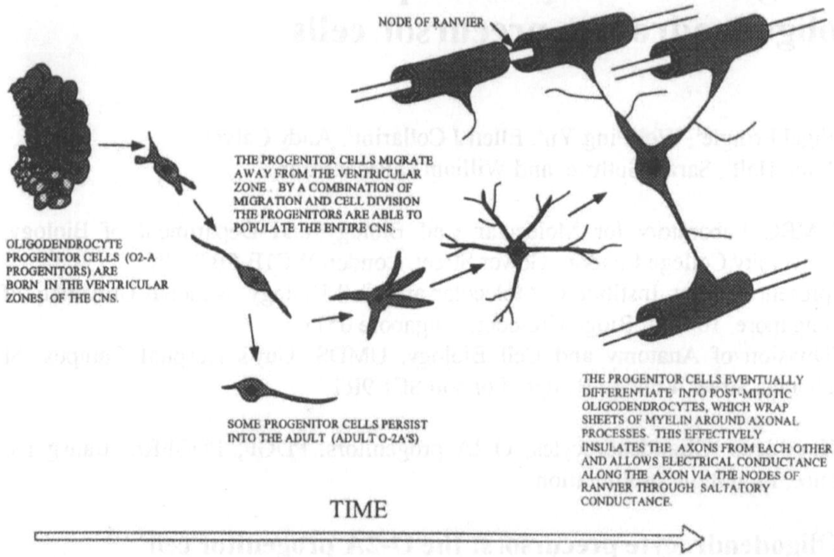

neurotrophic factor (CNTF) in collaboration with uncharacterized extracellular matrix components secreted by cultures of cortical astrocytes (12). Despite careful searching, there is still no definitive evidence for the existence of cells with the antigenic phenotype of type-2 astrocytes in vivo (13). Perhaps type-2 astrocytes will eventually turn up in a restricted region(s) of the CNS or under certain pathological conditions. For the present, however, we regard O-2A progenitors as dedicated oligodendrocyte progenitors in vivo, but retain the prefix O-2A as a reminder of their differentiation potential in vitro. Figure 1 shows a diagram of the life history of an oligodendrocyte.

O-2A progenitors migrate widely during normal development

O-2A progenitor cells are thought to be migratory cells both *in vitro* and *in vivo*. O-2A progenitors accumulate in the perinatal optic nerve in a wave starting at the brain end of the nerve and progressing toward the eye (14), suggesting that O-2A progenitors normally migrate into the developing optic nerve from the brain. This inferred migratory behaviour of O-2A progenitors has been confirmed in several ways. First, if enriched O-2A progenitors, tissue fragments containing O-2A progenitors or cell lines resembling O-2A progenitors are transplanted into normal, hypo-myelinated or de-myelinated hosts (usually neonatal or adult mice or rats), the transplanted progenitors and/or their differentiated progeny can be identified subsequently at a distance from the injection site (7,15,16,17,18,19,20,21 and 22). Second, if individual precursor cells in the ventricular zone of the embryonic chick spinal cord are infected with a recombinant retrovirus vector and the labelled progeny of those cells are visualized postnatally, some clones contain oligodendrocytes that are found dispersed widely from the site of infection (23). Third, if a segment of the caudal neural plate of an E2 chicken embryo is removed and replaced with homologous tissue from a quail embryo, and the chick-quail chimera is allowed to develop until the onset of myelination at E16-E18, then quail oligodendrocytes (identified with quail- and myelin-specific antibodies) can be found more than 2 mm away from the ends of the graft in both the rostral and caudal directions (24). The furthest-migrating cells appear selectively within the longitudinal nerve tracts, especially the dorso-lateral funiculus, indicating that axonal tracts provide a preferred substratum for O-2A progenitor migration. In contrast, quail neurons migrate hardly at all in the longitudinal directions. Together, these experiments prove that oligodendrocytes or, more likely, oligodendrocyte precursors migrate relatively large distances during normal development.

Oligodendrocyte precursors originate in a specialized region of the ventricular zone in the ventral half of the neural tube

It is clear that O-2A progenitors migrate widely through the CNS during embryonic and early postnatal development, but where do these cells come from? It has long been suspected that the oligodendrocyte lineage, like most neuronal cell lineages, might arise in the ventricular zones of the spinal cord and brain but it was not possible to demonstrate this by the classical tritiated thymidine pulse-chase methodology that has been so successful for mapping the origins of neuronal populations. This is because glial progenitor cells, unlike neurons, continue to divide after they leave the ventricular zones, thus diluting the radiolabel beyond the limits of autoradiographic detection. However, other kinds of experiments now provide clear evidence that the oligodendrocyte lineage originates in the ventricular zone. Since oligodendrocytes are found in gray and white matter in all regions of the CNS, with no obvious regard for position along either the rostro-caudal or dorso-ventral axes, one might have expected their precursor cells to arise in all parts of the ventricular germinal zones. However, contrary to expectations, it appears that O-2A progenitors originate in very restricted regions of the ventricular zone, in the ventral half of the spinal cord and brain. The evidence for this conclusion is summarized below.

Warf et al. (25) cultured cells from either the ventral or the dorsal halves of embryonic rat spinal cords and examined the ability of these two populations of cells to generate oligodendrocytes after an appropriate period of time in vitro. The outcome of this experiment depended on the age of the embryos from which the cultures were established. GC-positive oligodendrocytes developed in cultures of ventral spinal cord cells from E14 embryos, but did not develop in cultures of dorsal cells from E14 embryos. In contrast, oligodendrocytes developed in cultures of both ventral and dorsal spinal cord cells from E16 embryos. This suggested that oligodendrocyte precursor cells are present in the ventral, but not the dorsal half of the cord at E14, but spread to the dorsal half by E16. Around the same time in our laboratory, we were attempting to trace the origins of oligodendrocytes by in situ hybridization with a probe against the alpha subunit of the receptor for platelet-derived growth factor (PDGFRα), which we had previously suggested might be a useful marker for early stages of the oligodendrocyte lineage (25). We found narrow, bilateral longitudinal columns of PDGFRα-expressing cells in the ventral half of the E14 rat embryo (see figure 2) (26). Transverse sections revealed that these cells lie at the ventricular surface, 15% of the way from the ventral towards the dorsal extremity of the central canal (see figure 2, *panel a*). After E14, the PDGFRα^+ cells appeared to leave the ventricular surface and spread throughout the cross-section of the cord (see figure 2, *panel b*) (26). These data have now been confirmed by immunolabelling with an antibody against PDGFRα (27). Thus the behaviour of these cells in time and space matched very well the behaviour predicted for oligodendrocyte precursors from previous experiments (21) experiments. Identification of the ventral focus of PDGFRα-expressing cells as oligodendrocyte precursors has been strongly supported by labelling with other, more conventional, markers of the oligodendrocyte lineage. First, in situ hybridization with a probe against 2',3'-cyclic nucleotide 3'-phosphodiesterase (CNP), an abundant protein of myelinating oligodendrocytes that is also expressed precociously by proliferating rat O-2A progenitor cells in vitro (27), reveals a small focus of cells in the E14 rat spinal cord that precisely co-localizes to the PDGFRα^+ expression domain described above (28). Second, Ono *et al.* (29) showed that monoclonal antibody O4, which is known to label rat O-2A progenitors at a late stage of their stepwise progression towards oligodendrocyte differentiation, labels all stages of the chicken oligodendrocyte lineage including very early progenitors; immunolabelling of transverse sections of E6 chick spinal cord revealed a focus of O4$^+$ cells in the ventricular zone in the ventral half of the cord that corresponds closely to the PDGFRα^+, CNP$^+$ cells described in the rat. Thus, there are at least three plausible markers of early oligodendrocyte precursors that identify the same microdomain of the spinal cord ventricular zone.

We have further evidence that establishes beyond doubt that the PDGFRα^+ cells in the embryonic rat spinal cord are, indeed, oligodendrocyte precursors. We can establish cultures of >99% pure PDGFRα^+ cells from E17 rat spinal cords by immunoselection ("panning") with an antibody, raised in rabbits, against the extracellular domain of PDGFRα. When cultured in the presence of PDGF and basic fibroblast growth factor these cells display the characteristic morphology and antigenic phenotype (A2B5$^+$, NG2$^+$) expected for O-2A progenitors. When transferred to defined medium lacking growth factors, they all differentiate into GC$^+$ oligodendrocytes (authors unpublished observations).

The combined weight of evidence therefore supports the idea that spinal cord oligodendrocyte precursors originate at a precisely defined point in the ventral ventricular zone and disseminate from there to populate the rest of the spinal cord. This seems to be a conserved characteristic of vertebrates; analogous foci of migratory PDGFRα^+ oligodendrocyte precursors are expressed in the ventral ventricular zone in the spinal cords of embryonic mice, chickens and *Xenopus* at equivalent developmental stages (unpublished results from our laboratory).

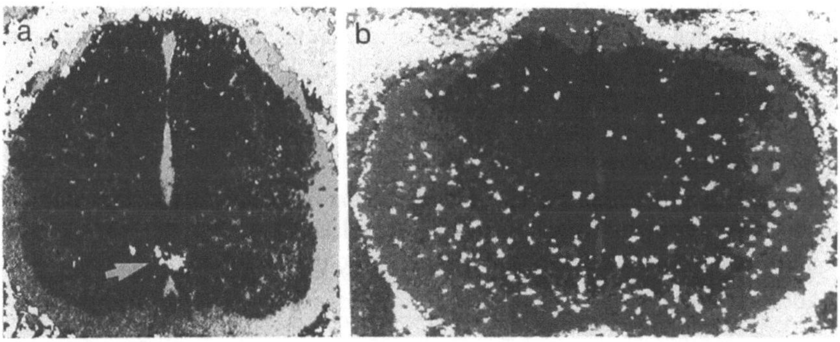

Figure 2. *In situ* hybridization of transverse sections of embryonic rat spinal cord with probes to PDGFRα. Following autoradiography, the sections were photographed under bright-field and dark-field microscope optics and the images superimposed by computer. *Panel a* shows an E14 rat thoracic spinal cord probed for PDGFRα. The white arrow indicates the focus of PDGFRα^+ oligodendrocyte precursor cells in the ventral ventricular zone. *Panel b* shows an E16 rat spinal cord section at E16, PDGFRα^+ oligodendrocyte progenitors are scattered throughout the ventral half of the rat spinal cord in both grey and white matter.

It should be emphasized that, while it is clear that all PDGFRα^+ cells in the late embryonic spinal cord can give rise to oligodendrocytes in vitro, we do not yet know if all oligodendrocytes develop from PDGFRα^+ cells. There might be an additional, unrecognized oligodendrocyte lineage(s) that does not express PDGFRα. It might be possible to address this possibility in future by removing all the PDGFRα^+ progenitors from mixed cultures of spinal cord cells, either by antibody-mediated complement lysis or negative immunoselection, to see whether oligodendrocytes still develop in these depleted cultures. Unfortunately, the anti-PDGFRα antibodies available at present are unable to fix complement (authors unpublished observations).

Control of glial cell numbers by PDGF in transgenic mice.

Oligodendrocyte precursor cells express the PDGFRα and synthesise DNA *in vitro* in response to PDGF. PDGF-A is expressed by neurons and astrocytes in the developing CNS which suggests that PDGF-A might normally stimulate proliferation and/or survival of oligodendrocyte lineage cells *in vivo*, but up to now there has been no direct evidence for this. We therefore examined the spinal cords of P14 PDGF-A chain knockout mice. We found very few (1% of wild type) of PDGFRα^+ oligodendrocyte progenitors and a corresponding decrease in the number of differentiated oligodendrocytes. Conversely when PDGF-A was over expressed in transgenic mice under control of the neuron-specific enolase (NSE) gene promoter, there were twice the normal number of PDGFRα^+ oligodendrocyte progenitor cells in the postnatal spinal cord. These experiments demonstrate that PDGF-A is important for development of the oligodendrocyte lineage *in vivo*, and suggest that PDGF-A is normally present in sub-saturating amounts in the developing CNS.

Acknowledgements
We would like to thank the UK Medical Research Council and the Multiple Sclerosis Society of Great Britain and Northern Ireland for their generous support of much of the research carried out and discussed in this article.

References

1. Raff, M.C. (1989). Glial cell diversification in the rat optic nerve. Science *243*, 1450-1455.
2. Richardson, W.D., Raff, M., and Noble, M. (1990). The oligodendrocyte-type-2 astrocyte lineage. Semin.Neurosci. *2*, 445-454.
3. Pfeiffer, S.E., Warrington, A.E., and Bansal, R. (1994). The oligodendrocyte and its many cellular processes. Trends Cell Biol. *3*, 191-197.
4. Raff, M.C., Miller, R.H., and Noble, M. (1983). A glial progenitor cell that develops in vitro into an astrocyte or an oligodendrocyte depending on the culture medium. Nature *303*, 390-396.
5. Eisenbarth, G.S., Walsh, F.S., and Nirenburg, M. (1979). Monoclonal anti-body to a plasma membrane antigen of neurons. Proc.Natn.Acad.Sci.U.S.A. *76*, 4913-4917.
6. Stallcup, W.B. and Beasley, L. (1987). Bipotential glial progenitor cells of the optic nerve express the NG2 proteoglycan. J.Neurosci. *7*, 2737-2744.
7. Tanabe, Y., Roelink, H., and Jessell, T.M. (1995). Induction of motor neurons by sonic hedgehog is independent of floor plate differentiation. Curr.Biol. *5*, 651-658.
7. Warrington, A.E., Barbarese, E., and Pfeiffer, S.E. (1993). Differential myelinogenic capacity of specific stages of theoligodendrocyte lineage upon transplantation into hypomyelinating hosts. J.Neurosci. *34*, 1-13.
8. Gard, A.L. and Pfeiffer, S.E. (1990). Two proliferative stages of the oligodendrocyte lineage (A2B5+O4- and O4+GalC-) under different mitogenic control. Neuron *5*, 615-625.

9. Sommer, I. and Schachner, M. (1981). Monoclonal antibodies (O1 to O4) to oligodendrocyte cell surfaces:an immunocytological study in the central nervous system. Dev.Biol. *83*, 311-327.
10. Raff, M.C., Mirsky, R., Fields, K.L., Lisak, R.P., Dorfman, S.H., Silberberg, D.H., Gregson, N.A., Liebowitz, S., and Kennedy, M. (1978). Galactocerebroside: a specific cell surface antigenic marker for oligodendrocytes in culture. Nature *274*, 813-816.
11. Temple, S. and Raff, M.C. (1985). Differentiation of a bipotential glial progenitor cell in single cell microculture. Nature *313*, 223-225.
12. Lillien, L.E. and Raff, M.C. (1990). Differentiation signals in the CNS: type-2 astrocyte development in vitroas a model system. Neuron *5*, 111-119.
13. Fulton, B.P., Burne, J.F., and Raff, M.C. (1991). Glial cells in the rat optic nerve. The search for the type-2 astrocyte. Annals N.Y.Acad.Sci. *633*, 27-34.
14. Small, R.K., Riddle, P., and Noble, M. (1987). Evidence for migration of oligodendrocyte-type-2 astrocyte progenitor cells into the developing rat optic nerve. Nature *328*, 155-157.
15. Espinosa de los Monteros, A., Bernard, R., Tiller, B., Rouget, P. and De Vellis, J. (1993a). Grafting of fast blue labelled glial cells into neonatal rat brain: differential survival and migration among cell types. Int.J.Dev.Neurosci. *11*, 625-639.
16. Espinosa de los Monteros, A. Zhang, M. and De Vellis, J. (1993b). O-2A progenitor cells transplanted into the neonatal rat brain develop into oligodendrocytes but not astrocytes. Proc.Natl.Acad.Sci.USA *90*, 50-54.
17. Grove, E.A., Williams, B.P., Li, D.Q., Hajihosseini, M., Friedrich, A., and Price, J. (1993). Multiple restricted lineages in the embryonic rat cerebral cortex. Development *117*, 553-561.
18. Trotter, J., Crang, A.J., Schachner, M. and Blakemore, W.F. (1993). Lines of glial precursor cells immortalised with a temperature-sensitive oncogene give rise to astrocytes and oligodendrocytes following transplantation into demyelinated lesions in the central nervous system. Glia *9*, 25-40.
19. Friedrich, V.L. and Lazzarini, R.A. (1993). Restricted migration of transplanted oligodendrocytes or their progenitors, revealed by transgenic marker M beta P. J.Neural.Transplant.Plast. *4*, 139-146.
20. Vignais, L., Nait-Oumesmar, B., Mellouk, F., Gout, O., Labourdette, G., Baron Van Evercooren, A. and Gumpel, M. (1993). Transplantation of oligodendrocyte precursors in the adult demyelinated spinal cord: migration and remyelination. Int.J.Dev.Neurosci. *11*, 603-612.
21. Warf, B.C., Fok-Seang, J., and Miller, R.H. (1991). Evidence for the ventral origin of oligodendrocyte precursors in the rat spinal cord. J.Neurosci. *11*, 2477-2488.
22. Tontsch, U., Archer, D.R., Dubois-Dalcq, M., and Duncan, I.D. (1994). Transplantation of an oligodendrocyte cell line leading to extensive myelination. Proc.Natl.Acad.Sci.USA *91*, 11616-11620.
23. Leber, S.M. and Sanes, J.R. (1995). Migratory paths of neurons and glia in the embryonic chick spinal cord. J.Neurosci. *15*, 1236-1248.

24. Pringle, N.P., Lumsden, A., Richardson, W,D and Guthrie, S. (1995). Analysis of oligodendrocyte precursor origins and migration patterns in the developing spinal cords of chick-quail chimeras. In preparation.
25. Pringle, N.P., Mudhar, H.S., Collarini, E.J., and Richardson, W.D. (1992). PDGF receptors in the CNS: during late neurogenesis, expression of PDGF alpha receptors appears to be restricted to glial cells of the oligodendrocyte lineage. Development *115*, 535-551.
26. Pringle, N.P. and Richardson, W.D. (1993). A singularity of PDGF alpha-receptor expression in the dorsoventral axis of the neural tube may define the origin of the oligodendrocyte lineage. Development *117*, 525-533.
27. Nishiyama, A., Lin, X.-H., Giese, N., Heldin, C.-H. and Stallcup, W.B. (1995), Co-localization of NG2 proteoglycan and PDGF α receptor on O2A progenitor cells in the developing rat brain. J.Neurosci.Res. in press.
28. Yu, W.-P., Collarini, E.J., Pringle, N.P., and Richardson, W.D. (1994). Embryonic expression of myelin genes: evidence for a focal source of oligodendrocyte precursors in the ventricular zone of the neural tube. Neuron *12*, 1353-1362.
29. Ono, K., Bansal, R., Payne, J., Rutishauser, U., and Miller, R.H. (1995). Early development and dispersal of oligodendrocyte precursors in the embryonic chick spinal cord. Development *121*, 1743-1754.

Cytokine and Growth Factor Synthesis in Human Astrocytes

Francesca Aloisi, Francesca Ceccherini Silberstein and Giulio Levi

Laboratory of Organ and System Pathophysiology,
Istituto Superiore di Sanità, Rome, ITALY

Abstract. Astrocytes are involved in a number of physiological and pathological processes of the CNS, in which cytokines may act both as stimulators and effectors of astrocyte functions. *In vitro* studies, including those presented here using highly enriched astrocyte cultures from the human embryonic brain, show that macrophage-derived cytokines (interleukin-1b and tumor necrosis factor-a) are very efficient in stimulating the synthesis of a large number of polypeptides displaying pro-inflammatory, anti-inflammatory or neurotrophic activities. The T cell-derived cytokine interferon-g can potentiate the stimulating effects of interleukin-1b and tumor necrosis factor-a on some inflammatory mediators, whereas transforming growth factor-b, a cytokine produced by neural and hematogenous cells, mainly enhances the production of growth factors active on neurons and oligodendrocytes. These *in vitro* data suggest that cytokines that are transiently expressed in the CNS during development or in pathological conditions can activate distinct functional programs in astrocytes, which may play a specific role in brain growth and in different brain diseases.

Keywords. Astrocytes, cytokines, inflammation, repair

1 Cytokines as modulators and effectors of astrocyte function

The multiplicity of functions that have been attributed to astrocytes on the basis of in situ morphological observations and analysis of *in vitro* systems indicate that this cell population, which is most probably highly heterogeneous (33), fulfills a very important role in brain development and in the maintenance of brain function and homeostasis. In nearly all neurological disorders, disruption of the structural and functional integrity of the brain leads to a prominent astrocyte reaction that is characterized by enhanced or de novo expression of a large panel of genes and proteins (11). The available experimental evidence suggests that cytokines may play a key role in this functional reprogramming, acting both as activators and mediators of astrocyte function (12). Cytokines are polypeptide

Molecular Signaling and Regulation in Glial Cells
A Key to Remyelination and Functional Repair

mediators, which act on multiple, often unrelated, cell types, and which serve as important regulators of a variety of cellular processes, including survival, proliferation, differentiation and migration. Synthesis and release of a wide range of cytokines by astrocytes has been largely documented *in vitro*, whereas, *in vivo*, cytokine expression in astrocytes appears to be generally restricted to the developmental period and/or to pathological situations (6, 11, 21). Some astrocyte-derived cytokines [e.g., neurotrophins, platelet-derived growth factor (PDGF), leukemia inhibitory factor (LIF), fibroblast growth factor] affect the growth, survival and differentiation of neurons and oligodendrocytes, whereas others [e.g., interleukin-6 (IL-6), colony-stimulating factors (CSFs), chemokines, transforming growth factor-β (TGF-β)] may be implicated in regulating growth and activation of lymphocytes and mononuclear phagocytes. Elucidation of the mechanisms controlling cytokine expression in astrocytes may help to better define the multiple roles of these cells in the normal and pathological brain.

Cultures highly enriched in astrocytes from the human embryonic brain have represented a useful experimental system to examine which are the relevant stimuli that trigger cytokine and growth factor production in astrocytes, and have permitted to unravel differences in the activation of astrocyte functional programs by cytokines that are transiently expressed in the brain following injury or inflammation.

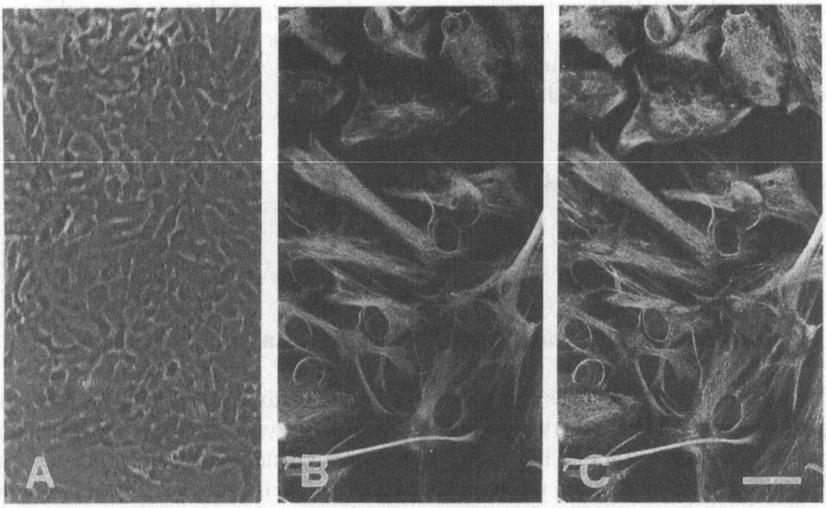

Fig. 1 Astrocyte-enriched subcultures from the human embryonic brain. **A:** Phase contrast micrograph of living cells. **B, C:** double immunofluorescence labeling with anti-GFAP polyclonal antibody (B) and anti-vimentin monoclonal antibody (C). Scale bar = 50 μm.

1.1 Astrocyte-enriched cultures derived from the human embryonic brain

In recent years, several laboratories, including our own, have developed culture conditions for maintaining and expanding *in vitro* highly enriched populations of astrocytes from the developing (embryonic, fetal) human CNS. Using human embryonic brain or spinal cord, we have established cultures which contain, after 2-3 *in vitro* passages, an elevated proportion (up to 90%) of highly proliferating glial fibrillary acidic protein$^+$ (GFAP$^+$) cells (1). Astrocyte-enriched subcultures also contain a certain proportion (8-18%) of cells with an undifferentiated phenotype (vimentin$^+$, neurofilament$^-$, GFAP$^-$), only rare contaminating fibroblasts (<2%) and neurons (<1%), but no detectable oligodendrocytes and macrophages/microglia. Due to the elevated degree of enrichment, high cellular yield, good survival after freezing and thawing, and, most importantly, lack of macrophages/microglia, the above cultures have represented a useful experimental tool for investigating cytokine synthesis and performing functional studies on astrocyte-T cell interactions (2, 3, 4, 20, 32).

1.2 Pro-inflammatory mediators

Various cytokines that are implicated in the recruitment and activation of lymphocytes and of cells of the monocyte/macrophage lineage have been shown to be transiently expressed in the cerebrospinal fluid or brain parenchyma in a number of neuropathological (inflammatory, demyelinating, infectious, and degenerative) diseases (6, 12). Previous work performed with rodent brain-derived cultures had shown that cytokines, neurotropic viruses, or lypopolisaccharide (LPS) are able to stimulate the astroglial production of IL-6, CSFs, IL-1 and TNF-α (11), suggesting that astrocytes may play a role in the amplification of the inflammatory process.

Human astrocyte cultures were examined for their ability to synthesize and secrete biologically significant amounts of IL-6 [a pleiotropic cytokine involved in B cell differentiation and T cell activation (30)], IL-8 [a mediator belonging to the large family of chemokines which acts as potent chemoattractant and activator for different leukocyte populations (23)], and CSFs [cytokines which affect migration and activation of mature granulocytes and macrophages/microglia (22)]. Synthesis of IL-6, IL-8, granulocyte-macrophage CSF (GM-CSF), macrophage CSF (M-CSF) and granulocyte CSF (G-CSF) was stimulated when astrocytes were exposed to IL-1β and TNF-α (2) (Table 1.1). Both at the messenger RNA and protein levels, IL-1β was a much more powerful inducer of all the above cytokines than TNF-α. The T cell-derived cytokine IFN-γ and TGF-β, a pleiotropic cytokine which is produced by different cell types including neural cells (8, 19), were ineffective in stimulating IL-6, IL-8 or CSF production; however, IFN-γ synergized with IL-1β and TNF-α in stimulating IL-6 and G-CSF, while TGF-β synergized with both cytokines only for IL-6 induction. Human astrocytes did not respond to LPS, a potent macrophage activator, with increased cytokine production. Other studies reported similar observations on M-CSF and IL-6 synthesis in cultures of purified human fetal astrocytes (14, 15).

Recently, Hurwitz et al. (13), reported that human astrocytes also produce monocyte chemoattractant protein-1, a member of the chemokine family, which specifically stimulates monocyte recruitment and activation.

Our studies in human astrocyte cultures failed to demonstrate any relevant synthesis of IL-1α, IL-1β or TNF-α protein, either in the absence or presence of inducing stimuli (IFN-γ, TNF-α, IL-1β, or superantigens), but provided evidence for elevated amounts of TNF-α specific transcripts already in basal culture conditions, suggesting the existence of a post-transcriptional blockade in the synthesis of this cytokine (2, 4). Although some limited TNF-α production has been reported in IL-1-stimulated human fetal astrocytes (14), and controversial data emerge from studies on rodent astrocytes (6), the overall available evidence supports the notion that astrocytes are not a major source of IL-1 and TNF-α. With the exception of one study describing TNF-α in astrocytes within active multiple sclerosis lesions (26), IL-1 and TNF-α immunoreactivities have been demonstrated mainly in activated macrophages/microglia in various neuropathological conditions (HIV encephalopathy, Alzheimer's disease, multiple sclerosis) (6, 12). Thus, macrophage-derived IL-1 and TNF-α may be indirectly involved in the recruitment and activation of lymphocytes and macrophages through the induction of pro-inflammatory cytokine production in astrocytes. Interestingly, GM-CSF immunoreactivity has been detected in activated astrocytes in multiple sclerosis lesions and in senile plaques of Alzheimer's disease (16), suggesting that the cytokine loops described *in vitro* may be functionally relevant at least in certain pathological conditions.

1.3 Anti-inflammatory mediators

Previous studies have shown that human astrocytes suppress antigen-dependent proliferation and cytokine production in T cells stimulated by antigen presenting cells from the peripheral blood (32), and that this inhibitory effect is mediated by soluble factors (at least in part by prostaglandins) (20). Search for soluble mediators with immunosuppressive activity revealed that, in agreement with previous work with human glioma cells and rat astrocytes (8), human astrocytes secrete both the latent and active form of TGF-β (mainly TGF-β_2) into the culture medium (20). Additional unpublished experiments indicated that cytokine (IL-1β, TNF-α or IFN-γ) treatments did not consistently affect the levels of secreted TGF-β (either latent or active). Although the inhibitory effect of astrocytes on T cell proliferation *in vitro* was not reverted by anti-TGF-β neutralizing antibodies, it is possible that, *in vivo*, astrocyte-derived TGF-β does play a role in inhibiting proliferation of T lymphocytes (31), and in downregulating proliferation and activation of macrophages/microglia (28). IL-10 [a potent inhibitor of cytokine synthesis in macrophages (9)], IL-4 [a cytokine suppressing microglial activation (7)], and IL-1 receptor antagonist [a naturally occurring anti-inflammatory mediator capable of counteracting IL-1 activity (10)], were not produced by cultured human astrocytes, even after cytokine stimulation (4, 20). We however found that astrocytes secreted soluble TNF receptors [the truncated forms of type I

(55 kDa) and type II (75 kDa) TNF receptors] (29), when they were stimulated by IL-1β or IL-1β/IFN-γ. It has been suggested that *in vivo* soluble TNF receptors may interfere with the ability of TNF-α and lymphotoxin to interact with their specific receptors, providing a protective mechanism to limit TNF degenerative effects. The above data suggest that astrocytes may play an active role in limiting the spread of inflammatory reactions within the CNS.

Table 1.1 Cytokine production in human astrocytes

Produced cytokines:	Stimulating cytokines
IL-6	IL-1β/IFN-γ = TNF-α/IFN-γ > IL-1β >> TNF-α
IL-8	IL-1β > TNF-α
M-CSF	IL-1β ^{2} TNF-α
GM-CSF	IL-1β > TNF-α
G-CSF	IL-1β/IFN-γ = TNF-α/IFN-γ > IL-1β >> TNF-α
TGF-β_2	none of the above cytokines
LIF	TGF-β_1 ^{2} IL-1β = TNF-α
PDGF AB	TGF-β_1/TNF-α ^{2} TGF-β_1/IL-1β > TGF-β_1 ^{2} TNF-α

1.4 Growth factors (LIF and PDGF)

During development and following experimental brain injury, astrocytes have been recognized as a source of cytokines that may play an important role in the growth, differentiation and survival of neuronal and glial cell populations (5, 25). Astroglial synthesis of nerve growth factor was previously shown to be stimulated by IL-1, TNF-α or TGF-β (17), suggesting a role for cytokine networks in regulating neurotrophic support in the CNS.

Our studies in human astrocyte cultures have focussed on the expression and regulation of two growth factors, LIF and PDGF, which were previously shown to promote growth, survival or differentiation of neurons and oligodendrocytes from the developing rat brain (5, 24, 27). Due to its chemiotactic and growth promoting activity on fibroblasts and endothelial cells, PDGF has also been implicated in wound healing after CNS injury (18). Expression of LIF specific transcripts and protein was consistently stimulated by TGF-β_1, TNF-α and IL-1β (3). TGF-β_1 and TNF-α, but not IL-1β, were also effective in stimulating the expression of PDGF B-, but not PDGF A-chain mRNA, and in enhancing the levels of secreted PDGF AB protein in astrocyte cultures; maximal PDGF stimulation was observed after combined citokine (TGF-β_1/TNF-α and TGF-β_1/IL-1β) treatments (Ceccherini Silberstein, De Simone, Levi and Aloisi, submitted). IFN-γ did not affect either PDGF or LIF production. These data suggest that resident brain and blood-derived cells could positively regulate LIF and PDGF synthesis via IL-1, TNF-α and TGF-β. Although these *in vitro* observations cannot be immediately extrapolated to the *in vivo* situation, it is possible that this regulatory mechanism is implicated in brain development as well as in brain protection and repair.

2 Effects of astrocyte-neuron interactions on cytokine synthesis

Due to their embryonic derivation and to *in vitro* manipulations, cultured astrocytes are likely to be more closely related to reactive rather than to normal astrocytes. To determine the influence of cell interactions on astrocyte function, we have compared cytokine production by human astrocytes in neuron/astrocyte co-cultures and after removal of neurons. In early primary cultures from the human embryonic brain (7-10 days *in vitro*), which contained mainly neuronal cells and immature glial/neuronal progenitors, IL-6, IL-8, CSFs and LIF were not induced in the presence of the most powerful stimulatory citokines (IL-1β, TNF-α or TGF-β_1) (2, 4). In late (3-4 week-old) primary neuronal-glial cultures, in which most glial precursors had acquired GFAP positivity, IL-1β, and to a lesser extent TNF-α, significantly increased IL-6 and IL-8 secretion, while only IL-1β enhanced the production of G-CSF and GM-CSF in a substantial way; in all cases, however, stimulated cytokine levels were much lower (one half to one third) than those detected in astrocyte-enriched subcultures, despite the fact that the primary cultures contained 4-6 fold more astrocytes (as determined by counting the number of GFAP$^+$ cells). Interestingly, LIF remained undetectable also in late neuronal/glial cultures even after cytokine treatments. These findings indicate that the capacity of astrocytes to produce a number of polypeptide mediators after cytokine stimulation is strongly reduced in the presence of neuronal cells, and suggest that neuronal-glial interactions may be implicated in determining the activation threshold of astrocytes to inflammatory cytokines. *In vivo*, prominent cytokine production by astrocytes could occur in pathological conditions as a consequence of alterations of neuronal-glial interactions combined with a local increase in the levels of inflammatory cytokines.

3 Concluding remarks

Examination of the molecular profile acquired by cultured astrocytes in response to cytokines (or combinations of cytokines) may be useful for understanding the functional consequences of astrocyte activation in various brain pathologies, in which distinct and transient expression of multiple cytokines is documented. As summarized in Table 1.1, IL-1β and TNF-α stimulate astrocytes to produce wide, largely overlapping sets of soluble mediators exhibiting immunoregolatory or neurotrophic activities. Since both IL-1β and TNF-α are produced by macrophages, it is likely that macrophage activation in response to brain injury provides the most relevant signals for stimulating astrocyte partecipation in host defense and tissue repair processes. The T cell-derived cytokine IFN-γ is by itself unable to induce astrocyte cytokine synthesis, but potentiates the effects of IL-1β and TNF-α on IL-6, G-CSF, and soluble TNF receptor production; this suggests that the simultaneous presence of activated macrophages and infiltrating T cells may enhance the regulatory influences of astrocytes on inflammatory and immune processes. TGF-β released by astrocytes and by other CNS or hematogenous cells, could be involved in the suppression of T cell and macrophage activation. In addition, TGF-β, via paracrine and/or autocrine mechanisms, could stimulate astrocytes to produce growth factors involved in wound healing and tissue repair.

Elucidation of the temporal and spatial patterns of cytokine expression within a given neuropathological condition may help to elucidate the contribution of astrocytes in different stages of the disease. This knowledge should help to predict possible mechanisms of action for cytokine-based therapeutical interventions that are aimed at inhibiting unwanted inflammatory reactions and/or at potentiating protective and repair processes.

References

1. Aloisi, F., Borsellino, G., Samoggia, P., Testa, U., Chelucci, C., Russo, G., Peschle, C. and Levi, G. (1992a) Astrocyte cultures from human embryonic brain: characterization and modulation of surface molecules by inflammatory cytokines. J. Neurosci. Res. 32, 494-506.
2. Aloisi, F., Carè, A., Borsellino, G., Gallo, P., Rosa, S., Bassani, A., Cabibbo, A., Testa, U., Levi, G. and Peschle, C. (1992b). Production of hemolymphopoietic cytokines (IL-6, IL-8, colony-stimulating factors) by normal human astrocytes in response to IL-1β and tumor necrosis factor-α. J. Immunol. 149, 2358-2366.
3. Aloisi, F., Rosa, S., Testa, U., Bonsi, P., Russo, G., Peschle, C., and Levi, G. (1994) Regulation of leukemia inhibitory factor synthesis in cultured human astrocytes. J. Immunol. 152, 5022-5031.
4. Aloisi, F., Borsellino, G., Carè, A., Testa, U., Gallo, P., Russo, G., Peschle, C. and Levi, G. (1995) Cytokine regulation of astrocyte function: in vitro studies using cells from the human brain. Int. J. Dev. Neurosci. 13, 265-274.
5. Barres, B.A. and Raff, M.C. (1994) Control of oligodendrocyte number in the developing rat optic nerve. Neuron 12, 935-942.
6. Benveniste, E.N. (1992) Inflammatory cytokines within the central nervous system: sources, function, and mechanism of action. Am. J. Physiol. 263, 1-14.
7. Chao, C.C., Molitor, T.W. and Hu, S. (1993) Neuroprotective role of IL-4 against activated microglia. J. Immunol. 151, 1473-1481.

8. Constam, D.B., Phillip, J., Malipiero, U.V., tenDijke, P., Schachner, M. and Fontana, A. (1992) Differential expression of transforming growth factor-β1, -β2, and -β3 by gliomastoma cells, astrocytes and microglia. J. Immunol. 148, 1404-1410.
9. de Waal Malefyt, R., Abrams, J., Bennett, B., Figdor, C.G. and de Vries, J.E. (1991) Interleukin-10 (IL-10) inhibits cytokine synthesis by human monocytes: an autoregolatory role of IL-10 produced by monocytes. J. Exp. Med. 174, 1209-1220.
10. Dinarello, C.A. and Thompson, R.C. (1991) Blocking IL-1: interleukin 1 receptor antagonists in vivo and in vitro. Immunol. Today 12, 404-410.
11. Eddleston, M. and Mucke, L. (1993) Molecular profile of reactive astrocytes-Implications for their role in neurologic disease. Neuroscience 54, 15-36.
12. Hopkins, S.J. and Rothwell, N.J. (1995) Cytokines and the nervous system I: expression and recognition. Trends Neurosci. 18, 83-88.
13. Hurwitz, A.A., Lyman, W.D. and Berman, J.W. (1995) Tumor necrosis factor and transforming growth factor β upregulate astrocyte expression of monocyte chemoattractant protein-1. J. Neuroimmunol. 57, 193-198.
14. Lee, S.C., Liu, W., Dickson, D.W., Brosnan, C.F. and Berman, J.W. (1993a) Cytokine production by human fetal microglia and astrocytes. J. Immunol. 150, 2659-2667.
15. Lee, S.C., Liu, W., Roth, P., Dickson, D.W., Berman, J.W. and Brosnan, C.F. (1993b) Macrophage colony-stimulating factor in human fetal astrocytes and microglia. J. Immunol. 150, 594-604.
16. Lee, S.C., Liu, W., Brosnan, C.F. and Dickson, D. W. (1994) GM-CSF promotes proliferation of human fetal and adult microglia in primary cultures. Glia 12, 309-318.
17. Lindholm, D., Hengerer, B., Zafra, F. and Thoenen, H. (1990) Transforming growth factor-β1 stimulates expression of nerve growth factor in the rat CNS. Neuroreport. 1, 9-12.
18. Lotan, M. and Schwartz, M. (1992) Postinjury changes in platelet-derived growth factor-like activity in fish and rat optic nerves. J. Neurochem. 58, 1637-1642.
19. Massague', J. (1990) The transforming growth factor-β family. Ann. Rev. Cell Biol. 6, 597-641.
20. Meinl, E., Aloisi, F., Ertl, B., Weber, F., de Waal Malefyt, R., Wekerle, H. and Holhfeld, R. (1994) Multiple Sclerosis: Immunomodulatory effects of human astrocytes on T cells. Brain 117, 1323-1332.
21. Merrill, J.E. (1992) Tumor necrosis factor-α, interleukin-1 and related cytokines in brain development: normal and pathological. Devl. Neurosci. 14, 1-10.
22. Nicola, N.A. (1989) Hemopoietic cell growth factors and their receptors. Ann. Rev. Biochem. 58, 45-77.
23. Oppenheim, J.J., Zachariae, C.O.C., Mukaida, N., and Matsushima, K. (1991) Properties of the novel pro- inflammatory supergene "intercrine" cytokine family. Ann. Rev. Immunol. 9, 616-648.

24. Patterson, P.H. and Nawa, H. (1993) Neuronal differentiation factors/cytokines and synaptic plasticity. Cell 72, 123-37.
25. Rudge, J.S. (1993) Astrocyte-derived neurotrophic factors. In S. Murphy (ed.), Astrocytes. Pharmacology and function, Academic Press, Inc., San Diego, pp. 267-305.
26. Selmaj, K., Raine, C.S., Cannella, B., and Brosnan, C.F. (1991) Identification of lymphotoxin and tumor necrosis factor in multiple sclerosis lesions. J. Clin. Invest. 87, 949-954.
27. Smits, A., Ballagi, A.E. and Funa, K. (1993) PDGF-BB exerts trophic activity on cultured GABA interneurons from the newborn rat cerebellum. Eur. J. Neurosci. 5, 986-994.
28. Suzumura, A., Sawada, M., Yamamoto, H. and Marunouchi, T. (1993) Transforming growth factor-β suppress activation and proliferation of microglia in vitro. J. Immunol. 151, 2150-2158.
29. Tartaglia, L.A. and Goeddel, D.V. (1992) Two TNF receptors. Immunol. Today 13, 151-153.
30. Van Snick, J. (1990) Interleukin-6: an overview. Ann. Rev. Immunol. 8, 253-278.
31. Wahl, S.M., Hunt, D.A., Wong, H.L., Dougherty, S., McCartney-Francis, N., Wahl, L.M. et al. (1988) Transforming growth factor-β is a potent immunosuppressive agent that inhibits IL-1 dependent lymphocyte proliferation. J. Immunol. 140, 3026-3032.
32. Weber, F., Meinl, E., Aloisi, F., Nevinny-Stickel, C., Albert, E., Wekerle, H. and Hohlfeld, R. (1994) Human astrocytes are lysed by antigen-specific, HLA-DR restricted T cells but fail to induce T cell proliferation. Brain 117, 59-69.
33. Wilkin, G.P., Marriott, D.R. and Cholewinski, A.J. (1990) Astrocyte heterogeneity. Trends Neurosci. 13, 43-46.

Mature Pig Astrocytes Do Not Respond to NGF as Oligodendrocytes Do

Hans H. Althaus, Tyede Schmidt-Schultz and Sabine Klöppner

AG Neural Regeneration, Max-Planck Institute for Exp. Medicine, Göttingen, Germany

Abstract: Recently, it was shown that NGF enhances the process regeneration of mature pig oligodendrocytes (OL) and that it induces a subset of mature OL to proliferate. This effect is mediated by TrkA, the NGF tyrosine kinase receptor. Astrocytes belong to the cells in the CNS, which synthesise and secrete NGF, however, it is unknown whether pig astrocytes can also respond to NGF. Firstly, we established the cultural requirements for mature pig astrocytes. During culturing three different types of $GFAP^+$ cells could be observed, of which the flat and the stellate type were predominant. NGF did neither influence the astroglial process formation nor did it alter the rate of the ^{3}H-thymidine incorporation. An in gel kinase assay revealed that a key enzyme of one of the intracellular signaling cascades, Erk1, is functioning in astrocytes, however, its activity cannot be enhanced by NGF.

Keywords: Astrocytes, NGF, MAP kinase, Oligodendrocytes

1 Introduction

Growth factors are known to play an important regulatory part in complex processes such as differentiation, survival, and regeneration. A subclass of growth factors represents the neurotrophin (NT) family. To this family belong NGF, BDNF,NT3, and NT4/5. Two types of NT receptors have been identified (18,30): the p75 receptor binds all NTs presently known (18,30), whereas to the other type, the Trk receptors, which exhibit tyrosine kinase activity and which are required for a biological response, a selective binding exists (25). Previously thought to affect only neurons it became clear that the NT and their receptors are more widespread (26). Members of the NT family have recently attracted attention in glial research. Binding of ^{125}I-NGF to glioma cell lines and to astrocytes has been shown (27). Moreover, the expression of TrkA,TrkB and TrC mRNA has been reported for rat astrocytes and oligodendrocytes (4,5,19). The results obtained, however, were not unequivocal (22). In addition, non signaling competent truncated forms of TrkB and TrkC have also been found in glial cells (6,16,23).

Recently, a clear morphological response to NGF was demonstrated for pig oligodendrocytes: NGF enhanced the process formation of cultured regenerating OL and induced a subset of mature OL to proliferate (3). Yet it is unclear, whether an autocrine and/ or a paracrine mechanism provides for NGF in vivo. Evidence for an NGF production in OL has been given (7,17), however,

Molecular Signaling and Regulation in Glial Cells
A Key to Remyelination and Functional Repair

astrocytes may be one of the major source for releasing NGF (29,31). In MS lesions certain properties of astrocytes such as forming a glial scar (12), secreting a number of cytokines (1), and being an antigen presenting cell (34) are of particular importance. Hence, it would be of interest to know, whether NGF contributes to the number of signaling molecules, to which astrocytes respond. Therefore, we isolated and cultured mature pig astrocytes and exposed them to NGF after several days in culture. We found that none of the parameters tested were significantly influenced by NGF.

2 Material & Methods

2.1 Cell culture of astrocytes

The gray matter of mature pig brains was dissected and minced in icecold hexose (5% glucose, 5% fructose)-Hepes (20 mM) buffered medium, pH 7.4, the tissue was gently sieved through nylon screens of 300-, 150-, and 70-µm pore size.The resulting cell-tissue suspension was layered onto a percoll gradient and centrifuged at 10.000xg for 30 min. The layer just above the residual erythrocytes was aspirated and washed in BME/F12 two times by centrifugating at 1000 rpm for 10 min. The final cellular pellet (cell yield: 2-3x10^6 cells/g gray matter) was seeded into poly-D-lysine coated 35 mm Ø Petri dishes or 96 wells plates. The BME/F12 medium was supplemented with 10 % fetal calf serum (fcs) and insulin (5 mg/L) and transferrin (10 mg/L). The cultures were incubated in an atmosphere of 95% air/5% CO_2. Mezlocillin (Bayer) was used as antibiotic (40 µg/ml).

2.2 Thymidine incorporation

Astrocytes cultured onto 96 wells plates for at least 7-8 days and up to 13 days under experimental conditions were exposed to NGF(2.5 S,Sigma and human recombinant, Boehringer) for 48 hrs, in the last 24 hrs 1µCi [^{3}H]-thymidine (Amersham) was also given. After another 48 hrs, the cells (+/- NGF) were washed in PBS several times and finally lysed in a lysis buffer. An aliquot was taken for protein determination and scintillation counting.

2.3 Immunocytochemistry

Immunofluorescent staining was performed by using the following antisera: anti-GFAP (DAKO), anti-A2B5 and anti-GC (Boehringer). The staining protocol was carried out as previously described (2).

2.4 MAP kinase in-gel kinase assay

MAP kinase activity was documented by using an in gel kinase assay, which was based on previously described protocols (14,24). Briefly, the astrocytes were treated with or without NGF, and the proteins resolved by SDS-PAGE. The SDS-polyacrylamide gels contain 0,5 mg /ml MBP. After electrophoresis, the gels were washed in de-and renaturating solutions as described ( ). The proteins were thereafter incubated in a reaction buffer containing 7.5 µCi/ml γ-[^{33}P] ATP (3000 Ci/mmol, Amersham) for 60 min. at 30° C, the reaction was terminated with washings by a solution containing 5% TCA and 1% $Na_4P_2O_7$. After washing, autoradiography was performed.

3 Results

3.1 Culture of astrocytes with or without NGF

Astroglial cells have lost their processes after the isolation procedure. Approximately after 7 days in culture (DIC) they start to form processes, the formation of processes was not enhanced by NGF as seen in oligodendroglial cultures. After 9-10 DIC the first plaques of flat astroglial cells were present. Morphologically, 3 types of astrocytes can be observed: the predominant type is the flat cell, which finally covers the whole dish; some cells exhibit the stellate type, which can be partly induced by exposing flat astrocytes to a phorbol ester such as TPA or by reducing the fcs to 1%; in addition, very few cells were present, smaller in diameter, ramified and reminiscent to oligodendrocytes. All these types were GFAP$^+$ (Fig. 1).

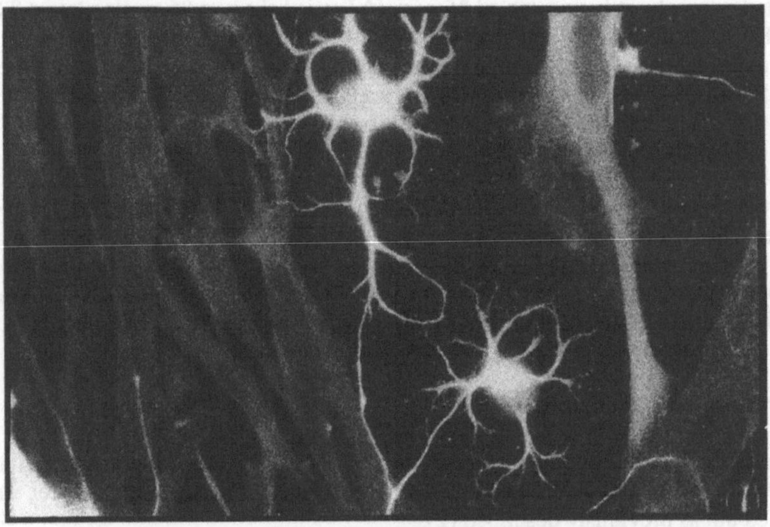

Fig. 1 Pig astrocytes after 14 days in culture: The predominant astroglial cell types are flat and stellate cells, GFAP staining

A2B5$^+$ cells, approximately 2-3 % of the initial cell population, can be observed. This cell type has never been detected in pig oligodendroglial cultures (dissected material: white matter of the corpus callosum and centrum semiovale). Under normal culture conditions only flat cells proliferate. NGF does not stimulate the rate of [^{3}H]-thymidine incorporation (Fig. 2).

4 Discussion

4.1 Astroglial cultures

Cultures of astrocytes, which were isolated from the gray matter of mature pig brains (avoiding the ventricular areas) revealed three $GFAP^+$ cell types. A flat type, corresponding to astrocyte type I (13), prevailed. Stellate cells, present under normal culture conditions, increased in number depending on the modifications of the culture medium. TPA as well as fcs reduction to 1% induced a certain percentage of the cells to become a more stellate morphology. In contrast, a fibrous morphology was not observed in rat astroglial cultures under serum reduced conditions unless NGF was added (20). However, previous reports showed that the morphology of cultured astrocytes may vary and is greatly dependent already on the type of media used (28).
Interestingly, a third $GFAP^+$ cell type was observed, which was smaller, ramified and could easily be mixed up with oligodendrocytes. It was already present after 1 week in culture (< 1% of the cell population) and did not change its morphology thereafter. Maybe, these cells represent a variant of fibrous astrocytes or a subpopulation of O-2A cells, which became $A2B5^-$. It differed from the $A2B5^+$ $GFAP^-$ cell type, which was also present. These cells emanated 3-4 processes, which gave them a different appearance to rat $O\text{-}2A^{adult}$ progenitors with their unipolar morphology (35).

4.2 Astroglial response to growth factors

Astrocytes synthesise and secrete various growth factors (1,31). Among these are, for example, bFGF and NGF. Concerning glial cells, a dual task of bFGF has been suggested: it might play a role in astroglial as well as in oligodendroglial growth and function (33). The significance of NGF in terms of rat glial cells is less clear. In contrast to pig oligodendrocytes, where NGF induces distinct effects (3), no comparable results have been reported for rat oligodendrocytes. This goes along with the, up to now, negative finding for TrkA in rat oligodendrocytes (22). However, two publications deal with NGF effects on cultured rat astrocytes. The one (36) reports on an increase of brain astrocytes when exposed to NGF suggesting that NGF either increased mitosis or enhanced survival of dividing cells. The other (20) describes a change in morphology under serum-free, NGF containing conditions. NGF did not stimulate astrocytic proliferation. In the first experimental design, a relatively high dosage of NGF (1-2µg/ml) is obvious. Hence, it is plausible that the effect observed was not transduced via the NGF high affinity receptor, TrkA, rather than with TrkB or TrkC, with which NGF interacts at higher concentrations (10). The morphological alteration observed in the second report seems to be different from the normal morphological change of flat to stellate astrocytes occurring under serum-free conditions (28). The change is only observed when NGF was added. The authors themselves discuss the possibility that this phenomenon does not represent differentiation rather than cell death. Interestingly, a recent report deals with the possibility that glial cells can undergo ceramide-dependent cell death under serum-deprived conditions (8). The signaling cascade for this event requires the p75 NGF receptor but not TrkA (11).

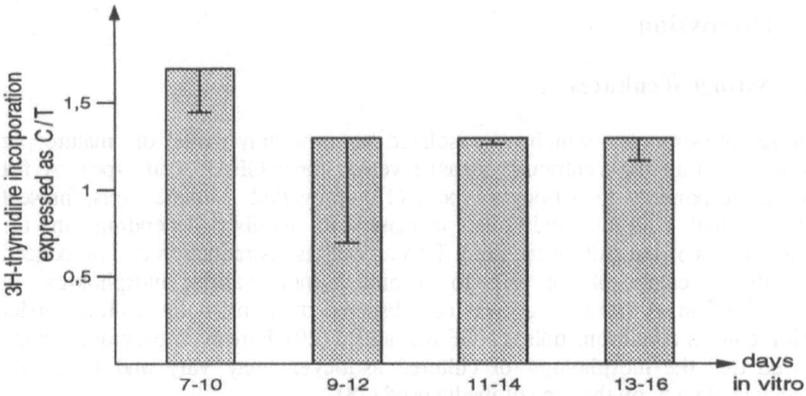

Fig.2 [^{3}H]-thymidine incorporation into astrocytes ± NGF expressed as control (C) versus NGF-treated cultures (T)

3.2 In-Gel kinase assay

By using antibodies against Erk1 and Erk2 (UBI-Biomol) two astroglial proteins of 44 kDa and 42 kDa were identified by Western blot analysis as Erk1 and Erk2. Treatment of astrocytes with NGF for 30 min. did not increase the Erk1 activity over control. In contrast, treatment with the protein kinase C activating phorbol ester TPA (30 min.) resulted in a marked increase of the Erk1 activity (Fig. 3).

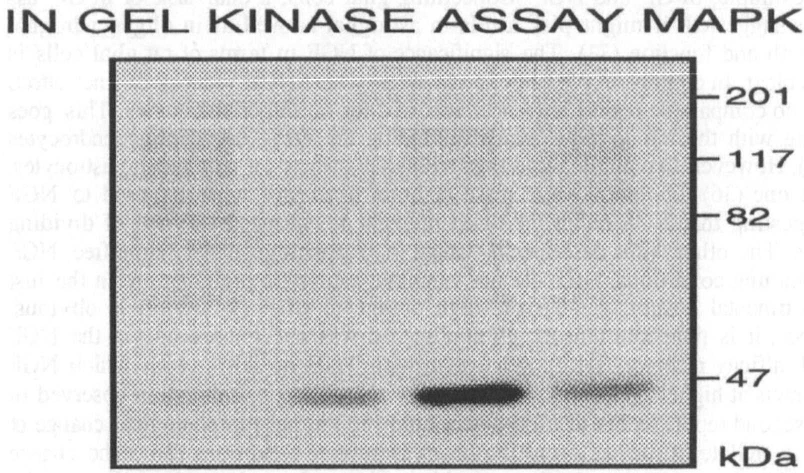

Fig.3 MAPK activity in astrocytes ± NGF and ± TPA as revealed by an in-gel kinase assay

In our astroglial cultur system, NGF does not induce a significant morphological change versus control conditions, at least for the period of observation. We have not tested NGF as to whether it influences astrocytes cultured serum-free. Furthermore, similar to the observation for rat astrocytes (20) no significant stimulation of the ^{3}H-thymidine incorporation could be detected.

4.3 Signal transduction by neurotrophins

Neurotrophins bind to the p75 receptor and to their respective tyrosine kinase receptors- the Trks (18). TrkB and TrkC either in full length or in truncated form have been shown to be expressed by astrocytes (5,6,16,22,23). In terms of TrkA the situation is less clear. No TrkA mRNA has been detected with Northern blots of astrocytes (22), however, RT-PCR provided evidence that astrocytes type I express TrkA (19). In addition, about an up-regulation of TrkA (and TrkB) in dorsal column astrocytes after injury was also reported (15). These authors suggest that their immunostainings might have picked up a truncated TrkA. However, a truncated TrkA has not yet been found in situ (18). Two months post-lesioning, an up-regulation of astroglial p75 and TrkA was observed in an experimental model, in which a progressive degeneration of target-deprived neurons in the rat thalamus occurred concomitant with a reactive gliosis. Interestingly, a full length TrkB was not found (21).

Unfortunately, no reports exist for astrocytes dealing with the signaling cascade downstream Trk. However, a part of the MAP kinase cascade, which is used by several growth factors, has been shown to be present and functioning: growth factors such as FGF, PDGF, and IGF, hormones, and phorbolesters such as TPA use this route (32). In our culture system, we could also show that the MAPK (p44) can be activated by TPA. On the other hand, no such activation was observed, when the cells were exposed to NGF. This suggests that NGF does not induce a TrkA dependent signaling path. Recent reports indicated that even the p75 receptor might be able to signal via a sphingomyelin cycle (11). Interestingly, neural cells display a differential sensitivity to this p75 directed path, which involves the SAPK cascade (8).

Conclusion:

We have cultured astrocytes isolated from mature pig brains. These cells were exposed to NGF. In the NGF concentrations used, no morphological response or an influence on astroglial proliferation was detected. Furthermore, NGF could not activate a component of the MAPK cascade, Erk1, which was otherwise activated by TPA. These results stand in contrast to those published for rat astrocytes. Hence, it is possible that not only oligodendrocytes but also astrocytes express receptors for growth factors species specifically.

References:

1. Aloisi F, Ceccherini Silberstein F, Levi G: Cytokine and growth factor synthesis in human astrocytes Proc. Ohrbeck (1996), this volume
2. Althaus HH, Schröter J, Spörri P, Schwartz P, Klöppner S, Rohmann A, Neuhoff V: Protein kinase C stimulation enhances the process formation of adult oligodendrocytes and induces proliferation J Neurosci Res 29 (1991) 481-489
3. Althaus HH, Klöppner S, Schmidt-Schultz T, Schwartz P: Nerve growth factor induces proliferation and enhances fiber regeneration in oligodendrocytes isolated from adult pig brain Neurosci Letters 135 (1992) 219-223
4. Althaus HH, Klöppner S, Kruska L, Heumann R: Neurotrophins affect cultured oligodendrocytes isolated from different species differently Proc 22nd Göttingen Neurobiol (1994) 122
5. Barres BA, Raff MC, Gaese F, Bartke I, Dechant G, Barde Y-A: A crucial role for neurotrophin-3 in oligodendrocyte development Nature 367 (1994) 371-375
6. Beck KD, Lamballe F, Klein R, Barbacid M, Schauwecker PE, McNeill TH, Finch CE, Hefti F, Day JR: Induction of noncatalytic trkB neurotrophin receptors during axonal sprouting in the adult hippocampus J Neurosci 13 (1993) 4001-4014
7. Byravan S, Foster LM, Phan T, Verity AN. Campagnoni AT: Murine oligodendroglial cells express nerve growth factor Proc Natl Acad Sci USA 91 (1994) 8812-8816
8. Casaccia-Bonnefil P, Aibal L, Chao MV: Central glial and neuronal populations display differential sensitivity to ceramide-dependent cell death J Neurosci Res 43 (1996) 382-389
9. Chao MV and Hempstead BL: p75 and Trk: A two-receptor system TINS 18 (1995) 321-326
10. Dechant G, Rodriguez-Tebar A, Barde Y_A: Neurotrophin receptors Progr Neurobiol 42 (1994) 347-352
11. Dobrowsky RT, Werner MH, Castellino AM, Chao MV, Hannun YA: Activation of the sphingomyelin cycle through the low-affinity neurotrophin receptor Science 265 (1994) 1596-1599
12. Eng L: Experimental models for astrocyte activation and fibrous gliosis In: HH Althaus&W Seifert (Eds.), Glial-Neuronal Communication in Development and Regeneration, NATO ASI Series, Springer, Heidelberg, 1987, pp 27-40
13. Fedoroff S: Prenatal ontogenesis of astrocytes In: S Fedoroff&A Vernadakis (Eds.) Astrocytes Vol 1, Academic Press, Orlando, 1986, pp 35-74
14. Ferrell JE and Martin GS: Identification of a 42-kilodalton phosphotyrosyl protein as a serine(threonine)protein kinase by renaturation Mol Cell Biol 10 (1990) 3020-3026
15. Foschini DR, Kestler AM, Egger MD. Crockett DP: The up-regulation of trkA and trkB in dorsal column astrocytes following dorsal rhizotomy Neurosci Letters 169 (1994) 21-24
16. Frisen J, Verge VMK, Fried K, Risling M, Persson H, Trotter J, Hökfelt T, Lindholm D: Characterization of glial trkB receptors: Differential response to injury in the central and peripheral nervous systems Proc Natl Acad Sci USA 90 (1993) 4971-4975
17. Gonzalez D, Dees WL, Hiney JK, Ojeda SR, Saneto RP: Expression of ?-nerve growth factor in cultured cells derived from the hypothalamus and cerebral cortex Brain Res 511 (1990) 249-258

18.Heumann R: Neurotrophin signalling Curr Opin Neurobiol 4 (1994) 668-679
19.Hutton LA, deVellis J, Perez-Polo JR: Expression of p75 ^{NGFR}, TrkA and TrkB mRNA in rat C6 glioma and type I astrocyte cultures J Neurosci Res 32 (1992) 375-383
20.Hutton LA,and Perez-Polo JR: In vitro glial responses to nerve growth factor J Neurosci Res 41 (1995) 185-196
21.Junier MP, Suzuki F, Onteniente B, Peschanski M: Target-deprived CNS neurons express the NGF gene while reactive glia around their axonal terminals contain low and high affinity NGF receptors Mol Brain Res 24 (1994) 247-260
22.Kumar S, Pene LA, deVellis JA: CNS glial cells express neurotrophin receptors whose levels are regulated by NGF Mol Brain Res 17 (1993) 163-168
23.Lamballe F, Tapley P, Barbacid M: trkC encodes multiple neurotrophin-3 receptors with distinct biological properties and substrate specificities EMBO J 12 (1993) 3083-3094
24.Leevers SJ and Marshall CJ: Activation of extracellular signal regulated kinase, Erk2, by p21 ras oncoprotein EMBO J 11 (1992) 569-574
25.Lindsay RM, Wiegand SJ, Altar CA, DiStefano PS: Neurotrophic factors: From molecule to man TINS 17 (1994) 182-190
26.Lomen-Hoerth C and Shooter EM: Widespread neurotrophin receptor expression in the immuns system and other nonneuronal rat tissues J Neurochem 64 (1995) 1780-1789
27.Marchetti D, Stach RW, Saneto R, deVellis J, Perez-Polo JR: Binding constants of soluble NGF-receptors in rat oligodendrocytes and astrocytes in culture Biochem Biophys Res Comm 147 (1987) 422-427
28.Michler-Stuke A: Putative autoregulative growth factor in conditioned medium of serum-free grown astrocytes In: HH Althaus&W Seifert (Eds.) Glial -Neuronal Communication in Development and Regeneration, NATO ASI Series, Springer, Heidelberg, 1987, pp 420-432
29.Norrgren G, Ebendal T, Belew M, Jacobson CO, Porath J: Release of nerve growth factor by human glial cells in culture Exp Cell Res 130 (1980) 31-39
30. Raffioni S,Bradshaw RA, Buxser SE, The receptor for nerve growth factor and other neurotrophins Annu Rev Biochem62 (1993) 823-850
31.Spranger M, Lindholm D, Bandtlow C, Heumann R, Gnahn H, Näjer-Noe M, Thoenen H: Regulation of nerve growth factor (NGF) synthesis in the rat central nervous system: comparison between the effects of interleukin-1 and various growth factors in astrocyte cultures and in vivo Europ L Neurosci 2 (1990) 69-76
32.Tournier C, Pomerance M, Gavaret JM,Pieere M: MAP kinase cascade in astrocytes Glia 10 (1994) 81-88
33.Vijayan VK, Lee Yl, Eng LE: Immunohistochemical localization of basic fibroblast growth factor in cultured rat astrocytes and oligodendrocytes Int J Dev Neurosci 11 (1993) 257-267
34.Weber F, Meinl E, Aloisi F, Nevinny-Stickel C, Albert E, Wekerle H, Hohlfeld R: Human astrocytes are only partially competent antigen presenting cells Brain 117 (1994) 59-69
35.Wren D, Wolswijk G, Noble M: In vitro analysis of the origin and maintenance of O-2A adult progenitor cells J Cell Biol 116 (1992) 167-176
36.Yokoyama M, Black IB, Dreyfus CF: NGF increases brain astrocyte number in culture Exp Neurol 124 (1993) 377-380

Endopeptidase-24.11/common acute lymphoblastic leukaemia antigen CD10 in Schwann cells: Evidence for a role in nerve development and regeneration

Rebecca Matsas and Stathis Meintanis

Department of Biochemistry, Hellenic Pastéur Institute, 127 Vas. Sofias Avenue, 115 21 Athens, Greece

Keywords. Surface molecule, Zn^{2+}-metalloprotease, regulation of peptide signals, Wallerian degeneration

1 Introduction

The regulation of peptide-induced responses in the nervous system is believed to occur through the action of cell-surface peptidases. A key enzyme in this process is endopeptidase-24.11 (E-24.11, EC 3.4.24.11, neprilysin), a 90kD cell surface glycoprotein which has the ability to hydrolyze a large number of physiologically active peptides. The enzyme is found in the nervous system and other peripheral organs and has been shown to be present in the surface membrane of many different cell types (for review see ref. 25). Molecular cloning and sequencing of E-24.11 cDNA from mouse, rat, rabbit and human revealed that it is a type II integral membrane protein which displays 90% identity in amino acid sequence between different species (8,12,31,33,34,48). These studies also demonstrated that E-24.11 is identical to the human common acute lymphoblastic leukemia antigen CD10 (31).

E-24.11 is an ectoenzyme with its catalytic domain projecting from the cell surface, well placed to hydrolyze peptide substrates released in its vicinity (for review see ref. 27). It belongs to the group of Zn^{2+}-metalloproteases, the active site of which shares the characteristic amino acid sequence HEXXH (where X stands for any amino acid) that constitutes a unique signature for this family of metalloproteases (18). Although E-24.11 was initially associated with the inactivation of the enkephalins and substance P (32,40), subsequent studies showed that the enzyme displays a broad *in vitro* specificity for a variety of regulatory peptides, such as other opioids and tachykinins (41), bradykinin and angiotensin II (15,41,51), atrial natriuretic factor (52), cholecystokinin (42), chemotactic peptide fMet-Leu-Phe (9). Furthermore, recent evidence suggests that such catalytic activity might be implicated in the *in vivo* biological regulation of the action of at least some of these peptides (16,35).

In the nervous system, E-24.11 is expressed by central, but not peripheral

Molecular Signaling and Regulation in Glial Cells
A Key to Remyelination and Functional Repair

neurons (39,43,54). In the central nervous system (CNS), the enzyme is found in restricted neuronal populations, most notably in the striatonigral pathway (4,43,45) and its physiological action has been associated in this tissue with the inactivation of the enkephalins and substance P (32,40).

In contrast to the intense interest in the study of E-24.11 in the central nervous system, little was known until recently, about its presence in the peripheral nervous system (PNS). In the course of mapping E-24.11 by immunocytochemistry in the pig CNS, it was observed for the first time that the enzyme, although present in the dorsal root ganglia and associated dorsal roots, it was not expressed by neuronal elements in these tissues (43). The E-24.11 immunoreactive cells were then tentatively identified as Schwann cells. This observation was subsequently verified in two independent studies performed on the sciatic nerve of the newborn pig (5,28).

Presence of E-24.11 on the surface of Schwann cells raised very intriguing implications concerning the role of the peptidase in the PNS. It is well established that Schwann cells have the unique ability to promote axonal growth during development and regeneration following injury. For this reason, the origin and development of the Schwann cell lineage has been extensively studied (22). It is now known that Schwann cells originate from a common precursor which gives rise to the immature Schwann cells of the developing peripheral nerves. These cells proliferate rapidly and then withdraw from the mitotic cycle to differentiate into the myelin-forming and the non-myelin-forming cells found in the adult nerves (55). Immature Schwann cells are directed to become myelin-forming by contact with large diameter axons, with which they form a one-to-one relationship (for review see ref. 44). Axonal contact specifying the myelin pathway also causes down-regulation of certain proteins present in the immature Schwann cells including N-CAM, A5E3, Ran-2, GAP43 and the NGF-receptor (10,20,21,37). However, these proteins persist in, and can be used as markers of, the mature non-myelin-forming Schwann cells (21). Interestingly, upon loss of normal axon-Schwann cell contact, these proteins become re-expressed by the entire Schwann cell population (10,11,17,20,21,38,53). This re-expression is likely to be important for nerve regeneration (23,36).

The establishment of Schwann cell markers, whose expression changes as Schwann cells differentiate is a very important step towards understanding the molecular mechanisms which underlie and regulate this process. We have therefore investigated the expression of E-24.11 during development of the rat sciatic nerve as well as during de- and regeneration of the adult nerve after injury. We have also attempted to define the factor(s) which regulate the expression of E-24.11 in these cells.

2 E-24.11 During Development of the Rat Sciatic Nerve

2.1 Prenatal Development

Expression of E-24.11 was studied by immunocytochemistry on enzymatically dissociated cells from sciatic nerves (7) which were maintained in culture for 3h. This was done to minimize any changes that might occur as a result of removal of the cells from axonal contact. Therefore, the immunostaining obtained should mirror closely that obtained *in vivo*. Presence of E-24.11 in these and all subsequent experiments described here was demonstrated using the anti-endopeptidase-specific monoclonal antibody 23B11 (2). This antibody was initially raised against the rabbit endopeptidase but has been subsequently shown to recognize effectively the enzyme from pig (28) and rat (29). The epitope for this antibody lies within the intracellular part of E-24.11 (2). It is therefore necessary to open the cells before its application so that the corresponding epitope becomes accessible for immunocytochemical detection.

E-24.11 immunoreactivity was first detectable at embryonic day 18 continuously rising up to five days postnatal (29). The type of cells expressing E-24.11 in cultures from 5-day old animals was identified by immunofluorescence double labeling with antibodies to E-24.11 and S-100 protein which was used as a Schwann cell marker (44). It was thus revealed that E-24.11 always co-localized with S-100 protein and was expressed by all Schwann cells at this age (29). Cells with fibroblastic morphology were always E-24.11 negative.

2.2 Postnatal Development

In order to identify the type of cells expressing E-24.11 and determine its distribution with respect to myelin-forming and non-myelin-forming Schwann cells during postnatal development, we used teased nerve preparations of the sciatic nerve. The teased nerves allow a clear distinction to be made between myelinated and unmyelinated fibres by use of phase-contrast optics.

First, the sciatic nerve from 5-day old rats was examined by immunofluorescence double labeling with E-24.11 and S-100 antibodies to specifically identify Schwann cells (Mirsky and Jessen, 1990). In these experiments it was shown that all the S-100-positive cells were also E-24.11-positive, in accordance with our data using cultured Schwann cells. It was thus revealed that, at this age, both unmyelinated and early myelinating fibres were immunostained by the antibody to E-24.11 (29).

In 18-day old rats, staining with the antibody to E-24.11 persisted in non-myelin-forming Schwann cells as well as in those forming the myelin sheath (29). However, by 31 days postnatal, labeling of myelin-forming Schwann cells was weak and appeared in distinct focal concentrations along the fibre, probably

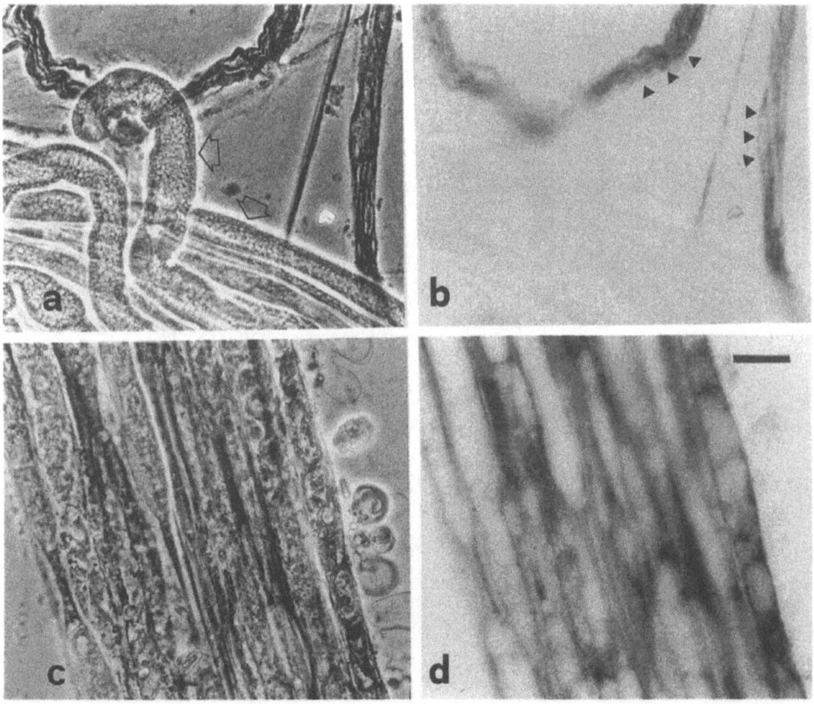

Fig. 1. Immunoperoxidase staining for E-24.11 in a teased fibre preparation of normal (a,b) and axotomized (c,d) adult rat sciatic nerve, 14 days after injury. Phase-contrast (left) and brightfield (right) optics were used to visualize E-24.11. The arrowheads in (a) illustrate two E-24.11-positive unmyelinated fibres in the normal nerve. Note the absence of E-24.11 staining in the myelinated fibres (thick arrows) of the normal nerve. In contrast, the myelinated fibres undergoing Wallerian degeneration in the distal part of the axotomized nerve are E-24.11-positive. Bar=20 μm.

corresponding to Schmidt-Lantermann incisures (29). This immunostaining was no longer detectable in the myelinated fibres of the adult animal (Fig. 1a,b and Fig. 2a-c). In contrast, non-myelin-forming Schwann cells exhibited E-24.11 immunoreactivity throughout development and in the adult (Fig.1a-c).

3 E-24.11 After Injury of the Adult Rat Sciatic Nerve

3.1 Nerve transection

To investigate how the expression of E-24.11 is regulated in Schwann cells after loss of axonal contact, we cut the sciatic nerve of adult rats and inhibited

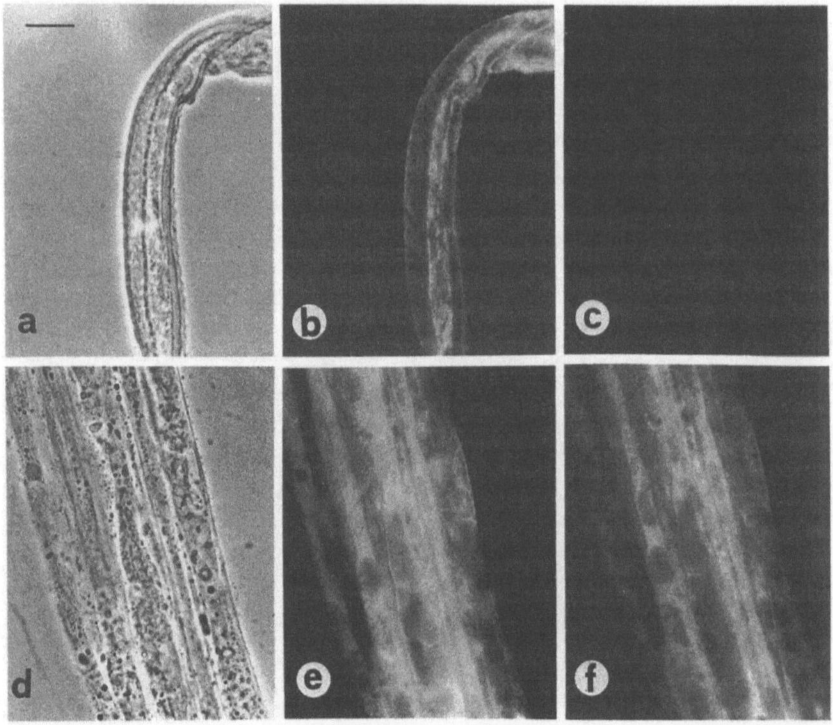

Fig. 2. Double-label immunofluorescence using antibodies to E-24.11 and S-100 protein in a teased fibre preparation of normal (a-c) and axotomized (d-f) adult rat sciatic nerve, 14 days post cut. Phase-contrast (a,d); rhodamine optics (b,e) and fluorescein optics (c,f) to visualize S-100 protein and E-24.11, respectively. Note absence of E-24.11 staining in the S-100-positive myelinated fibres of the normal nerve (a-c). In contrast, both antigens give an identical staining pattern in the degenerating nerve (d-f). Bar=20 μm.

regeneration by deflection of the proximal stump. Teased fibres of sciatic nerve were then immunostained in order to monitor the expression of E-24.11 in Schwann cells undergoing Wallerian degeneration. Four days after the transection most myelinated fibres in the distal segment of the sciatic nerve showed typical changes of Wallerian degeneration by breaking up into linear rows of myelin ovoids. At the same time the Schwann cells of these fibres containing myelin debris began to express E-24.11. The staining reached a maximum 14 days after injury (Fig. 1c,d). At this time, the denervated Schwann cell columns, which were readily identifiable either by phase-contrast optics or by double labeling with the antibody to S-100 protein (Fig. 2d-f), exhibited very strong E-24.11 immunoreactivity. No other cells present in the distal part of the transected nerve, appart from Schwann cells, exhibited E-24.11-immunoreactivity. In particular, absence of E-24.11 expression by macrophages

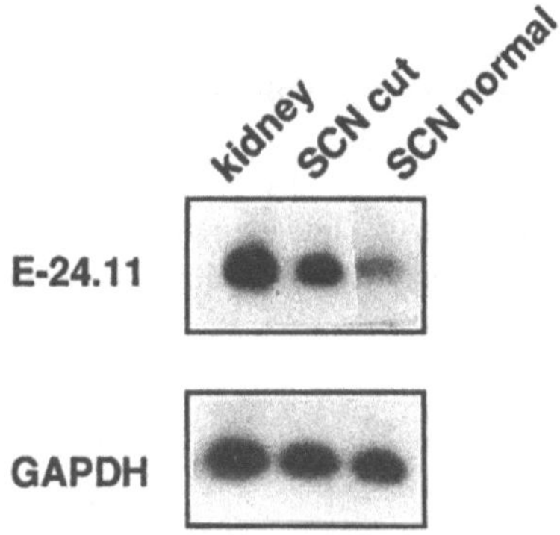

Fig. 3. PCR analysis of E-24.11 mRNA from rat kidney, sciatic nerve (SCN normal) and the distal segment of the transected sciatic nerve (SCN cut), 14 days after injury. Specific amplification of E-24.11 cDNA was performed and PCR products were hybridized with the E-24.11 probe (upper panel). A 468 bp band corresponding to the E-24.11 PCR products is apparent. GAPDH hybridization serves as control for loaded RNA (lower panel). The 190 bp band is shown.

invading at the site of lesion, was demonstrated in double labeling experiments in which the macrophage-specific marker OX-42 was used to identify these cells (not shown).

To confirm the molecular identity of the 23B11 immunoreactivity detected in the denervated distal segment of the sciatic nerve after transection, we performed RT-PCR analysis. RNA was isolated (3) from the distal part of the transected nerve 2 weeks after the cut, and from the contralateral non-injured sciatic nerve. RNA was also isolated from rat kidney which represents a rich source for E-24.11 (25) and served as positive control in these experiments (Fig.3). Specific amplification of E-24.11 cDNA was carried out using E-24.11-specific primers (30), and PCR products were hybridized with a rat full-length cDNA probe (6). With the help of this method we were able to confirm that E-24.11 is differentially expressed in normal and axotomized nerve (Fig. 3). Although PCR analysis is not suitable for RNA quantitation, it was interesting to note that the hybridization signal obtained for E-24.11 in the distal segment of the transected nerve was significantly higher than that obtained for the contralateral non-injured control nerve.

To further quantify the increase in E-24.11 expression after transection, we measured the enzyme activity of membrane preparations made from the distal part of the transected nerve 2 weeks after the cut, and from the contralateral non-injured sciatic nerve. E-24.11 activity was measured by a two-stage assay adopted for use on 96-well microtitration plates by using succinyl-Ala-Ala-Leu-4-nitroanilide as substrate supplemented with *Streptomyces griseus* aminopeptidase, as described (19). E-24.11 cleaves the peptide bond between the second alanine residue and leucine, whilst the exogenously added aminopeptidase liberates leucine. The proteolytic activity of E-24.11 was measured by following the increase in absorbance at 405nm, due to the release of 4-nitroaniline. The activity of the various fractions assayed is shown in Table 1. In all cases the activity measured could be completely inhibited by 1 μM phosphoramidon which is an E-24.11-specific inhibitor (40). It was thus shown that there is an approximately eight-fold increase in the levels of E-24.11 activity in the transected versus the control nerve. Interestingly, this increase corresponds to the endopeptidase levels previously determined in the developing sciatic nerve of the early postnatal period (29).

Table 1. Hydrolysis of suc-Ala-Ala-Leu-4-nitroanilide by E-24.11 in membrane fractions

Tissue	specific activity (nmol.min.mg^{-1} of protein)	% phosphoramidon inhibition
Sciatic nerve, normal	3.2±0.1	>99
Sciatic nerve, cut *	22.4±0.9	>99
kidney microvilli	290.0±10	>99

Sciatic nerve and kidney membrane fractions were prepared as described (29).
* Membranes were prepared from the distal part of the nerve 2 weeks after transection.

3.2 Nerve Crush

To determine whether E-24.11 expression could again be repressed in Schwann cells as a result of axonal regeneration, teased fibres of the distal segment were immunostained for E-24.11 after sciatic nerve crush. This procedure causes Wallerian degeneration of the fibres distal to the crush but preserves the continuity of the endoneurium, which allows subsequent axonal regeneration. In agreement with the results obtained after nerve transection, the myelinated fibres undergoing Wallerian degeneration showed up-regulation of E-24.11 (Fig. 4a,b). Subsequently, E-24.11 expression was suppressed in the newly myelinated fibres of the reinnervated distal segment three months after crush injury (Fig. 4c,d). These results suggest that the expression of E-24.11 by Schwann cells depends on their contact with the axon and seems to be regulated by axon-Schwann cell signals.

4 Regulation of E-24.11 expression in Schwann cells

To determine the factors which regulate E-24.11 expression in Schwann cells, we investigated the ability of nerve growth factor (NGF) and tumour necrosis

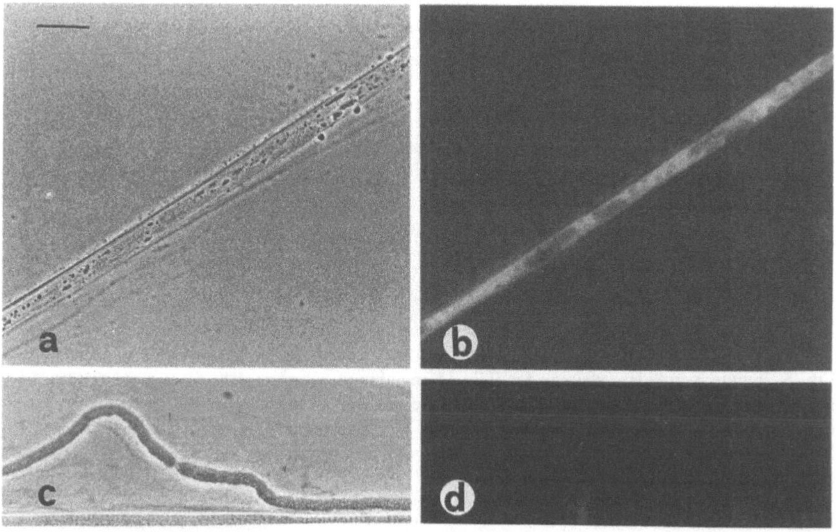

Fig. 4. E-24.11 immunofluorescence in a teased fibre preparation of the adult rat sciatic nerve after crush injury. Phase-contrast (left) and fluorescein (right) optics were used to visualize E-24.11, 14 days (a,b) and three months (c,d) after nerve crush. Bar=20 μm.

factor (TNF) to modify the endopeptidase levels in cultured Schwann cells. Both of these factors are known to be present in large amounts in the developing and regenerating peripheral nerves (17,46). For this purpose, Schwann cells were cultured (7) in the presence of either NGF (100ng/ml) or TNF (10ng/ml) which were added in the culture medium at the time of plating. The expression of E-24.11 was initially monitored by imunocytochemistry on treated and non-treated cells. An increase in E-24.11 immunoreactivity was evident in the NGF (Fig. 5) but not in the TNF treated cells, maintained for 1 or 3 days *in vitro.* This increase could be blocked by antibodies to NGF added to the cultures at the same time as NGF (Fig. 5).

To further analyze the stimulatory effect of NGF on E-24.11 expression, we performed RT-PCR analysis as described above for the degenerating nerve. That is, RNA was extracted from Schwann cells cultured for one day in the presence or absence of NGF and the generated PCR products were identified by Southern blot analysis using the specific E-24.11 probe. It was thus revealed that the hybridization signal was much higher in the NGF-treated than in the non-treated cells (Fig. 6). On the other hand, GAPDH (14) PCR products which were generated under identical conditions yielded a lower hybridization signal in the NGF-treated as compared to the non-treated cells (Fig. 6). We conclude that NGF may represent one of the axonal signals which regulates the expression of E-24.11 in Schwann cells.

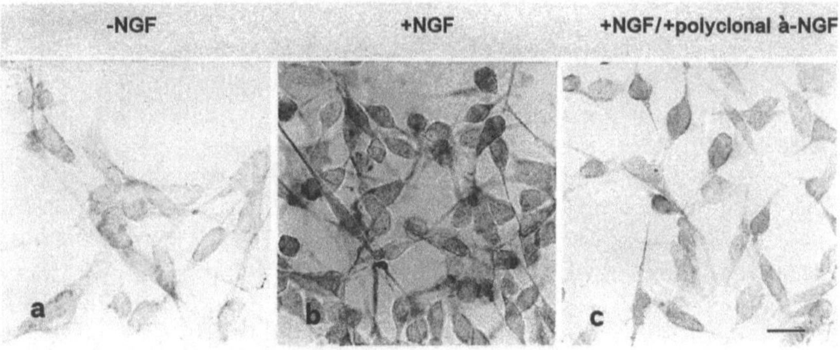

Fig. 5. Immunoperoxidase staining for E-24.11 in cultured Schwann cells from postnatal day five nerves, maintained for 1 day *in vitro*. Cells were cultured in the absence (a) or in the presence (b,c) of NGF. A neutralizing anti-NGF antibody was also included in (c). Note the stronger E-24.11 immunoreactivity in the presence of NGF which is blocked by the action of the anti-NGF antibody. Bar=20 μm.

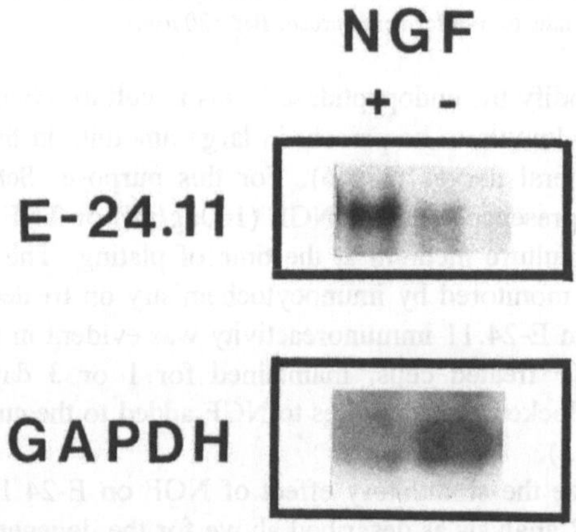

Fig. 6. PCR analysis of E-24.11 mRNA from cultured Schwann cells prepared from postnatal day five nerves, maintained for 1 day *in vitro*. Cells were cultured in the presence (+) or absence (-) of NGF. Specific amplification of E-24.11 cDNA was performed and PCR products were hybridized with the E-24.11 probe (upper panel). A 468 bp band corresponding to the E-24.11 PCR products is apparent. GAPDH hybridization serves as control for loaded RNA (lower panel). The 190 bp band is shown.

5 Discussion

The data presented here on the expression and distribution of E-24.11 in the peripheral nervous system have unraveled a potential, previously unrecognized role for this enzyme in nerve development and regeneration. The expression of E-24.11 in the sciatic nerve follows a particular pattern in time and space. That is, the enzyme is present in all neonatal and early postnatal Schwann cells and in mature non-myelin-forming Schwann cells, but is down-regulated in mature myelin-forming cells (29). Moreover, we have obtained evidence that the expression of E-24.11 is induced *in vivo* after lesion of the adult rat sciatic nerve. Collectively, our results demonstrate that E-24.11 belongs to the group of the so called Schwann cell differentiation antigens, such as the low affinity NGF receptor, growth-associated protein GAP-43 and the cell adhesion molecules NCAM and L1 (for references, see the Introduction). Interestingly, all these molecules have a similar to the endopeptidase expression pattern during development and also become re-expressed by the entire population of Schwann cells after nerve injury. Therefore they are considered to be critical for nerve regeneration. Upon completion of axonal regeneration, as during nerve maturation, these molecules are repressed in myelinated fibres but remain associated with the unmyelinated fibres in the adult animal.

The profile of E-24.11 expression during peripheral nerve development and regeneration has been studied both by immunocytochemical means and by measuring in each case enzyme activity (29,30 and this study). Both methods yielded identical results demonstrating that during development as well as during nerve repair there is a similar down-regulation of E-24.11 activity. This observation is relevant to any hypotheses put forward for the possible role of E-24.11 in the peripheral nervous system. Hence, we propose that our data support the notion that E-24.11 may participate in mechanisms underlying nerve development and repair. Alternatively, it is also plausible to assume that E-24.11 may be specifically associated with the Schwann cell response to injury and in particular with the proliferation and migration of these cells, events which take place during both nerve development and repair. Of interest, it has been previously shown that the low-affinity NGF receptor is involved in Schwann cell migration (1) and accordingly, it is possible to hypothesize a similar role for E-24.11. To test these alternative hypotheses we are currently attempting to develop an *in vivo* model by which to perturb the action of E-24.11 in the PNS and monitor possible deviations from physiological development. To this end, we have also generated oligoclonal antibodies against synthetic peptides modelled form the active site of E-24.11 (49) and intend to use them in such a system. It is worthwile mentioning that an involvement of E-24.11 in ontogenetic events, such as the craniofacial development, has been previously reported (50).

The mechanism by which E-24.11 may participate in nerve development and regeneration awaits and merits clarification. However, previous studies have shown that E-24.11 functions in many systems to regulate peptide-induced responses, in both vertebrate and invertebrate species. For example, E-24.11 has been implicated in the inactivation of opioid peptides and substance P in neurons in the brain (32,40), of the chemotactic peptide fMet-Leu-Phe on polymorphonuclear granulocytes (9), and of a variety of regulatory peptides on the surface of the proximal tubule epithelial cells of the kidney (51,52). E-24.11 has also been shown to be involved in peptide-mediated inflammatory responses and in T cell activation and the regulation of interleukin-2 production (35). It is possible that E-24.11 has a similar regulatory role with respect to peptide-induced responses that control axonal growth in the developing and regenerating peripheral nerves. In this context, it is interesting to consider that a large number of neuropeptides, including the enkephalins and substance P, have been identified in the peripheral nervous system, some of which may act as neurotransmitters whilst others may have trophic effects (24). In the CNS, and in particular in the striatonigral pathway, there are convincing arguments that E-24.11 regulates peptide signals, especially those evoked by the enkephalins and substance P. It is therefore tempting to speculate that one of the physiological roles of E-24.11 during nerve development and regeneration is associated with the regulation of the action of these or other neuropeptides.

Of particular interest is our finding reported here, that Schwann cells respond to NGF by increased surface expression of E-24.11. This means that Schwann cells are directed by NGF to produce a quantitative change in a physiologically significant parameter. A similar observation has been made before for the cell adhesion molecule L1 (47). This function should presumably be mediated through the p75 low-affinity NGF receptor present on the Schwann cell surface (23). The role of this receptor in signal transduction has remained largely undefined for many years. Recently, it was reported that NGF can activate in T9 glioma cells the sphingomyelin cycle, generating the putative second messenger ceramide (13). This effect was clearly shown to be mediated through the p75 low-affinity NGF receptor. Whether a similar cascade of events takes place in Schwann cells as well, remains to be established. In this respect the E-24.11 model which we have analyzed should prove to be an invaluable tool and should assist in dissecting the molecular and cellular mechanisms of peripheral nerve development and regeneration.

Acknowledgments. We wish to thank P. Crine, L.B. Hersh and S. Blumberg for their generous gift of the antibody to the endopeptidase, the rat E-24.11 cDNA probe and the substrate suc-Ala-Ala-Leu-4-nitroanilide, respectively. This work was supported by the Greek General Secretariat of Research and Technology and the EU grants Biotechnology BIO2-CT930326 and Biomedicine and Health BMH-CT941378.

References

1. Anton, E.S., Weskamp, G., Reichardt, L.F. and Matthew, W.D. (1994) Nerve growth factor and its low-affinity receptor promote Schwann cell migration. Proc. Natl. Acad.Sci. USA 91, 2795-2799.
2. Aubry, M., Crine, P., Fortin, S., Legrimellec, C. Venien, C. and Zollinger, M. (1988) Monoclonal antibodies as probes for the transmembrane structure of neutral endopeptidase-24.11 (enkephalinase). Biochim. Biophys. Acta 967, 56-64.
3. Auffray, C. and Rougeon, F. (1980) Purification of mouse immunoglobulin heavy-chain messenger RNAs from total myeloma tumour RNA. Eur. J. Biochem. 107, 303-314.
4. Barnes, K., Matsas, R., Hooper, N. M., Turner, A.J. and Kenny, A.J. (1988) Endopeptidase-24.11 is striosomally ordered in pig brain and in constrast to aminopeptidase N and peptidyl dipeptidase A (angiotensin converting enzyme) is a marker for a set of striatal efferent fibres. Neuroscience 27, 799-817.
5. Barnes, K., Bourne, A., Cook, P.A., Turner, A.J. and Kenny, A.J. (1991) Membrane peptidases is the peripheral nervous system of the pig their localization by immunohistochemistry at light and electron microscopic levels. Neuroscience 44, 245-261.
6. Bateman, R.C., Jackson, D., Slaughter, C.A., Unnithan, S., Chai, Y.G., Moomaw, C. and Hersh, L.B. (1989) Identification of the active site arginine in rat neutral endopeptidase 24.11 (enkephalinase) as arginine 102 and analysis of a glutamine 102 mutant. J. Biol. Chem. 264, 6151-6157.
7. Brockes, J.P., Fields, K.L. and Raff, M.C. (1979) Studies on cultured rat Schwann cells. Establishment of purified populations from cultures of peripheral nerve. Brain Res. 165, 105-118.
8. Chen, C.Y., Salles, G. Seldin, M.F., Kister, A.E., Reinherz, E. and Shipp, M.A. (1992) Murine common acute lymphoblastic leukemia antigen (CD10 neutral endopeptidase-24.11). Molecular characterization, chromosomal localization and modelling of the active site. J. Immunol. 148, 2817-2825.
9. Connelly, J.C., Skidgel, R.A., Schulz, W.W., Johnson, A.R. and Erdos, E.G. (1985). Neutral endopeptidase 24.11 in human neutrophils: cleavage of chemotactic peptide. Proc. Natl. Acad. Sci. USA 82, 8737-8741.
10. Curtis, R., Stewart, H.J.S., Hall, S.M., Wilkin, G.P., Mirsky, R. and Jessen, K.R. (1992) GAP-43 is expressed by non-myelin-forming Schwann cells of the peripheral nervous system. J. Cell Biol. 116, 1455-1464.
11. Daniloff, J.K., Levi, G., Grumet, G., Rieger, F. and Edelman, G.M. (1986) Altered expression of neuronal cell adhesion molecules induced by nerve injury and repair. J. Cell. Biol. 103, 929-945.
12. Devault, A., Lazure, C., Nault, C., Le Moual, H., Seidah, N.G., Chretien, M., Kahn, P., Powell, J., Mallet, J., Baumont, A., Roques, B., Crine, P. and Boileau, G. (1987) Amino acid sequence of rabbit kidney neutral endopeptidase-24.11 (enkephalinase) deduced from a complementary DNA. EMBO J. 6, 1317-1322.

13. Dobrowsky, R.T., Werner, M.H., Castellino, A.M., Chao, M.V. and Hannun, Y.A. (1994) Activation of the sphingomyelin cycle through the low-affinity neurotrophin receptor. Science 265, 1596-1599.

14. Fort, P., Marty, L., Piechaczyk, M., Sabrouly, S.E., Dani, C., Janteur, P. and Blanchard, J.M. (1985) Various rat adult tissues express only one major mRNA species from the glutaraldehyde-3-phosphate-dehydrogenase multigenic family. Nucleic Acids Res. 13, 1431-1442.

15. Gafford, T.J., Skidgel, R.A., Erdos, E.G. and Hersh, L.B. (1983) Human kidney enkephalinase: a neutral metallo-endopeptidase that cleaves active peptides. Biochemistry 22, 3265-3271.

16. Gros, C., Souque, A., Schwartz, J.C., Duchier, J., Cournot, A., Baumer, P and Leconte. J.M. (1989) Protection of atrial natriuretic factor against degradation: diurectic and natriuretic responses after in vivo inhibition of enkephalinase (EC 3.4.24.11) by acetorphan. Proc. Natl. Acad. Sci. USA 86, 7580-7584.

17. Heumann, R., Lindholm, D., Bandtlow, C., Meyer, M., Radeke, M.J., Misko,T.P., Shooter, E. and Thoenen, H. (1987) Differential regulation of nerve growth factor (NGF) and NGF-receptor mRNA in the rat sciatic nerve during development, degeneration and regeneration: role of macrophages. Proc. Natl. Acad. Sci. USA 84, 8735-8739.

18. Hooper, N.M. (1994) Families of zinc metalloproteases. FEBS Lett. 354, 1-6.

19. Indig, F.E., Ben-Meir, D., Spungin, A. And Blumberg S. (1989) Investigation of neutral endopeptidases (EC 3.4.24.11) and of neutral proteinases (EC 3.4.24.4) using a new sensitive two-stage enzymatic reaction. FEBS Lett. 255, 237-240.

20. Jessen, K.R., Mirsky, R. and Morgan, L. (1987) Myelinated but not unmyelinated axons reversibly down-regulate N-CAM in Schwann cells. J. Neurocytol. 16, 681-688.

21. Jessen, K.R., Morgan, L., Stewart, H.T.S. and Mirsky, R. (1990) Three markers of adult non-myelin forming Schwann cells 217c (Ran-1), A5E3 and GFAP: development and regulation by neuron-Schwann cell interactions. Development 109, 91-103.

22. Jessen, K.R., Brennan, A., Morgan, L., Mirsky, R., Kent, A., Hashimoto, Y. and Gavrilovic J. (1994) The Schwann cell precursor and its fate: A study of cell death and differentiation during gliogenesis in rat embryonic nerves. Neuron 12, 509-527.

23. Johnson, E.M., Taniuchi, M. and Di Stefano, P.S. (1988) Expression and possible function of nerve growth factor receptors on Schwann cells. Trends Neurosci. 11, 299-303.

24. Ju, G., Hokfelt, T., Brodin, E., Fahrenkrug, J., Fischer, J.A., Frey, P., Elde, P.R. and Brown, J.C. (1987) Primary sensory neurons of the rat showing calcitonin gene-related peptide immunoreactivity and their relation to substance P-, somatostatin-, galanin-, vasoactive intestinal polypeptide- and cholecystokinin-immunoreactive ganglion cells. Cell Tissue Res. 247, 417-431.

25. Kenny, A.J. (1986) Cell surface peptidases are neither peptide- nor organ-specific. Trends Biochem. Sci. 11, 40-42.
26. Kenny, A.J. and Bourne, A. (1991) Cellular reorganization of membrane peptidases in Wallerian degeneration of pig peripheral nerve J. Neurocytol. 20, 875-885.
27. Kenny, A. J., Stephenson, S.L. and Turner, A.J. (1987) Cell surface peptidases. In Kenny, A.J. and Turner, A.J. (eds), Mammalian Ectoenzymes, Elsevier, Amsterdam, pp. 169-210.
28. Kioussi, C. and Matsas, R. (1991) Endopeptidase-24.11, a cell-surface peptidase of central nervous system neurons, is expressed by Schwann cells in the pig peripheral nervous system. J. Neurochem. 57, 431-440.
29. Kioussi, C., Crine, P. and Matsas, R. (1992) Endopeptidase-24.11 is suppressed in myelin-forming but not in non-myelin-forming Schwann cells during development of the rat sciatic nerve. Neuroscience 50, 69-83.
30. Kioussi, C., Mamalaki, A., Jessen, K.R., Hersh, L.B., Mirsky, R. and Matsas, R. (1995) Expression of endopeptidase-24.11 (common acute lymphoblastic leukaemia antigen CD10) in the sciatic nerve of the adult rat after lesion and during regeneration. Eur. J. Neurosci. 7, 951-961.
31. Letarte, M., Vera, S., Tran, R., Addis, J.B., Omizuka, J.R., Quackenbush, E.J., Jangenel, C.V. and Mclnnis, R.R. (1988) Common acute lymphocytic leukemia antigen is identical to neutral endopeptidase. J. Exp. Med., 168, 1247-1253.
32. Malfroy, B., Swertz, J. P., Guyon, A., Roques, B.P. and Schwartz, J.C. (1978) High-affinity enkephalin-degrading peptidase in brain is increased after morphine. Nature 276, 523-526.
33. Malfroy, B., Schofield, P.R., Kuang, W.G., Seeburg, P.H., Mason, A.J. and Henzel, W.J. (1987) Molecular cloning and amino acid sequence of rat enkephalinase. Biochem. Biophys. Res. Commun. 144, 59-66.
34. Malfroy, B., Kuang, W.J., Seeburg, P.H., Mason, A.J. and Schofield, P.R. (1988) Molecular cloning and aminoacid sequence of human enkephalinase (neutral endopeptidase) FEBS Lett. 229, 206-210.
35. Mari, B., Checler, F., Ponzio, G., Peyron, J.F., Manie, S., Farahifar, D., Rossi, B. and Auberger, P. (1992) Jurkat T cells express a functional neutral endopeptidase activity (CALLA) involved in T cell activation EMBO J. 11, 3875-3885.
36. Martini, R. (1994) Expression and functional roles of neural cell surface molecules and extracellular matrix components during development and regeneration of peripheral nerves. J. Neurocytol. 23, 1-28.
37. Martini, R. and Schachner, M. (1986) Immunoelectron microscopic localization of neural cell adhesion molecules (L1, NCAM and MAG) and their shared carbohydrate epitope and myelin basic protein in developing sciatic nerve. J. Cell Biol. 103, 2439-2448.
38. Martini R. and Schachner, M. (1988) Immunoelectron microscopic localization of neural cell adhesion molecules (L1, N-CAM and myelin associated glycoprotein) in regenerating adult mouse sciatic nerve. J. Cell Biol., 106, 1735-1746.

39. Matsas, R. and Kenny, A.J. (1989) Immunocytochemical localization of endopeptidase-24.11 in cultured neurons from pig striatum. Neuroscience, 31, 237-246.

40. Matsas, R., Fulcher, I.S., Kenny, A.J. and Turner, A.J. (1983) Substance P and [leu]enkephalin are hydrolyzed by an enzyme in pig caudate synaptic membranes that is identical with the endopeptidase of kidney microvilli. Proc. Natl. Acad. Sci. USA 80, 3111-3115.

41. Matsas, R., Kenny, A.J. and Turner, A.J. (1984a) The metabolism of neuropeptides. The hydrolysis of peptides including enkephalins, tachykinins and their analogues, by endopeptidase-24.11. Biochem. J. 223, 433-440.

42. Matsas, R., Kenny, A.J. and Turner, A.J. (1984b) Endopeptidase-24.11 and aminopeptidase activity in brain synaptic membranes are jointly responsible for the hydrolysis of cholecystokinin octapeptide. FEBS Lett. 175, 124-128.

43. Matsas, R., Turner, A.J. and Kenny, A.J. (1986) Immunocytochemical localization of endopeptidase-24.11 (enkephalinase) in the pig nervous system. Neuroscience 18, 991-1012.

44. Mirsky, R. and Jessen, K.R. (1990) Schwann cell development and the regulation of myelination. Semin. Neurosci. 2, 423-435.

45. Pollard, H., Bouthenet, M.L., Moreau, J., Souil, E., Verroust, P., Ronco, P. and Schwarz, J.C. (1989) Detailed immunoautoradiographic mapping of enkephalinase (EC 3.4.24.11) in rat central nervous system: comparison with enkephalins and substance P. Neuroscience 30, 339-376.

46. Schneider-Schaulies, J., Kirchhoff, F., Archelos, J. and Schachner, M. (1991) Down-regulation of myelin-associated glycoprotein on Schwann cells by interferon-α and tumour necrosis factor-α affects neurite outgrowth. Neuron 7, 995-1005.

47. Seilheimer, B. and Schachner M. (1987) Regulation of neural cell adhesion molecule expression on cultured mouse Schwann cells by nerve growth factor. EMBO J. 6, 1611-1616.

48. Shipp, M.A., Richardson, N.E., Sayre, P.H., Brown, N.R., Masteller, E.L., Clayton, L.K., Ritz, J. and Reinherz, E.L. (1988) Molecular cloning of the common acute lymphoblastic leukemia antigen (CALLA) identifies a type II integral membrane protein. Proc. Natl. Acad. Sci. USA 85, 4819-4823.

49. Soteriadou, K.P., Tzinia, A.K., Panou-Pomonis E., Tsikaris, V., Sakarellos-Daitsioti, M., Sakarellos, C., Papapoulou, Y. and Matsas, R. (1995) Antigenicity and conformational analysis of the zinc binding site of two zinc-metalloproteases: *Leishmania* gp63 and mammalian endopeptidase-24.11. Biochem. J. in press.

50. Spencer-Dene, B., Thorogood, P., Nair, S., Kenny, A.J., Harris, M and Henderson B. (1994) Distribution and a putative role for, the cell-surface neutral metallo-endopeptidases during mammalian craniofacial development. Development 120, 3213-3226.

51. Stephenson, S.L. and Kenny, A.J. (1987a) Metabolism of neuropeptides: hydrolysis of the angiotensins, bradykinin, substance P and oxytocin by pig kidney microvillar membranes. Biochem. J. 241, 237-243.
52. Stephenson, S.L. and Kenny, A.J. (1987b). The hydrolysis of a human atrial natriuretic peptide by pig kidney microvillar membranes is initiated by endopeptidase-24.11. Biochem. J. 243, 183-187.
53. Taniuchi, M., Clark, H.B. and Johnson, E.M. (1986) Induction of nerve growth factor receptor in Schwann cells after axotomy. Proc. Natl. Acad. Sci. USA 83, 4094-4098.
54. Waksman, G. Hamel, E., Delay-Goyet, P. and Roques, B.P. (1986) Neuronal localization of the neutral endopeptidase enkephalinase in rat brain revealed by lesions and autoradiography. EMBO J. 5, 3163-3166.
55. Webster, H. De F. and Favilla, J.T. (1984) Development of peripheral nerve fibres. In Dyck, P.J., Thomas, P.K., Lambert, E.H. and Bunge, R.P. (eds), Peripheral Neuropathy, W.B. Saunders, Philadelphia, pp.329-359.

Regulation of Oligodendrocyte Development by the LIF Cytokine Subfamily: Studies of the Neonatal Lineage Cultured from Rat Cerebrum

Anthony Gard, Wesley Williams II and Martha Burrell

Department of Structural and Cellular Biology
University of South Alabama College of Medicine
Mobile, AL 36688, USA

Keywords. astrocyte, ciliary neurotrophic factor, leukemia inhibitory factor, oligodendroblast, oligodendrocyte, oligodendrogliotrophic factor, oncostatin-M, platelet-derived growth factor

1 Introduction

We are investigating the differentiation potential of oligodendrocyte precursor cells and their dependence on trophic factors during defineable stages leading to CNS myeli-nation. Our studies concern the lineage response to a growing cytokine subfamily currently represented by ciliary neurotrophic factor (CNTF), leukemia inhibitory factor (LIF) and oncostatin-M, all of which are related by activation of a common transmembrane receptor heterodimer, signal transducers gp130 and LIFR (41). Previous in vitro studies have shown that CNTF and LIF can have opposing effects on oligodendrocyte development (Figure. 1.1). They can divert bipotential glial proge-

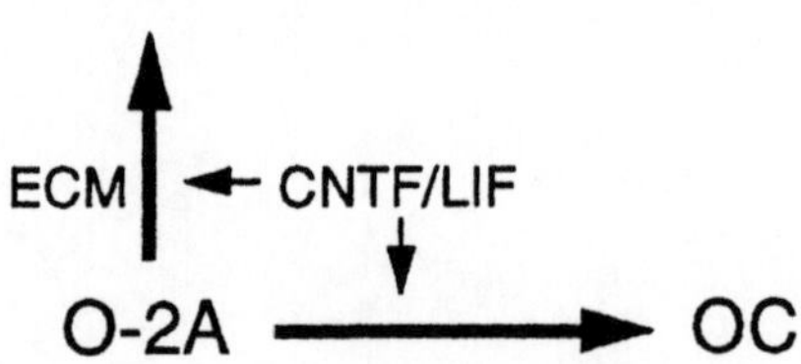

Fig. 1.1. Pleiotropic effects of CNTF and LIF on oligodendrocyte progenitors considered as a single O-2A lineage compartment. Accordingly, contact with an ECM component determines whether these cytokines function to induce type-2 astrocyte (A) formation or promote oligodendrocyte (OC) development.

Molecular Signaling and Regulation in Glial Cells
A Key to Remyelination and Functional Repair

glial progenitors (O-2A) described in cultures of neonatal optic nerve (reviewed in 34) to an astrocytic phenotype if presented in combination with an extracellular matrix (ECM) component of cultured meningeal or bovine aortic endothelial cells (26,29). Alterna-tively, in the absence of prepared ECM, they can promote development into oligodendrocytes as cues enhancing cell survival (5,29) and differentiation (29).

This brief report describes our ongoing examination of factors modulating oligodendrocyte development modeled through in vitro studies of the neonatal lineage in rat cerebrum. We show how sampling progenitors at successive, phenotypically distinct stages from tissue by antibody capture (15) has enabled the identification of two functionally distinct progenitor intervals within the O-2A lineage, an early migratory, bipotential progenitor which responds to CNTF, LIF and OS-M as astrocyte-inducing agents, and a late progenitor committed to oligodendrocyte differentiation instead (18). Further, we review evidence for soluble factors enabling oligodendroblasts to survive terminal differentiation and maturation and how the search for such agents in our culture system has identified astrocytes as a specific and highly efficacious source of activity consisting of at least two components, platelet derived growth factor (PDGF) and a LIF-like protein (14).

2 Differentiation Potential

Earlier studies of cultured optic nerve glia described fetal bovine serum (FBS) as the source of proteinaceous activity inducing O-2A progenitors to become type-2 astrocytes (35). The search for a neural equivalent of the serum stimulus led to a line of evidence suggesting a polypeptide similar, if not identical, to CNTF drives the formation of a second, functionally distinct wave of astrocytes during gliogenesis (reviewed in 25). Notably, induction by CNTF differed from FBS in several respects. First, CNTF was observed to be the less effective stimulus, inducing with increasing postnatal age the expression of glial fibrillary acidic protein (GFAP) in a progressively smaller subpopulation of FBS-responsive O-2A progenitors (19). Second, for O-2A progenitors grown apart from other cell types, astrocyte induction by CNTF, unlike FBS, was restricted to cells grown on a substratum-bound ECM component of cultured meningeal cells types (26) traced to vascular endothelium (26,29). Third, attempts to characterize the astrocyte-inducing molecule in FBS and similar activity in human plasma identified multiple polypeptides by gel filtration analysis, none of which corresponded to CNTF in size (24). These observations cautioned that lineage bipotentiality under regulation by CNTF in vivo could be overestimated by serum responsiveness in culture.

Resurgent evidence for the existence of separate astrocyte and oligodendrocyte lineages in the developing rat CNS comes from histologic investigations demonstrating type-2 astrocytes defined by in vitro criteria fail to arise in vivo (13,39). While fate studies of subventricular zone precursor cells examined by retroviral tracing also found no evidence of a *major* bipotential O-2A pathway operative in gliogenesis, a subpopulation of oligodendrocytes and astrocytes was indeed concluded to derive from a common presursor according to a temporally and spatially restricted developmental program (22,23).

2.1 Properties of Oligodendrocyte Progenitors in Vivo

Supplanted by the O-2A hypothesis, the concept of a committed oligodendroblast arose from electron microscopic and thymidine autoradiographic studies of postnatal cerebrum and optic nerve (2, reviewed in 39). Classified by subtle ultrastructural features, this progenitor was distinguished from earlier glial precursors as being newly post-migratory and completing the proliferative phase of myelinogenesis (20). Given the reduced role of a bipotential O-2A pathway now foreseen in gliogenesis, antibodies to cell surface components [A2B5, GD3, O4, galactocerebroside (GalC)] should be considered as defining two successive progenitor stages (A2B5/G_{D3}^{+}O4^{-} and O4^{+}GalC^{-}) within the rat oligodendrocyte lineage leading to post-mitotic terminal differentiation (Figure 2.1). In developing rat brain, both stages are proliferative as determined by bromodeoxyuridine incorporation (16,46). Platelet-derived growth factor

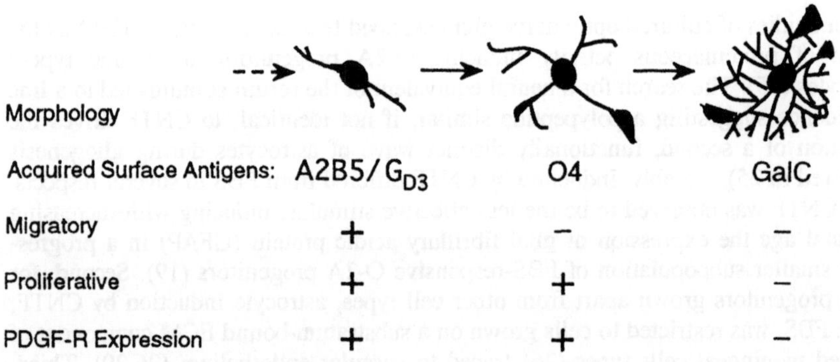

Fig. 2.1. In vivo properties of oligodendrocytes at three developmental stages preceding their elaboration of myelin in postnatal rat brain. For further subdivision based on additional surface antigens and in vitro studies, see Pfeiffer et al. (32).

(PDGF) receptor-alpha (PDGF-R) expression, a specific marker for the neonatal oligodendrocyte lineage (33), also characterizes both progenitor phenotypes before ceasing as GalC expression begins (8). Key differences among the two stages include process-bearing morphology and migratory capacity upon transplantation (46,47), suggesting that an O4^{+}GalC^{-} phenotype signifies a post-migratory oligodendroblast, at least in rat (31). As a surface marker for live cells, O4 immu-noreactivity, unlike A2B5/G_{D3} immunoreactivity, shows restriction to the oligodendrocyte lineage (40). However, in vitro studies indicated that $O4^+$ cells become astrocytic in response to FBS, thus behaving as O-2A progenitors (43).

2.2 Bipotential Oligodendrocyte Progenitors and Oligodendroblasts Redefined by OS-M/LIF/CNTF Responsiveness

To determine whether the aformentioned progenitor phenotypes respond to defined and serum inducers of astrocyte gene expression alike, we compared their differenti-ation potential in developmental stage-specific primary cultures prepared by immunopanning (18). The results demonstrated that progenitors having acquired O4 anti-gen in postnatal cerebrum no longer respond in culture to the astrocyte-inducing effect of CNTF and meningeal cell ECM previously described for O-2A cells (26); instead, they differentiate into myelinogenic oligodendrocytes regardless of whether CNTF is present in defined, hormone-supplemented medium (14,17,19).

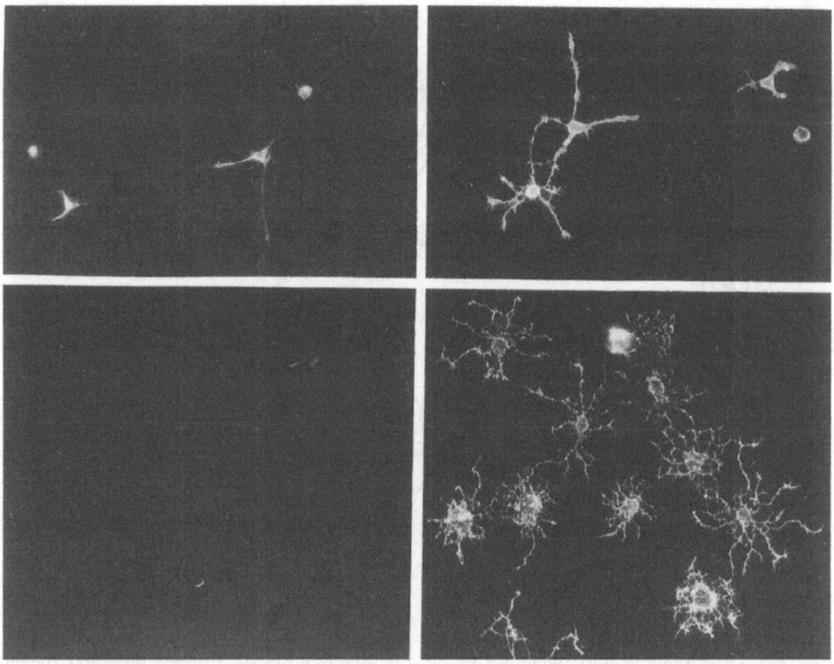

Fig. 2.2. Oligodendrocyte progenitors sampled from postnatal rat cerebrum as $G_{D3}{}^+O4^-$ (upper panel) and $O4^+GalC^-$ (lower panel) populations are shown by immuno-staining 2 days after primary culture on a substratum of meningeal ECM in defined medium containing 10 ng/ml OS-M. Whereas approximately 50% of the cultured $G_{D3}{}^+O4^-$ population (G_{D3}, upper left) is $GFAP^+$ (upper right), representing a 10-fold increase over

controls without added cytokine (not shown), seeded $O4^+GalC^-$ progenitors (O4, lower left) remain $GFAP^-$ (lower right) and have begun to differentiate morphologically into oligodendrocytes.

Astrocyte induction is restricted to the preceding $A2B5/G_{D3}{}^+O4^-$ stage, suggesting that lineage decisions made by bipotential progenitors detectable in postnatally developing rat brain (22,23) occur as cells approach their migratory destination, a conclusion supported by retroviral lineage tracing (23).

Like CNTF and LIF (19,29), OS-M functions interchangeably with these cytokines as an astrocyte-inducer targeting pre-$O4^+$progenitors (Figure 2.2) and also requires ECM for activity in primary cultures (19). By comparison, evidence of reduced or negligible GFAP induction by interleukin-6 (19,21) suggests that signal transduction activated by gp130 homodimerization alone is insufficient for astro-cyte induction. Importantly, these studies imply a role for multiple gp130 ligands in gliogenesis.

The capacity of FBS to defeat the oligodendrocyte differentiation program realized by $O4^+GalC^-$ progenitors challenged with CNTF (19; Figure 2.3) could impact on our understanding of remyelination. It may signify the potential for similar astrocyte-inducing activity in human plasma (24) to work more aggressively through a non-CNTF receptor-mediated mechanism contributing to

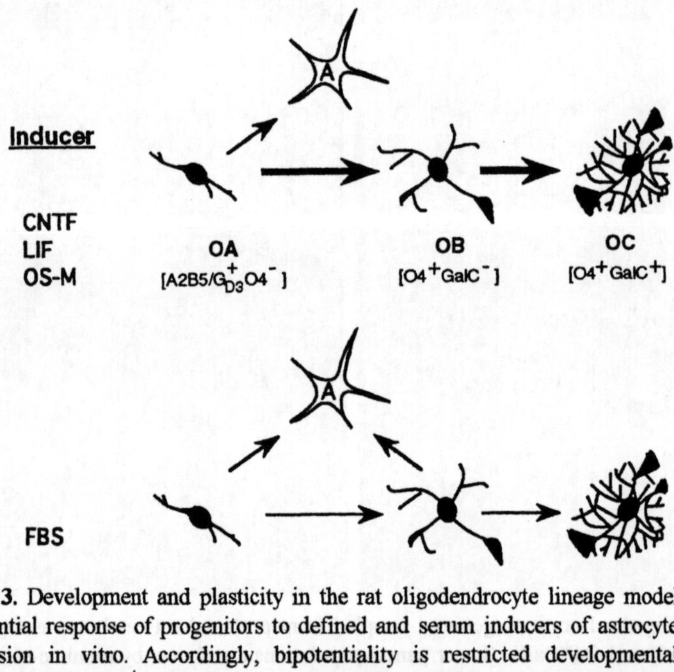

Fig. 2.3. Development and plasticity in the rat oligodendrocyte lineage modeled by the differential response of progenitors to defined and serum inducers of astrocyte (A) gene expression in vitro. Accordingly, bipotentiality is restricted developmentally (upper sequence) to pre-$O4^+$ cells (OA) and induced by CNTF or another neurally expressed ligand sharing the gp130/LIFR transducer heterodimer. By contrast, lineage commitment

evident in $O4^{+}GalC^{-}$ oligodendroblasts under these conditions is over-ridden by serum factors targeting both progenitor stages.

reactive gliosis. The possibility that adult oligodendrocyte progenitors normally destined for remyelination are diverted to an astroglial phenotype by humoral factors would seem to merit closer scrutiny in lesions where blood-brain barrier damage occurs.

3 Oligodendrogliotrophic Factors

The regulation of neuroglial survival has only recently gained attention in the myelin field through the discovery of extensive programmed cell death occurring in the oligodendrocyte lineage (3). Immunoaffinity-purified populations of oligoden-drocyte precursor cells from rat optic nerve and cerebrum rapidly die under basal, defined culture conditions on polycationic substrata (3,5,14,15,17) by a cyclo-heximide-inhibitable mechanism indicative of apoptosis (3). A number of cyto-kines with in vitro activity are known. These include the insulin-like growth factors, PDGF, neurotrophin-3 (NT-3), CNTF, LIF and interleukin-6 (reviewed in 6). When tested individually, the efficacy of these agents is low. PDGF supports oligodendrocyte progenitors but not oligodendrocytes (3) which cease to express the corresponding receptors (8). The remaining cytokines can extend, but not sustain, lineage survival through terminal differentiation (5). Survival can be prolonged, however, if a combination of cytokines represented by a gp130 ligand (LIF, CNTF or interleukin-6), NT-3 and insulin-like growth factor-1 is included in the medium. This has been interpreted to reflect the lineage requirement of multiple trophic signals acting in concert (5).

Environmental signals of oligodendrocyte survival operating in vivo remain unknown. Increasing the concentration of exogenous insulin-like growth factor, PDGF or CNTF can at least transiently reduce programmed death in rat optic nerve (5,6). A role for axons in this regard is supported by increased oligodendrocyte death observed in optic nerve following transection (4,6). On the other hand, astrocytes are also known to promote oligodendrocyte growth in culture (30) where they express a plethora of diffusible cytokines (37), including PDGF (36) and several neurotrophins. Astrocyte replenishment appears to be essential for remyelination by oligodendrocytes transplanted into hosts depleted of macroglia by ethidium bromide (11,12).

3.1 A Powerful Oligodendrocyte Stimulus Released by Cultured Astrocytes

Defined culture media which promote the survival of oligodendrocyte lineage cells arising in mixed primary cultures do not support oligodendroblasts purified from postnatal rat cerebrum by immunopanning (15,17). Working under the hypothesis that diffusible oligodendrogliotrophic factors were missing from this system, we conditioned the defined medium with a number of other neural and non-neural cell types and tested it for survival-promoting activity. The results of our investigation (14) indicate that astrocytes specifically release into this medium a powerful trophic stimulus which is neither mitogenic for the starting population nor does it prevent their maturation into oligodendrocytes sustainable for weeks of study in a myelinogenic state.

Immunoabsorption of the astrocyte-conditioned medium (ACM) with antibodies specific for known astrocyte-derived cytokines indicated that activity was speci-fically neutralized with polylonal LIF antibodies and reduced by treatment with monoclonal PDGF antibodies. In accordance with these findings, cultured astro-cytes have previously been shown to express and secrete LIF (1). However, further analysis of the stimulus in ACM suggests the action of a LIF-like protein distinct from LIF itself. ACM is >4-fold more efficacious than recombinant rat CNTF or murine LIF, tested alone or in combination with human recombinant NT-3, and promotes survival additively with these agents.

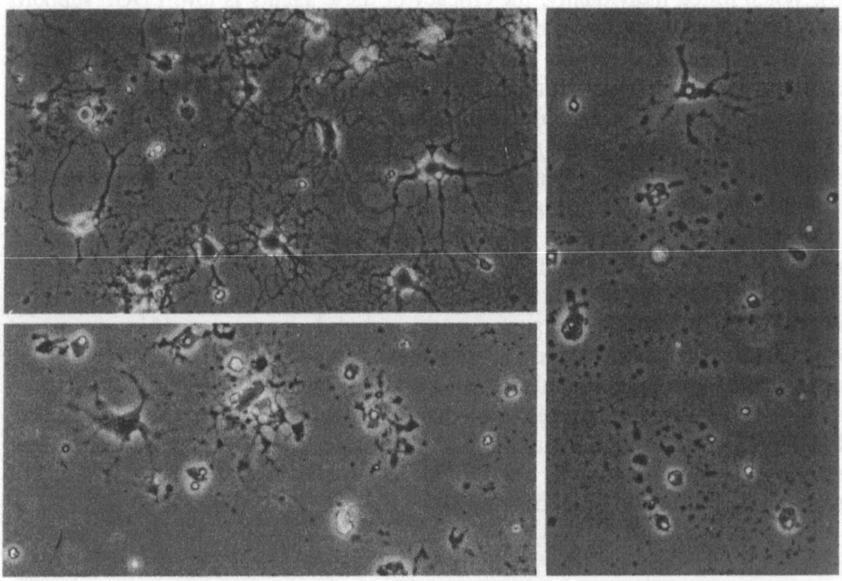

Fig. 3.1. Phase contrast micrographs of cultured oligodendroblasts surviving differentiation 7 days after seeding in defined medium conditioned by cultured astrocytes (upper left panel) and degenerating in the same basally defined formulation to which 10 ng/ml human recombinant OS-M (lower left panel) or no supplement (right panel) was added.

Because the survival activity partitions by ultrafiltration to 10-30 kDa, a size range smaller than rat LIF (43 kDa), we examined the effect of recombinant human OS-M (28 kDa), a structural homologue of LIF (28), on oligodendroblast survival.

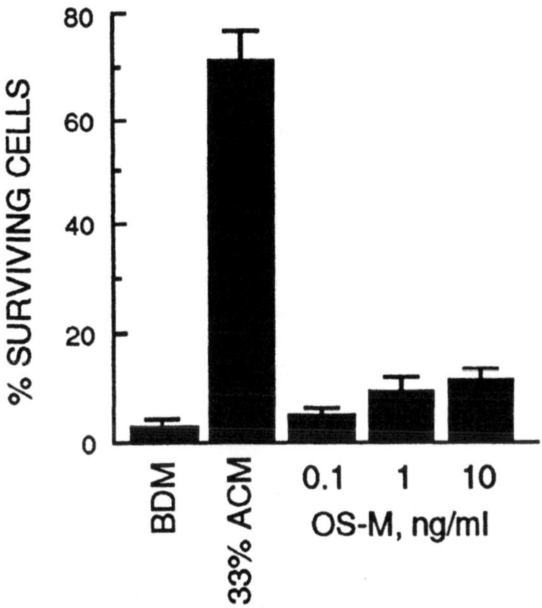

Fig. 3.2. Oligodendrocyte survival quantified in 7-day-old cultures as a percentage of oligodendroblasts attached to the substratum at the time of seeding in defined medium (BDM) containing 33% (v/v) astrocyte-conditioned BDM (ACM) or OS-M at the indicated concentrations.

Seven days after seeding, OS-M applied applied at concentrations shown to maximally induce GFAP expression in pre-$O4^+$ progenitors (18) effected only modest survival by comparison to ACM (Figures 3.1 and 3.2). While these preliminary studies do not exclude the possibility of LIF itself or OS-M (the murine OS-M gene has not been reported) contributing to the effect of ACM, they stimulate the search for additional gp130 receptor ligands with survival-promoting activity as purification of the astroglial factor(s) commences.

4 Concluding Thoughts on Remyelination

Remyelination of the CNS is acheivable through transplantation of either dissociated neonatal oligodendrocyte progenitor cells or immortalized oligodendrocyte cell lines (9,11,42). Studies of cell lines, in particular CG-4, have revealed that both oligodendrocytes and astrocytes are generated from donor populations grafted into glia-depleted hosts (12,44). This could reflect the maintenance of a develop-mental bipotential mechanism persisting in adulthood or the consequences of specific pathology (11). Our findings introduce the possibility that manipulation of donor progenitor stage as well as host factors could determine these lineage decisions, an advantage in situations when astrocyte development is undesired. Interestingly, CG-4 cells propogated in vitro include $A2B5/G_{D3}^{+}O4^{-}$ and $O4^{+}GalC^{-}$ subpopulations (21), the former of which could account for astrocytes generated upon transplantation. Consideration should also be given to the possi-bility that transformed cells are more sensitive than non-tranformed progenitors to astrocyte-inducing ligands of the gp130 family; growth on an artificially prepared ECM is not required for induction in CG-4 (21). Although bipotentiality was not evident in a study of progenitors expanded from primary cultures of rat cerebrum and transplanted into neonatal rat brain (9), a rigorous comparison of the fate $G_{D3}^{+}O4^{-}$ and $O4^{+}GalC^{-}$ progenitors supbopulations in this regard awaits further study.

We can draw on a number of studies and our own to summarize the role of astrocyte-derived growth factors promoting rat oligodendrocyte development in culture (Figure 4.1).

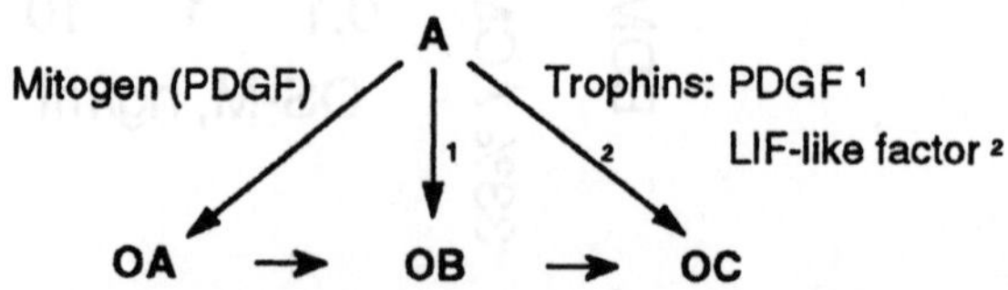

Fig. 4.1. Astroglial control of oligodendrocyte development during lineage progres-sion by factors released in serumless culture. Signaling changes from mitogenic stimulation of pre-$O4^{+}$ progenitors (16,36) to trophic support of oligodendroblasts by PDGF and oligodendrocytes by LIF-like protein (14).

Given the important function which astrocytes appear to have in oligodendrocyte regeneration and remyelination (12), the mitogenic and oligodendrogliotrophic signals they provide in murine culture should provide the seed for further investigations of human astrocyte-oligodendrocyte lineage cell interactions, factors targeting adult human progenitors (2,38) and the changing nature of macroglial interactions possibly brought about by reactive gliosis.

We appreciate the important contributions to this work of Sheila White for manu-script processing and Carlos Monteiro for imaging. These studies were supported by Shannon Award and NIH Grant NS29648 from NINDS.

References

1. Aloisi, F., Rosa, S., Testa, U., Bonsi, P., Russo, G., Peschle, C. and Levi, G. (1994) Regulation of leukemia inhibitory factor synthesis in cultured human astrocytes. J. Immunol. 152, 5022-5031.

2. Armstrong, R. C., Dorn, H. H., Kufta, C. V., Friedman, E. and Dubois-Dalcq, M. E. (1992) Pre-oligodendrocytes from adult human CNS. J Neurosci. 12, 1538-1547.
3. Barres, B. A., Hart, I. K., Coles, H. S. R., Burne, J. F., Voyvodic, J. T., Richardson, W. D. and Raff, M. C. (1992) Cell death and control of cell survival in the oligodendrocyte lineage. Cell 70, 31-46.
4. Barres, B. A., Jacobsen, M. D., Schmid, R., Sendtner, M. and Raff, M. C. (1993) Does oligodendrocyte survival depend on axons? Curr. Biol. 3, 489-497.
5. Barres, B. A., Schmid, R., Sendtner, M. and Raff, M. C. (1993) Multiple extracellular signals are required for long-term oligodendrocyte survival. Development 118, 283-295.
6. Barres, B. A. and Raff, M. D. (1994) Control of oligodendrocyte number in the developing rat optic nerve. Neuron 12, 935-942.
7. Blakemore, W. F. and Crang, A. J. (1989) The relationship between type-1 astrocytes, Schwann cells and oligodendrocytes following transplantation of glial cell cultures into demyelinating lesions in the adult rat spinal cord. J. Neurocytol. 18, 519-528.
8. Ellison, J. A. and De Vellis, J. (1993) Platelet-derived growth factor receptor is expressed by cells in the early oligodendrocyte lineage. J. Neurosci. Res. 37, 116-128.
9. Espinosa de los Monteros, A., Zhang, M. and De Vellis, J. (1993) O2A progenitor cells transplanted into the neonatal rat brain develop into oligodendrocytes but not astrocytes. Proc. Natl. Acad. Sci. USA 90, 50-54.
10. Franklin, R. J. M., Bayley, S. A., Milner, R., ffrench-Constant, C. and Blakemore, W. F. (1995) Differentiation of the O-2A progenitor cell line CG-4 into oligodendrocytes and astrocytes following transplantation into glia-deficient areas of CNS white matter. Glia 13, 39-44.
11. Franklin, R. J. M. and Blakemore, W. F. (1995) Glial-cell transplantation and plasticity in the O-2A lineage-implications for CNS repair. Trends Neurosci. 18, 151-156.
12. Franklin, R. J. M., Crang, A. J. and Blakemore, W. F. (1991) Transplanted type-1 astrocytes facilitate myelin repair of demyelinated lesions by host oligodendrocytes in adult rat spinal cord. J. Neurocytol. 20, 420-430.
13. Fulton, B. P., Burne, J. F. and Raff, M. C. (1992) Visualization of O-2A progenitors in developing and adult rat optic nerve by quisqualate-stimulated cobalt uptake. J. Neurosci. 12, 4816-4833.
14. Gard, A. L., Burrell, M. R., Pfeiffer, S. E., Rudge, J. S. and Williams II, W. C. (1995) Astroglial control of oligodendrocyte survival mediated by PDGF and leukemia inhibitory factor-like protein. Development 121, 2187-2197.
15. Gard, A. L. and Pfeiffer, S. E. (1989) Oligodendrocyte progenitors isolated directly from developing telencephalon at a specific phenotypic stage: myelinogenic potential in a defined environment. Development 106, 119-132.

16. Gard, A. L. and Pfeiffer, S. E. (1990) Two proliferative stages of the oligodendrocyte lineage (A2B5$^+$O4$^-$) and O4$^+$GalC$^-$) under different mitogenic control. Neuron 5, 615-625.
17. Gard, A. L. and Pfeiffer, S. E. (1993) Glial cell mitogens bFGF and PDGF differentially regulate development of O4$^+$GalC$^-$ oligodendrocyte progeni-tors. Dev. Biol., 159, 618-630.
18. Gard, A. L., Williams II, W. C. and Burrell, M. R. (1995) Oligodendroblasts distinguished from O-2A glial progenitors by surface phenotype (O4$^+$GalC$^-$) and response to cytokines using signal transducer LIFRb Dev. Biol. 167, 596-608.
19. Hughes, S. M. and Raff, M. C. (1987) An inducer protein may control the timing of fate switching in a bipotential glial progenitor cell in the rat optic nerve. Development 101, 157-167.
20. Imamoto, K., Paterson, J. A. and Leblond, C. (1978) Radioautographic investigation of gliogenesis in the corpus callosum of young rats. J. Comp. Neurol. 180, 115-138.
21. Kahn, M. A. and De Vellis, J. (1994) Regulation of an oligodendrocyte progenitor cell line by the interleukin-6 family of cytokines. Glia 12,87-98.
22. Levison, S. W., Chuang, C., Bernetta, J. and Goldman, J. E. (1993) The migrational patterns and developmental fates of glial precursors in the rat subventricular zone are temporally regulated. Development 119, 611-622.
23. Levison, S. W. and Goldman, J. E. (1993) Both oligodendrocytes and astrocytes develop from progenitors in the subventricular zone of postnatal rat forebrain. Neuron 10, 201-212.
24. Levison, S. W. and McCarthy, K. D. (1991) Characterization and partial puri-fication of AIM: A plasma protein that induces rat cerebral type 2 astro-glia from bipotential glial progenitors. J. Neurochem. 57, 782-794.
25. Lillien, L. E. and Raff, M. C. (1990). Analysis of the cell-cell interactions that control type-2 astrocyte development in vitro. Neuron 4, 525-534.
26. Lillien, L. E., Sendtner, M. and Raff, M. C. (1990) Extracellular matrix-asso-ciated molecules collaborate with ciliary neurotrophic factor to induce type-2 astrocyte development. J. Cell Biol. 111, 636-644.
27. Louis, J. -C., Magal, E., Takayama, S. and Varon, S. (1993) CNTF protection of oligodendrocytes against natural and tumor necrosis factor-induced death. Science 259, 689-692.
28. Malik, N., Haugen, H. S., Modrell, B., Shoyab, M. and Clegg, C. H. (1995) Developmental abnormalities in mice transgenic for bovine oncostatin M. Molec. Cell Biol. 15, 2349-2358.
29. Mayer, M., Bhakoo, K. and Noble, M. (1994) Ciliary neurotrophic factor and leukemia inhibitory factor promote the generation, maturation and survival of oligodendrocytes in vitro. Development 120, 143-153.
30. Noble, M. and Murray, K. (1984) Purified astrocytes promote the in vitro division of a bipotential glial progenitor cell. EMBO J. 3, 2243-2247.

31. Ono, K., Bansal, R., Payne, J., Rutishauser, U. and Miller, R. H. (1995) Early development and dispersal of oligodendrocyte precursors in the embryonic chick spinal cord. Development 121, 1743-1754.
32. Pfeiffer, S. E., Warrington, A. E. and Bansal, R. (1993) The oligodendrocyte and its many cellular processes. Trends Cell Biol. 3, 191-197.
33. Pringle, N., Mudhar, H. S., Collarini, E. J. and Richardson, W. D. (1992) PDGF receptors in the rat CNS: during late neurogenesis, PDGF alpha-receptor expression appears to be restricted to glial cells of the oligodendrocyte lineage. Development 115, 535-551.
34. Raff, M. C. (1989) Glial cell diversification in the rat optic nerve. Science 243, 1509-1524.
35. Raff, M. C., Miller, R. H. and Noble, M. (1983) A glial progenitor that develops in vitro into an aastrocyte or an oligodendrocyte depending on the culture medium. Nature 303, 389-396.
36. Richardson, W. D., N. Pringle, M. J. Mosley, B. Westermark and Dubois-Dalcq, M. (1988) A role for platelet-derived growth factor in normal glio-genesis in the central nervous system. Cell 53, 309-319.
37. Rudge, J. S. (1993) Astrocyte-derived neurotrophic factors. In S. Murphy (ed.), Astrocytes: Pharmacology and Function, Academic Press, San Diego, pp. 267-305.
38. Scolding, N. J., Rayner, P. J., Sussman, J., Shaw, C. and Compston, D. A. S. (1995) A proliferative adult human oligodendrocyte progenitor. NeuroReport 6, 441-445.
39. Skoff, R. P. and Knapp, P. E. (1991) Division of astroblasts and oligodendroblasts in postnatal rodent brain: Evidence for separate astrocyte and oli-godendrocyte lineages. Glia 4, 165-174.
40. Sommer, I. and Schachner, M. (1981) Monoclonal antibodies (O1 and O4) to oligodendrocyte cell surfaces: an immunocytochemical study in the central nervous system. Dev. Biol. 83, 311-327.
41. Stahl, N. and Yancopoulos, G. D. (1994) The tripartite CNTF receptor complex: Activation and signaling involves components shared with other cy-tokines. J. Neurobiol. 25, 1454-1466.
42. Tontsch, U., Archer, D. A., Dubois-Dalcq, M. and Duncan, I. (1994) Transplantation of an oligodendrocyte cell line leading to extensive myelination. Proc. Natl. Acad. Sci. USA. 91, 11616-11620.
43. Trotter, J. and Schachner, M. (1989) Cells positive for the O4 surface antigen isolated by cell sorting are able to differentiate into astrocytes or oligoden-drocytes. Dev. Brain Res. 46, 115-122.
44. Trotter, J., Crang, A. J., Schachner, M. and Blakemore, W. F. (1993) Lines of glial precursor cells immortalised with a temperature-senstitive oncogene give rise to astrocytes and oligodendrocytes following transplantation into demyelinated lesions in the central nervous system. Glia 9, 25-40.
45. Ware, C. B., Horowitx, M. C., Renshaw, B. R., Hunt, J. S., Liggitt, D., Koblar, S. A., Gliniak, B. C., McKenna,H. J., Papayannopoulo, T., Thoma, B., Cheng, L., Donovan, P. J., Peschon, J. J., Bartlett, P. F.,

Willis, C. R., Wright, B. D., Carpenter, M. K., Davison, B. L. and Gearing, D. P. (1995) Targeted disruption of the low-affinity leukemia inhibitory factor receptor gene causes placental, skeletal, neural and metabolic defects and results in perinatal death.

46. Warrington, A. E. and Pfeiffer, S. E. (1992) Morphology, proliferation and differentiation of oligodendrocytes in unfixed postnatal rat cerebellum. J. Neurosci. Res. 33, 338-353.
47. Warrington, A. E., Barbarese, E. and Pfeiffer, S. E. (1993) Differential myeli-nogenic capacity of specific developmental stages of the oligodendrocyte lineage upon transplantation into hypomyelinating hosts. J. Neurosci. Res. 34, 1-13.
48. Wesselingh, S. L., Gough, N. M., Kinlay-Jones, J. J. and McDonald, P. L. (1990) Detection of cytokine mRNA in astrocyte culture using the polymerase chain reaction. Lymphokine Res. 9, 177-185.

HIV-1 Env and Glutamate Induction of NO, IL1, and TNF

Jean E. Merrill[1], Bruce D. Howard[2], and Nigel T. Maidment[3]

Departments of Neurology[1], Biological Chemistry[2], and Psychiatry and Biobehavioral Sciences[3], UCLA School of Medicine, Los Angeles, CA 90024 USA

Keywords. HIV-1, gp160, gp120, gp41, glutamate, IL1, TNF, and NO

1 Introduction.

Several studies which have focused on central nervous system (CNS) AIDS have implicated HIV-1 envelope (env) proteins in glutamate- and calcium-induced damage of neurons, most likely through nitric oxide (NO) induction (1). We have proposed that white matter damage, both to oligodendrocytes and myelin, could be mediated by NO, IL1 and TNFα in CNS AIDS (2) and multiple sclerosis (MS) (3). Since axonal damage and Wallerian degeneration are also seen in MS (4), glutamate might be elevated in the CNS of MS patients, aggravating the pathological condition created by the inflammatory immune component. Previously, we have demonstrated that substance P (SP) induction of cytokines in glial cells was calcium ($[Ca^{++}]_i$)-dependent (5) and that glutamate induced an increase in $[Ca^{++}]_i$ in glial cells (6). We have also previously shown that HIV-1 env proteins induce IL1, TNFα, and NO in rat and human glial cultures (7,8). In this study, we demonstrate that glutamate directly induces IL1, TNFα, and NO production in rat and human glial cell cultures. Furthermore, we show that HIV-1 env protein induction of these proinflammatory agents is glutamate-dependent. Additionally, gp160 and gp41 treatment of glia leads to an increased efflux and/or decreased uptake of aspartate/glutamate. We conclude that HIV-1 may alter the function of the glutamate transporter leading to an accumulation of extracellular glutamate and triggering of cytokine and NO production through the quisqualate receptor.

2 Materials and Methods

2.1 Rodent and Human Glial Cell Cultures.

Production of mixed rodent and human glial cell cultures by this laboratory has been described extensively elsewhere (7,8). Rodent glial cell cultures were either purified microglia or both astrocytes and microglia (75%:25%). Human glial cell cultures were comprised of 50% microglia and 50% astrocytes as previously

Molecular Signaling and Regulation in Glial Cells
A Key to Remyelination and Functional Repair

described (7,8).

2.1.1 Reagents for Glial Cell Treatment.

We determined empirically that glial cells produced cytokines and NO maximally in response to 100uM glutamate (Sigma). Kynurenic acid (KA, 3mM, Sigma) and CNQX (5uM, Cambridge Research Biochemicals) were titrated to optimal concentrations and used as inhibitors of the glial cell quisqualate receptor. APV (100uM, Cambridge Research Biochemicals) and MK801 (12uM, gift of Dr. Claude Wasterlain, UCLA) were also titrated and used as the negative receptor control since glia do not have NMDA receptors. These reagents were added at the beginning of the experiment when HIV-1 or glutamate were added.

The requirment for calcium in NO or cytokine production was determined as previously described (5), by chelation of extracellular calcium with 1mM EGTA (Sigma) and intracellular calcium with 10uM dibromoBAPTA/AM (Sigma). The role of TNFα in the induction, or TGFβ in the inhibition, of NO production was tesed by blocking TNF alpha with anti- murine TNFα (1:100, R&D Systems) or addition of porcine TGFβ 1+2 (1ng/ml, R&D Systems) at the initiation of the experiment.

NO production was blocked either by degradation of the substrate with arginase (5U/ml, Sigma) or the competitive inhibition of substrate for enzyme using the arginine analogue N-monomethyl-L- arginine (MM-L-A, 1mM, Sigma) or its negative enantiomer control, MM-D-A added at time 0.

Heat-inactivated, gradient-purified virus (HIV-1 JRFL, 200ng/ml, gift of Dr. Irvin Chen, UCLA) or recombinant env proteins (gp160, gp120, gp41, 4ug/ml, ARRRP, NIH) were used to stimulate glial cultures. Gag p24 was the control HIV-1 protein also used at 4ug/ml. The Mock control was supernatant derived from uninfected T cells; it was put through the gradient purification step and used in the same way as virus-containing supernatants (7,8).

2.2 IL1, TNFα, and NO Assays.

Both IL1 and TNFα were assessed using bioassays as previously described (5). NO was determined as total NO (NO_x^-) by the method of boiling vanadium reduction of nitrite and nitrate (9).

2.3 ^{3}H-D-Aspartate Efflux from Glia.

Short term aspartate efflux was determined by labeling cells for 1hr at 37°C with D-[^{3}H]-aspartate (2uCi/ml,NEN). The cells were washed with PBS and incubated for 15 min at 37°C with env proteins and then rinsed with medium and lysed with 0.1% Triton X-100. Cell lysates were counted for radioactivity using a Beckman

LS7000 Scintillation Counter for retained intracellular D-aspartate. The long term exposure of cells to env proteins consisted of a 24hr exposure followed by a 1hr uptake of ^{3}H-D-aspartate. Cells were rinsed with medium and incubated 15 min. Cells were rinsed again, lysed and counted as above.

2.4 HPLC Determination of Amino Acids.

An HPLC fluorometric detection procedure was used for quantitative amino acid analysis of brain dialysates, as modified from Larsen and West (10). Medium alone was used as the basal measurement of amino acids. Cells were either untreated or exposed to env (or gag) proteins for 3 or 24hr. Supernatants were harvested and amino acids derivatized with O-phthaldehyde (OPA) prior to automatized injection into the HPLC apparatus (Gilson). The OPA-amino acid adducts were resolved on a reversed phase 2X150mm column (Hypersil, 5um,C18) using a linear gradient of acetonitrile from 5-50% sodium acetate (50mM,pH 5.6) as aqueous solvent.

3 Results

3.1 Induction of NO in rat glial cell cultures by glutamate

Glutamate induces NO production in a time and dose dependent fashion as shown in Table 1. The production is inhibitable if the substrate is degraded by arginase or the enzyme is inactivated by the competitive substrate analogue, but not by its enantiomer. Like LPS induction of NO, the production is inhibited by TGFβ. NO is not inhibited by anti-TNFα, suggesting that glutamate induction of NO is not TNF-mediated, as it is in LPS induction (9). The 50% reduction of NO by TGFβ suggests that IL1 may be involved in the induction (9).

3.2 Induction of cytokines in rat glial cell cultures by glutamate and HIV-1

Table 2 demonstrates that glutamate induces IL1 and TNFα at the same concentration as it does for NO induction. The induction of both cytokines is through the quisqualate but not the NMDA receptor, since KA and CNQX but not APV or MK801 inhibit the induction. Compared to the mock control, HIV-1 also induces both cytokines and may do so via the increase in glutamate since KA and CNQX inhibit this induction as well.

3.3 Induction of NO, IL1, and TNFα in human glial cell cultures by glutamate and HIV-1

Table 1 **Induction of Nitric Oxide in Rat Glial Cell Cultures by Glutamate**

*Treatment of cells	Total NO Production (μM) 6hr	24hr	48hr
0	0±0.5	5.2±4.1	5.8±4.0
glutamate (μM)			
1	7.3	8.0±7.5	9.9±2.0
10	6.9	9.1±3.1	9.4±1.8
100	8.9	13.6±2.5	14.3±2.5
100 + ^{N}MM-L-A	-	4.2±2.0	8.0±1.0
100 + ^{N}MM-D-A	-	13.4±1.3	15.5±5.1
100 + Arginase	-	4.4±0.5	8.5±1.3
100 + TGFβ	-	6.9±1.7	7.7±0.2
100 + Anti TNFα	-	14.2±1.1	17.1±0.0

*Mean ± SD of 3 exps at 24 and 48 hr; 10^5 microglia/well; Anti TNFα 1:100; TGFβ, 1ng/ml; Arginase, 5U/ml; ^{N}MM-L-A and ^{N}MM-D-A: N-Monomethyl-L or D-arginine, 1 mM.

- = Not done

Table 2 Induction of Cytokines in Rat Glial Cell Cultures by Glutamate and HIV-1

Mixed Glial Cell Treatment*	IL1+ Units of Activity	TNFα+ (pM)
glutamate (μM)		
0	5.0± 4.5	0.8±0.6
100	77.5±27.0	9.6±0.1
100 + KA	6.7± 6.0	7.2±3.1
100 + CNQX	10.0± 8.1	6.3±1.8
100 + APV	137.6±35.2	13.9±3.8
100 + MK801	99.3±22.6	12.3±7.2
$HIV\text{-}1_{JR\text{-}FL}$	87.5± 5.4	10.50±3.1
HIV-1 + KA	5.1± 5.0	4.80±2.0
HIV-1 + CNQX	40.7± 3.7	4.29±1.1
HIV-1 + MK801	109.7±30.3	9.40±1.8
MOCK	0.0±0.0	1.00±1.0
MOCK + KA	0.0±0.0	1.30±0.8

*Mean ± SD of 4 exps.: 10^5 glia/well; ^{N}MM-L-A, Arginase as in Table 1. Kynurenic acid KA, 3mM; CNQX, 5μM; APV, 100μM; MK801, 12μM; $HIV\text{-}1_{JR\text{-}FL}$, 200 ng/ml.
+Measured at 24 hours after exposure to stimuli or other reagents.

As shown in Table 3, glutamate and HIV-1 induce NO and cytokines through the quisqualate receptor in human glial cell cultures as these two stimuli did in rodent glial cultures. Interestingly, since EGTA and BAPTA/AM were inhibitory, glutamate induction of IL1 and TNFα, but not that of NO, was calcium-dependent. These findings support the data in Table 1 showing that TNFα is not contributing to NO production in these glutamate-induced cultures. It also resembles data previously reported from this laboratory in which the neurotransmitter SP induced IL1 in a calcium-dependent manner (5).

3.4 Effect of HIV-1 env proteins on amino acid metabolism in rat glial cell cultures

Recent studies have suggested that arachidonic acid or gp120 can inhibit the glutamate transporter (11,12). If this happens, there would be an accumulation of extracellular glutamate that might activate increased intracellular calcium and give rise to cell damage. We tested the hypothesis that HIV-1 env proteins might affect extracellular glutamate as shown in Table 4. As determined by HPLC, there were no changes at 3 hr in those amino acids being released (aspartate, asparagine, taurine, GABA) or taken up (glutamate) by glia exposed to any HIV-1 proteins. At 24 hr, asparagine's release was blocked by gp160, gp120, and gp41, but not by p24. Aspartate's release was enhanced by gp41 and gp160. Glutamate's uptake was blocked by gp160, gp41, and to a lesser (but still significant extent) by gp120. Thus, both aspartate and glutamate accumulated in greater concentrations in the extracellular medium. Taurine's release was enhanced by gp41 but GABA's release was not significantly altered by any of the recombinant HIV-1 proteins. Serine was also released by glial cells in vitro, but its levels in glial supernatants was unaffected by HIV-1 env (data not shown). Cells did not take up or release glutamine, threonine, glycine, or tyrosine in culture whether env proteins were present or not (data not shown).

3.5 Assessment of ^{3}H-D-aspartate content in glial cells exposed to HIV-1 env proteins

As a different approach to assessing the glutamate transporter function, efflux of incorporated ^{3}H-D-aspartate from glial cells was assessed. Cells were allowed to take up the labeled D-aspartate, a non-metabolizable substitute for glutamate on the sodium-dependent glutamate transporter, and the acute (15 min) effects of the HIV-env proteins on acute transporter activity (15 min) were assessed. As shown in Table 5, both gp160 and gp41 increased the efflux of D-aspartate from glia, while gp120 had no such acute effect. To assess the effects of HIV-1 chronic exposure on acute uptake and release, cells were first exposed to env proteins for 24 hr and then allowed to take up labeled D-aspartate for 1 hr. Treated and untreated cells took up the same amount of label, suggesting the transporter still functioned normally in its short term response of uptake of aspartate/glutamate. Unlike the

Table 3 Induction of Nitric Oxide and Cytokines in Human Glial Cultures by Glutamate and HIV-1

Treatment of Cells	NO_x^- (μM)	IL1 (Units of activity)	TNFα (pM)
glutamate (μM)			
0	0.5	0	0
100	17.2	130	7.7
100 + CNQX	4.4	0	1.0
100 + MK801	15.5	75	12.3
100 + EGTA + BAPTA/AM	13.5	32	0
HIV-1_{JR-FL}	19.2	115	-
HIV-1 + CNQX	8.0	0	-
HIV-1 + MK801	19.7	97	-

Concentrations of reagents as in Table 2; Tissue from 10 y.o. pediatric patient. Cells with no glutamate but with CNQX, MK801, or BAPTA/AM plus EGTA gave no response. Culture supernatants harvested at 48 hours.

Table 4 Effect of HIV-1 Envelope Proteins on Amino Acid Metabolism in Rat Glial Cell Cultures

Amino Acid		Asp	Glu	Asn	Tau	GABA
			(nmoles/ml in culture supernatants)			
Medium alone		8.3±2.0	12.8±1.0	2.8±0.8	1.5±0.2	0.8±0.1
Unstimulated Cells						
3hr		7.8±0.3	8.5±0.8	54.5±2.5	29.5±0.8	1.0±0.05
24hr		17.8±1.5	3.8±0.5	111.3±9.5	30.5±5.8	9.5±2.5
Stimulated Cells						
3hr	p24	7.5±2.5	8.0±0.8	47.5±4.0	28.5±2.5	0.5±0.3
	gp41	5.5±0.3	8.5±0.7	50.8±2.3	33.0±3.0	1.0±0.1
	gp160	4.5±0.3	7.0±0.7	45.8±2.3	27.5±3.0	1.0±0.1
	gp120	3.5±1.5	8.3±0.8	57.8±0.8	34.0±8.5	0.8±0.3
24hr	p24	17.0±1.8	4.5±0.5	111.3±5.3	34.5±3.5	11.5±2.3
	gp41	**25.3±2.5**	**9.0±0.3**	**87.0±3.8**	**45.3±0.3**	10.8±1.3
	gp160	**24.8±2.0**	**7.0±0.1**	**84.3±3.0**	36.8±2.8	8.0±5.0
	gp120	19.0±1.5	**6.5±0.5**	**74.8±6.0**	29.8±3.3	9.8±1.0

Means ± SD of 5 exps. Bold type indicates significance $P \leq 0.01$ compared to unstimulated cells at same time point. Asp, aspartate; glu, glutamate; asn, asparagine; tau, taurine; GABA, gamma aminobutyric acid. Recombinant HIV-1 proteins were used at 4ug/ml/10^6 cells.

Table 5

Assessment of ^{3}H-D-aspartate Content in Cells Exposed to HIV-1 Env Proteins

Cell Associated D-aspartate (% radioisotope content in untreated control)

Exposure to env protein	15 min*	24 hr**
gp120	101.7±4.5	96.8±3.6
gp160	58.3±23.2	98.7±5.0
gp41	69.9±20.5	95.5±4.5

Mean ± SD shown
*Average of 3 experiments of duplicate or triplicate wells
^{3}H-D-aspartate radioisotope content (cpm) in untreated control cultures = 239,863 ± 136 471
**One experiment of triplicate wells.
^{3}H-D-aspartate radioisotope content (cpm) in untreated control cultures = 220,267 ± 15,312

experiment on acute env exposure, chronic env exposure had no effect on transporter-mediated efflux of D-aspartate as measured for a short (15 min) interval. Given that the long term function of the transporter is altered during exposure to HIV-1 (Table 4), these data suggest that HIV-1 induces two events, one acute and the other long term, in which the glutamate transporter works to actively transport aspartate/glutamate out of the cell and is impaired in the (re)uptake of glutamate.

4 Discussion

The data presented here point to the possibility that HIV-1, cytokines, NO, and glutamate interact with each other to produce pathology in the CNS. HIV-1 env proteins appear to have produced both an acute and chronic effect on the glutamate transporter, which suggests that two different mechanisms are involved. The enhanced level of glutamate in the extracellular medium might then induce IL1, TNF, and NO in glial cells. These proinflammatory agents might indirectly cause grey matter (neurons) and white matter (myelin, oligodendrocytes) damage. The implications of this in CNS AIDS have been discussed elsewhere (1,2). If axonal damage in the CNS of MS patients (4) gives rise to extracellular glutamate, a similar cascade of events in white matter leading to demyelination might occur. In vitro culture systems, in vivo analysis of the experimental model for MS (EAE), and immunohistochemistry of lesions in MS brain tissue have provided evidence for roles for NO, IL1, and TNFα in the pathology of oligodendrocyte cell death, demyelination, and/or clinical findings (3,9,13,14).

Several articles have suggested that free radicals of oxygen or arachidonic acid directly interfere with the glutamate transporter in astrocytes (11,15). Mixed reports have appeared suggesting that gp120 may (12) or may not (16,17) inhibit the glutamate transporter in glial cells resulting in increased extracellular glutamate. Ciardi and Meldolesi have shown that gp120 increases intracellular calcium in astrocytes, although this is not through the glutamate receptor (18).

We and others have shown that oligodendrocytes are exquisitely sensitive to NO, glutamate, and oxygen free radicals in vitro (3,9,19-22). It is therefore possible that glutamate, if enhanced in the extracellular environment could lead to a rise in intracellular calcium in oligodendrocytes. In oligodendrocytes, calcium might activate PLA2, arachidonate metabolism, and more $[Ca^{++}]i$ (23) or free radicals (24). Cell damage or death might occur through several pathways including (1) Ca^{++}-dependent degradation of mRNAs for myelin proteins (25), (2) glutathione depletion and oxygen free radical attack (15,21-24), or (3) NO-mediated damage to DNA or ferrosulfur-containing enzymes (3,9). From our experiments, glutamate directly induces NO suggesting that, in addition to cytokines as stimuli in CNS inflammatory disorders such as MS and CNS AIDS, neurotransmitters may also play a role. Knowing this, our focus might be toward several metabolic pathways as targets for therapeutic intervention in the treatment of these diseases.

5 References

1. Lipton, S.A. (1994) *Devel. Neurosci.* **16**, 145-151.
2. Merrill, J.E., and Chen, I.S.Y. (1991) *FASEB J.* **5**, 2391-2397.
3. Mitrovic, B., St. Pierre, B.A., MacKenzie-Graham, A.J., and Merrill, J.E. (1994) *Ann. N.Y. Acad. Sci.* **738**, 436-446.
4. Prineas, J.W. (1985) *in* Handbook of Clinical Neurology (Vinken, P.J., Bruyn, G.W., and Klawans, H.L., Eds.), pp213-257, Elsevier Science Publishers, Amsterdam.
5. Martin, F.C., Charles, A.C., Sanderson, M.J., and Merrill, J.E. (1992) *Brain Res.* **599**, 13-18.
6. Charles, A.C., Merrill, J.E., Dirksen, E.R., and Sanderson, M.J. (1991) *Neuron* **6**, 983-992.
7. Koka, P., He, K., Camerini, D., Tran, T., Yashar, S.S., and Merrill, J.E. (1995) *J. Neuroimmunol.* **57**, 179-191.
8. Koka, P., He, K., Zack, J.A., Kitchen, S., Peacock, W., Fried, I., Tran, T., Yashar, S.S., and Merrill, J.E. (1995) *J. Exp. Med.*, in press.
9. Merrill, J.E., Ignarro, L.J., Sherman, M.P., Melinek, J., and Lane, T.E. (1993) *J. Immunol.* **151**, 2132-2141.
10. Larsen, B.R. and West, F.G., (1981) *J. Chromat. Sci.* **19**, 259-265.
11. Volterra, A., Trotti, D., Casutti, P., Tromba, C., Salvaggio, A., Melcangi, R.C., and Racagni, G. (1992) *J. Neurochem.* **59**, 600-606.
12. Benos, D.J., Hahn, B.H., Bubien, J.K., Ghosh, S.K., Mashburn, N.A., Chaikin, M.A., Shaw, G.M., Benveniste, E.N. (1994) *Proc. Natl. Acad. Sci.* **91**, 494-498.
13. Cross, A.H., Misko, T.P., Lin, R.F., Hickey, W.F., Trotter, J.L. and Tilton, R.G., (1994) *J. Clin Invest.* **93**, 2684-2690.
14. Bo, L., Dawson, T.M., Wesselingh, S., Mork, S., Choi, S., Kong, P.A., Hanley, D., and Trapp, B.D. (1994) *Ann. Neurol.* **36**, 778-786.
15. Piani, D., Frei, K., Pfister, H-W., and Fontana, A. (1993) *J. Neuro-Immunol.* **48**, 99-104.
16. Pulliam, L., Herndier, B.G., Tang, N.M., and McGrath, M.S. (1991) *J. Clin Invest.* **87**, 503-512.
17. Patrizio, M., and Levi, G. *Neurosci. Lett.* **178**, 184-189.
18. Ciardi, A. and Meldolesi, J. (1993) *Eur. J. Neurosci.* **5**, 1711-1718.
19. Mitrovic, B., Ignarro, L.J., Montestruque, S., Smoll, A., and Merrill, J.E. (1994) *Neurosci.* **61**, 578-585.
20. Mitrovic, B., Ignarro, L.J., Vinters, H., Aker, S., Uittenbogaart, C., Schmid, I., and Merrill, J.E. (1995) *Neurosci.* **65**, 531-539.
21. Oka, A., Belliveau, M.J., Rosenberg, P.A., and Volpe, J.J. (1993) *J. Neurosci.* **13**, 1441-1453.
22. Griot, C., Vendervelde, M., Richard, A., Peterhans, E., and Stoker, R. (1990) *Free Rad. Res. Comm.* **11**, 181-193.
23. Soliven, B., Takeda, M., Shandy, T., and Nelson, D.J. (1993) *Cell*

Physiol. **33**, C632-C640.
24. Coyle, J.T. and Puttfracken, P. (1993) *Science* **262**, 689-695.
25. Shirazi, Y., Rus, H.G., Macklin, W.B. and Shin, M.L. (1993) *J. Neuro-Immunol.* **150**, 4581-4590.

Protein Kinase C Stimulation Induces Astrocyte and Oligodendrocyte Proliferation, Demyelination and Enhanced Remyelination in Aggregating Rat Brain Cell Cultures

Jean-Marie Matthieu[1], Sandrine Pouly[2] and Paul Honegger[2]

[1]Laboratoire de neurochimie, Département de pédiatrie, Centre hospitalier universitaire vaudois, Lausanne, Switzerland.
[2]Institut de physiologie, Université de Lausanne, Lausanne, Switzerland.

Abstract. The protein kinase C-activating tumor-promoter, mezerein induces cell proliferation in differentiated aggregating rat brain cell cultures. This effect was accompanied by a drastic loss of myelin constituents - like myelin basic protein and 2',3'-cyclic nucleotide 3'-phosphohydrolase - indicating that mature oligodendrocytes entering cell division dedifferentiate and lose their myelin sheath. This phase of proliferation was followed by remyelination which eventually led to higher concentrations of myelin constituents than in cultures which were not treated with mezerein. While mezerein had no effect on choline acetyltransferase, a specific neuronal marker, it induced a two-fold increase of glutamine synthetase, an astrocytic marker. These results confirm that mature oligodendrocytes and astrocytes are able to proliferate. Protein kinase-C plays a central role in stimulating oligodendrocytes as well as astrocytes. This effect on astrocytes could impair affect the process of remyelination.

Keywords. Astrocytes, cell proliferation, demyelination, mezerein, oligodendrocytes, protein kinase C, remyelination

Introduction

In vivo, demyelinating diseases, such as multiple sclerosis in humans, frequently cause irreversible loss of function and for many years such diseases were considered irreversible. Experimental work performed in the early sixties brought some hope of remyelination [6, 7]. Since these pioneering studies, a growing body of evidence indicates that remyelination occurs, although these repair attempts are not very efficient. Many factors are held responsible for this unsuccessful repair, one of them is the limited number of oligodendrocytes available for remyelination.

Dissociated fetal rat telencephalon cells under constant rotation in a chemically defined medium form highly differentiated three-dimensional aggregates [12]. Cells differentiate on a time scale similar to that in the developing brain. Cellular proliferation and differentiation is enhanced in these cultures by hormones and growth factors [1-3, 10, 11, 16,]. Myelinated axons are present and myelin fractions can be isolated which are chemically similar to those in normal brain tissue [19]. Myelin

Molecular Signaling and Regulation in Glial Cells
A Key to Remyelination and Functional Repair

basic protein (MBP) content correlates with the extent of myelination in the murine CNS as assessed by electron microscopy [24]. 2',3'-Cyclic nucleotide 3'-phosphohydrolase (CNP) activity, being localized predominantly in oligodendrocytes as opposed to myelin, is a useful marker for mature oligodendrocytes. Glutamine synthetase (GS) serves as an astrocyte marker.

In the presence of complement, demyelination can be induced in aggregate cultures by antibodies directed against myelin oligodendrocyte glycoprotein (MOG) without loss of oligodendrocytes [14]. Recently, these results were confirmed and various cytokines increased the degree of demyelination obtained with MOG antibodies [17]. Following demyelination, spontaneous remyelination occurs in aggregate cultures and full recovery is completed after 7 days [20]. Using aggregating brain cell cultures we are currently studying strategies to manipulate and improve remyelination. In the study presented in this paper we used mezerein, a non-phorbol ester tumor promoter which activates specifically protein kinase C to induce cell proliferation and subsequent remyelination by mature oligodendrocytes. This work has been published in a condensed form [22].

Materials and methods

Cell cultures

Serum-free aggregating cell cultures were prepared from mechanically dissociated fetal (16 days of gestation) rat telencephalon as described previously in detail [13]. The cultures were maintained in DeLong flasks under constant gyratory agitation (70-80 r.p.m.) at 37°C and in an atmosphere of 10% CO_2/90% humidified air. The chemically defined culture medium was supplemented with nutritional factors, vitamins, trace elements, and hormones [13]. At day 20, cultures were subdivided and their medium supplemented with albumin-bound lipids (Albumax, GIBCO, final concentration 0.1%) [26]. Cultures were treated with a single dose of mezerein (100 nM) at day 26 when oligodendrocytes reached the highest degree of differentiation [13]. Three and 19 days after treatment, cultures were labelled for 18 hours with [^{14}C]-thymidine (30 nCi/ml) and then harvested for biochemical analyses.

Analytical procedures

For biochemical analyses, aggregates were washed twice with 5 ml of ice-cold phosphate-buffered saline and homogenized in 0.5 ml of 2 mM potassium phosphate containing 1 mM EDTA (pH 6.8), using glass-glass homogenizers. The homogenates were divided into aliquots for the different assays and stored at -80°C. The incorporation of [^{14}C]-thymidine into a trichloroacetic acid-precipitable fraction was determined as described previously [16]. DNA content of the cells was measured by fluorimetry [9]. The activity of 2',3'-cyclic nucleotide 3'-phosphohydrolase (CNP; EC 3.1.4.37) was determined according to Kurihara and Tsukada (1967) [15]. Myelin basic protein (MBP) was measured by radioimmunoassay [8]. Glutamine synthetase (GS) [21], choline acetyltransferase (ChAT) [23, 25] and total protein [18] were the other parameters analyzed in this study. All experiments were repeated at least twice.

Statistical significance was assessed using the Student's t-test. Differences with $p<0.05$ were considered as significant.

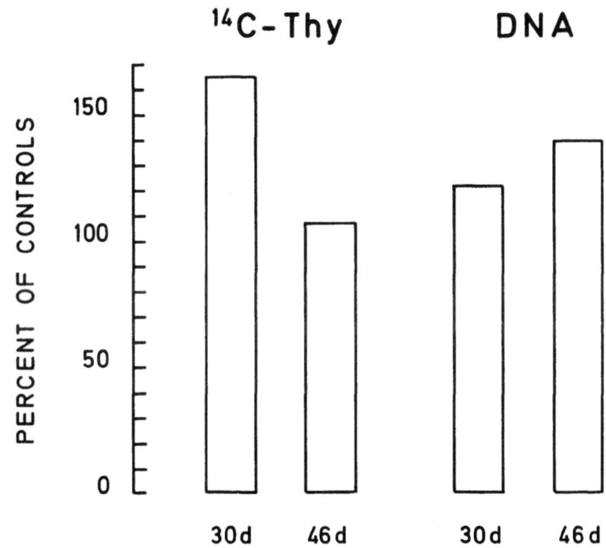

Fig. 1 [^{14}C]-thymidine incorporation (^{14}C-Thy) and DNA content of aggregating brain cell culture at day 30 (30d) and at day 46 (46d) in culture. Results are expressed in percentage of untreated cultures of the same age. They are means of 6 separate cultures. The differences observed were highly significant ($p<0.001$ for [^{14}C]-thymidine and $p<0.01$ for DNA).

Results

The mitogenic capacity of the tumor promoter mezerein was investigated in highly differentiated brain cell aggregates treated at day 26 in vitro with a single dose (100 nM) by measuring the incorporation of [^{14}C]-thymidine into a macromolecular fraction as well as the DNA content of the cultures at day 30 (i.e. 4 days after stimulation) and at day 46 (i.e. 20 days after stimulation). At day 30, thymidine incorporation was significantly increased (Fig.1) indicating cell proliferation in the cultures. This response became attenuated at day 46. Cell proliferation was confirmed by an increase of the DNA content of 20 and 40 percent at 30 and 46 days, respectively (Fig. 1).

During the time of active cell proliferation induced by mezerein, the expression of myelin markers decreased dramatically by 35 and 65 percent for CNP and MBP, respectively (Fig. 2). This phase of demyelination was followed by remyelination and at 46 days, cultures treated with mezerein showed very high levels of both myelin markers which exceeded the controls by 40 and 25 percent for CNP and MBP, respectively (Fig. 2).

In contrast to the myelin markers, GS, a marker for astrocytes was not affected shortly after mezerein treatment (Fig. 3). Nevertheless, 16 days later, a drastic increase of GS was observed, whereas the activity of ChAT, a neuron-specific marker was not different from control cultures at 30 and 46 days (Fig. 3).

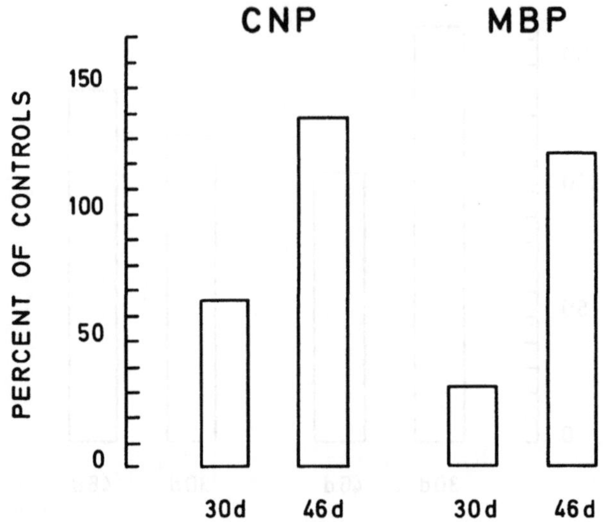

Fig. 2. Oligodendrocyte (CNP: 2', 3'-cyclic nucleotide 3'-phosphohydrolase) and myelin (MBP : myelin basic protein) markers at day 30 (30d) and at day 46 (46d) in culture. Results are expressed in percentage of untreated cultures of the same age. They are means of 6 separate cultures. The decrease observed at day 30 was significant at the $p<0.001$ and $p<0.01$ level, for CNP and MBP, respectively. The increase observed for both parameters between 30 and 46 days was highly significant.

Discussion

In aggregating brain cell cultures myelination begins between days 14 and 16 and reaches a maximum around day 26 [13]. The accumulation of MBP in the aggregates correlates with the formation of myelin sheaths around axons. In the mature cultures, within 4 days, as much as 60 % of the myelin content can be lost following the addition of antibodies against MOG in the presence of complement [14-17]. But, full remyelination occurs within a week [20]. The present study shows that in these highly differentiated cultures, tumor promoters induce demyelination and cell proliferation suggesting that mature oligodendrocytes loose contact with axons, dedifferentiate, and enter a proliferative phase since an increase in silver grains was observed over oligodendrocytes after [^{3}H]-thymidine incorporation (Pouly et al., in preparation). Althaus et al. (1991) [4] using oligodendrocyte cultures prepared from adult pig brains, reported that a small population of differentiated oligodendrocytes proliferate when stimulated with the tumor promoter phorbol 12-myristate 13-acetate (PMA).

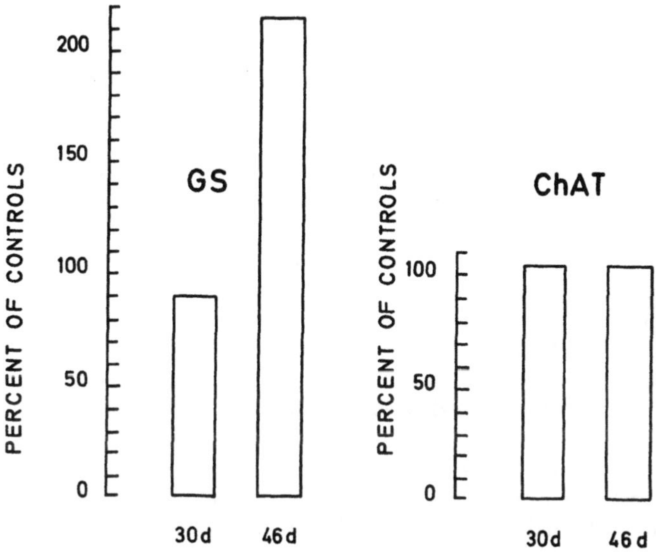

Fig. 3. The astrocytic marker glutamine synthetase (GS) activity doubled by day 46. A neuronal marker choline acetyltransferase (ChAT) remained unaffected by the treatment with mezerein. Results are expressed in percentage of untreated cultures of the same age. They are means of 6 separate cultures.

Furthermore, Asotra and Macklin (1993) [5] showed that long-term exposure of cultured oligodendrocytes to phorbol esters down-regulated the expression of myelin genes. In our study, mezerein, a non-phorbol ester specific activator of protein kinase C, induced demyelination and proliferation of oligodendrocytes. The loss of MBP content and CNP activity was reversed within 2 to 3 weeks, indicating that remyelination occurred. At the same time, the rate of cell proliferation returned to control levels. Work in progress in our laboratories (Pouly et al., in preparation) indicates that growth factors such as basic fibroblast growth factor can enhance remyelination. While mezerein had no effect on the highly differentiated neurons, it caused a significant increase of glutamine synthetase activity at day 46. This astrocytic response did not seem to affect remyelination in aggregate cell cultures, whereas in the CNS in vivo, reactive or proliferating astrocytes may impede remyelination. Therefore, selective proliferation and differentiation of oligodendrocytes will be crucial for promoting efficient remyelination in the CNS. The three-dimensional aggregating brain cell cultures represent a useful model to tackle this problem.

Acknowledgments

This work was supported by the Swiss National Science Foundation (grants 31.30787.91 and 31.37292.93) and the Multiple Sclerosis Society of Switzerland.

References

1. Almazan, G., Honegger, P. and Matthieu, J.-M. (1985) Triiodothyronine stimulation of oligodendroglial differentiation and myelination. Develop. Neurosci., 7, 45-54.
2. Almazan, G., Honegger, P., and Matthieu, J.-M. and Guentert-Lauber, B. (1985) Epidermal growth factor and bovine growth hormone stimulate differentiation and myelination of brain cell aggregates in culture. Develop. Brain Res. 21, 257-264.
3. Almazan, G., Honegger, P., Du Pasquier, P. and Matthieu, J.-M. (1986) Dexamethasone stimulates the biochemical differentiation of fetal forebrain cells in reaggrating cultures. Develop. Neurosci. 8, 14-23.
4. Althaus, H.H., Schroter, J., Spoerri, P., Schwartz, P., Kloppner, S., Rohmann, A. and Neuhoff, V. (1991) Protein kinase C stimulation enhances the process formation of adult oligodendrocytes and induces proliferation. J. Neurosci. Res. 29, 481-489.
5. Asotra, K. and Macklin, W.B. (1993) Protein kinase C activity modulates myelin gene expression in enriched oligodendrocytes. J. Neurosci. Res. 34, 571-588.
6. Bornstein, M.B. and Appel, S.H. (1961) The application of tissue culture to the study of experimental all enceph. I. Pattern of demyelination. J. Neuropath. Exp. Neurol. 209, 141-157.
7. Bunge, M.B., Bunge, R.B. and Ris, H. (1961) Ultrastructural study of remyelination in an experimental lesion in adult cat spinal cord. J. Biophys. Biochem. Cytol. 10, 67-94.
8. Bürgisser, P. (1983) Mise au point d'un dosage radioimmunologique de la protéine basique de la myéline dans le liquide céphalo-rachidien et dans le tissu nerveux (M.D. thesis). Université de Lausanne, Lausanne, Switzerland.
9. Downs, T.R. and Wilfinger, W.W. (1983) Fluorometric quantification of DNA in cells and tissues. Analyt. Biochem. 131, 538-547.
10. Guentert-Lauber, B. and Honegger, P. (1985) Responsiveness of astrocytes in serum-free aggregate cultures to epidermal growth factor : dependence on the cell cycle and the epidermal growth factor concentration. Dev. Neurosci. 7, 286-295.
11. Honegger, P. and Guentert-Lauber, B. (1983) Epidermal growth factor (EGF) stimulation of cultured brain cells. I. Enhancement of the developmental increase in glial enzymatic activity. Dev. Brain Res. 11, 245-251.
12. Honegger, P. (1985) Biochemical differentiation in serum-free aggregating brain cell cultures. In J.E. Bottenstein, and G. Sato (eds.), Cell culture in the Neurosciences, Plenum Press, New York, pp. 223-243.
13. Honegger, P. and Matthieu, J.-M. (1990) Aggregating brain cell cultures : a model to study myelination and demyelination. In G. Jeserich, H.H. Althaus and T.V. Waehneldt (eds.), Cellular and Molecular Biology of Myelination, NATO ASI Series Volume H 43, Springer, Berlin, pp. 155-170.
14. Kerlero de Rosbo, N., Honegger, P., Lassman, H. and Matthieu, J.-M. (1990) Demyelination induced in aggregating brain cell cultures by a monoclonal antibody against myelin/oligodendrocyte glycoprotein. J. Neurochem. 55, 583-587.

15. Kurihara, T. and Tsukada, Y. (1967) The regional and subcellular distribution of 2', 3'-cyclic nucleotide 3'-phosphohydrolase in the central nervous system. J. Neurochem. 14, 1167-1174.
16. Lenoir, D. and Honegger, P. (1983) Insulin-like growth factor I (IGFI) stimulates DNA synthesis in fetal rat brain cell cultures. Dev. Brain Res. 7, 205-213.
17. Loughlin, A.J., Honegger, P., Woodroofe, M.N., Comte, V., Matthieu, J.-M. and Cuzner, M.L. (1994) Myelin basic protein content of aggregating rat brain cell cultures treated with cytokines and/or demyelinating antibody : Effects of macrophage enrichment. J. Neurosci. Res. 37, 647-653.
18. Lowry, O.H., Rosebrough, N.J., Farr, A.L and Randall, R.J. (1951) Protein measurement with the Folin phenal reagent. J. Biol. Chem. 193, 265-275.
19. Matthieu, J.-M., Honegger, P., Favrod, P., Gautier, E. and Dolivo, M. (1979) Biochemical characterization of a myelin fraction isolated from rat brain aggregating cell cultures. J. Neurochem. 32, 869-881.
20. Matthieu, J.-M., Comte, V., Tosic, M. and Honegger, P. (1992) Myelin gene expression during demyelination and remyelination in aggregating brain cell cultures. J. Neuroimmunol. 40, 231-234.
21. Pishak, M.R. and Phillips, A.T. (1979) A modified radioisotopic assay for measuring glutamine synthetase activity in tissue extracts. Analyt. Biochem. 92, 82-88.
22. Pouly, S., Matthieu, J.-M. and Honegger, P. (1994) Mature oligodendrocytes in three-dimensional brain cell culture respond to protein kinase C stimulation by dedifferentiation, proliferation and remyelination. Schweiz. Archiv Neurol. Psy. 145, 27-29.
23. Schrier, B.K. and Shuster, L. (1967) Simplified radiochemical assay for choline acetyltransferase. J. Neurochem. 14, 977-985.
24. Shine, H.D., Readhead, C., Popko, B., Hood, L. and Sidman, R.L. (1992) Morphometric analysis of normal, mutant, and transgenic CNS : correlation of myelin basic protein expression to myelinogenesis. J. Neurochem. 58, 342-349.
25. Wilson, S.H., Schrier, B.K., Farber, J.-L. Thompson, E.J., Rosenberg, R.N., Blumme, A.J. and Nirenberg, M.W. (1972) Markers for gene expression in cultured cells from the nervous system. J. Biol. Chem. 247, 3159-3169.
26. Zurich, M.-G., Matthieu, J.-M. and Honegger, P. (1993) Improvement of culture conditions for aggregating brain cells to study remyelination. Schweiz. Arch. Neurol. Psy. 144, 228-232.

Monosialogangliosides and Nerve Growth Factor Stimulate Synthesis of Glycoproteins in Glioma Cells

Ephraim Yavin

Department of Neurobiology, The Weizmann Institute of Science, 76100 Rehovot, Israel

Keywords. Glycoconjugates, gangliosides, C6 glioma, extracellular matrix, growth factors, oligodendroglia

1 Glycoconjugates in Neurogenesis

The development of the neural cell phenotype is influenced by the genetic information contained within the nucleus, as well as by time-designated epigenetic contributions such as local modulation of the environment and cellular interactions with other developing cells and their cell surface signaling or secreted molecules. Differentiation is frequently accompanied by acquisition and modulation of cell surface constituents which permit both homotypic and heterotypic interactions between cells (5, 8, 20). Among these molecular constituents, glycoproteins (Gp)[1] are believed to provide the specificity required for cell-cell and cell-substrate recognition and to participate in the development of the neural phenotype.

In addition to Gp, the nervous system is enriched in complex gangliosides (Gg) that reside predominantly in the outer leaflet of the plasma membrane. Already during intrauterine life the developing brain shows enhanced formation of complex Gg (27). This period in the rat corresponds to the neuronal growth spurt and predates gliogenesis. In dissociated cells from the fetal rat cerebral cortex (26) cells extend neurites and form synapses in serum-free medium without hormonal supplements (32). They synthesize complex polysialo-Gg (27, 28) presumably as part of a repertoire which may enable them to interact with other neurons or with glia cells. Evidence in favor of the possibility that molecules on the surface of one cells interact with the oligosaccharide portion of Gg on the surface of another cells has been recently reviewed (19).

---

[1] **Abbreviations used:** CNS – central nervous system; DMEM – Dulbecco's modified Eagle medium; ECM – extracellular matrix; EGF – epidermal growth factor; FGF – fibroblast growth factor; Gg – ganglioside; Gp – glycoprotein; GD_{1a} – Gal[3-aNeuAc]b-3GalNacb-4Gal[3-aNeuAc]b-4G1c-Cer; GT_{1b} – Gal[3-aNeuAc]b-3GalNacb-4Gal[3-aNeuAc-aNeuAc]b-4G1c-Cer; GM_1 – Galb-3Gal Nacb-4Gal[3-aNeuAc]b-4G1c-Cer; GM_2 – Gal Nacb-4Gal[3-aNeuAc]b-4G1c-Cer; GM_3 – Gal[3-aNeuAc]b-4G1c-Cer; MW, molecular weight; NGF, nerve growth factor; SDS-PAGE, sodium dodecyl sulfate-polyacrylamide gel electrophoresis; TCA, tricholoroacetic acid.

2 Neurotrophic Influences of Exogenous Gangliosides on Cell Morphology

2.1 Effects on Neurons

Artificial supplements of Gg have been reported to induce marked morphological changes in a variety of cell lines and primary cultures (14). A combination of GM_1 and nerve growth factor (NGF)-stimulated neuritogenesis in PC12 cells (4), while GQ_{1b} stimulated neuritogenesis in a human cell line (22).

When SB21B1 neurohybrid cells were subjected to relatively high concentrations of a bovine brain Gg mixture, most cell bodies rounded up and assumed a neuron-like appearance with fibers extending from nearly every cell (18). The fiber network was maintained as long as Gg were present in the culture medium. Unlike the SB21B1 cells, rat pheochromocytoma PC12 cells were unable to extend neurites when artificially supplemented with GM_1 or GD_{1a}.

2.2 Effects on Neuroglia

Although there are a few reports that Gg affect the properties of astroglia in culture (3, 7, 21) much less attention has been directed towards the effects of Gg on neuroglia morphology than that of neurons. In a previous study we demonstrated that isolated bovine oligodendrocytes from the adult brain grown in serum-free medium and supplemented for one day with 50 mg/ml of GM_3, had significantly longer processes with more branching than control cells (33). After 50 days in culture treatment for one day with GM_3 resulted in a thickening of the processes and the appearance of intricate branches. The effect of GM_3 was also evident on the process formation in the O-2A lineage enriched from primary cultures of the neonatal rat brain (34).

3 Ganglioside-Modulated Gene Expression

In addition to their effect on cell morphology, Gg have been shown to affect the expression of a variety of cellular activities, possibly at the translational and posttranslational levels. These include stimulation of adenylate cyclase (25), inhibition of protein kinase C (9) activation of Na^+, K^+ ATPase (11) stimulation of phosphorylation (15), epidermal growth factor (EGF)-stimulated EGF receptor autophosphorylation (6) activation of the high affinity trkA receptor (16) and suppression of interferon-g-inducible expression of the major histocompatibility class I and class II mRNA levels in astrocytes (12).

The fact that Gg can modulate the cellular cytoarchitecture, prompted us to investigate their effect on expression of the cytoskeletal genes, tubulin and actin. In the neurohybrid SB21B1 clone after two weeks in the presence of Gg, tubulin mRNA species were greatly elevated, whereas few or no changes in the actin mRNA levels were detected (18). The elevated level of the tubulin message expression depended on the continuous presence of Gg, and returned to the basal level when Gg were removed. PC12 cells also responded to Gg supplements and displayed a significant increase in tubulin mRNA after one day (18). GD1a and GM_1, but not asialo-GM_1, were most effective and growth factors such as NGF and insulin induced elevation of tubulin message even more rapidly. We concluded that: (i) certain Gg selectively affect the expression of cytoskeletal gene products; (ii) amplification of such gene products occurs at distinct periods and may require the presence of the Gg; and (iii) Gg-enhanced increase in tubulin mRNA gene expression does not necessarily predict neurite outgrowth. In a subsequent study we examined *in vivo* the effect of Gg on central nervous system (CNS) lesions and found that a single intraperitoneal injection of GM_1 resulted in a substantial increase in the total RNA levels in the *substantia nigra* , but not in the striatum or cerebellum of the lesioned animals (30).

4 Ganglioside-Stimulated Release of Glycoproteins

Many surface Gp are released into the extracellular milieu. Possible mechanisms for this release include the shedding of membrane fragments, the active release of particles enriched with extracellular matrix (ECM) constituents and/or, perhaps most interesting, the purposeful release during normal neurogenesis or nerve regeneration. For example, the compounds released in the conditioned medium from C6 glioma cells has been a widely used biological source for putative growth factors in studying various aspects of survival and differentiation of nerve cells (2, 24). Many of the secreted substances in the medium are glycoconjugates which constitute the ECM. Secretion of glycoconjugates is a normal physiological process, not a pathologic consequence of cell damage or death (1, 17, 29).

4.1 Release of Glycoproteins

We have used C6 glioma cells to study the role of Gg and other extrinsic signals in regulating synthesis, secretion and assembly of ECM components by glia cells (31). When grown in monolayer cultures for a period of 24 h in the absence of serum or growth supplements, C6 cells incorporate D-[^{14}C]glucosamine into cellular metabolites, of which, most are Gp. A substantial fraction of the incorporated radioisotope (19.8%) was released and subsequently recovered in the medium after 24 h at 37°C as TCA insoluble material (Fig. 1). Addition of the growth factors b-fibroblast growth factor (b-FGF) and EGF each at 20 ng/ml, and insulin (10 ng/ml) had little effect on the amount of precipitable labeled material released into the culture medium. However, NGF at 20 ng/ml tripled the relative amount of the released material. Substances such as hydrocortisone (0.01 mM), dibutyryl cAMP (1 mM) and TPA (0.05 mM), which affect cell growth did not stimulate release. The Gg GM_3, GM_2 and GM_1, added at 50 mg/ml concentration, stimulated the release of TCA-labeled material into the culture medium by 24 h almost 3-fold while GD_{1a} and GT_{1b} had no effect.

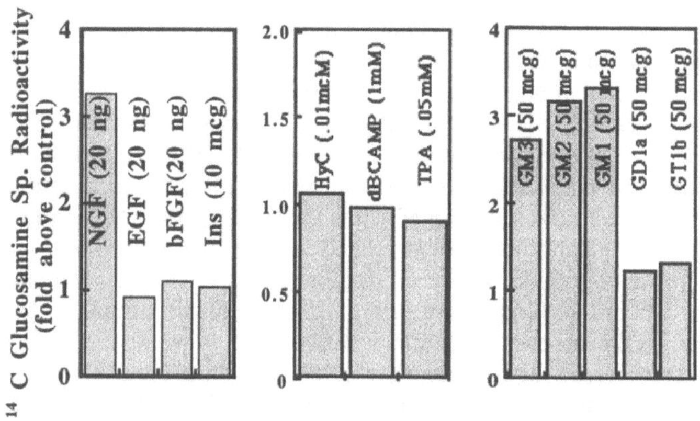

Fig. 1. Release of D-[^{14}C] glucosamine labeled Gp in C6 glioma cells in the presence of various agents. Cells were seeded in 24-well culture dishes and after 16 h the medium was removed and cells rinsed twice with serum-free medium. Each well was then supplemented with 0.2 ml Dulbecco's modified Eagle medium (DMEM) containing various additives at designated concentrations (given in brackets) in the presence of D-[^{14}C] glucosamine (2.5 mCi/ml). Incubation was carried out for 24 h. Trichloroacetic acid (TCA)-precipitable radioactivity in the medium was determined as described elsewhere (31). Values obtained from 4–6 culture wells (0.2 mg protein/well) are expressed as fold over control.

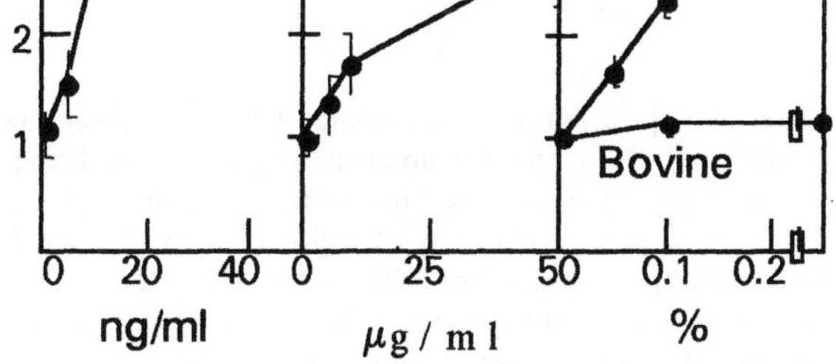

Fig. 2. Effect of serum and growth factors on Gp release. Experimental conditions were similar to those described in Fig. 1. NGF (50 ng/ml), GM_3 (50 mg/ml) or serum (%) were added for 24 h. Values expressed as arbitrary units were obtained from triplicate cultures from two separate experiments.

The effect of NGF and GM_1 concentration on release of labeled Gp is depicted in Fig. 2. Evidently 50 ng/ml NGF or 50 mg/ml GM_3 caused a nearly 4- and 3-fold increase, respectively, in the amount of released Gp. Addition of horse serum (0.5%) but not bovine serum, is as effective as NGF in enhancing the amount of Gp release. NGF-stimulated and Gg-enhanced release of Gp has been documented in PC12 pheochromocytoma cells (17, 29). Although the NGF receptors are less abundant in C6 glioma cells (23) they were sufficient to mediate Gp synthesis and secretion,which is the first defined neurotrophic factor-stimulated biochemical process demonstrated in these glia-derived cells

The fluorographic profile of medium-released Gp after stimulation of cells with NGF and GM_3 is shown in Fig. 3. Both agents stimulated release in cells grown in the absence of serum. Addition of horse serum or a defined medium (31) increased the basal levels of released Gp and the effect of GM_3 and NGF were less prominent. This suggests that serum and a combination of growth factors, as in the M2 medium, potentiate release.

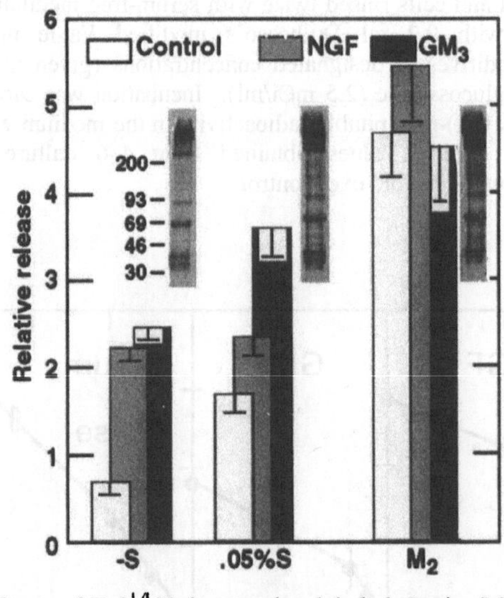

Fig. 3. Release of D-[^{14}C]glucosamine labeled Gp in C6 glioma cells in the presence of NGF and GM_3 and the absence or presence of horse serum or artificial medium. Experimental conditions were similar to these described in Fig. 1. Cells were incubated with D-[^{14}C] glucosamine (2.5 mCi/ml) in the presence of either GM_3, (50 mg/ml) or NGF (50 ng/ml). Medium was collected after 48 h, the Gp were TCA-precipitated, subjected to 7.5–15% sodium dodecyl sulfate-polyacrylamide gel electrophoresis (SDS-PAGE) and fluorography. Fluorographs in the insets are from cultures treated with NGF. The left size of the fluorograph depicts the MW markers in kDa.

The fluorographs in the inset of Fig. 3 illustrate radioactive bands in the region of molecular weights (MW) greater than 250 kDa, and also at 220 and 69 kDa. Radioactively labeled Gp were also detected in the 31–92.5 kDa area. These lower MW bands were mainly seen during incubations of longer duration (48 h). These labeled polypeptides could be degradation products of Gp or delayed products related to the NGF or Gg response.

GM_3 but not GT_{1b} Gg were effective in releasing two major Gp at 220 and 69 kDa as shown in Fig. 4. At all conditions studied, a 270 kDa band was detected in the culture medium, partly comprised of a high Mr protein complex released by the C6 cells, independent of the stimulus.

The appearance of radioactively labeled sialyl residues in the released Gp after NGF or GM_3 stimulation was compared to the cell-associated Gp. As shown in Fig. 5, the Gp labeled profile is fairly different between the medium and the cells. In the presence of sialidase, marked reductions in the amount and Mr size of the 220, 93, 69 kDa Gp was noticed. A drastic reduction in the amount of a low MW Gp (about 38 kDa) accompanied by the formation of a lower Mr compound was seen. The effect of sialidase on the cell-associated Gp was only moderate.

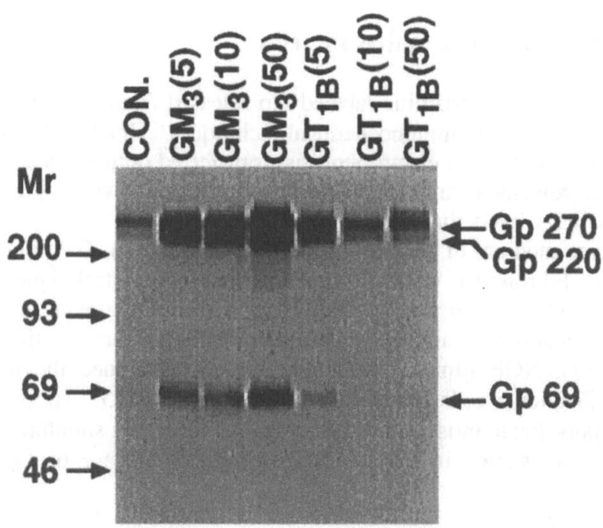

Fig. 4. Fluorograph of D-[^{14}C] glucosamine labeled Gp released from C6 cells upon exposure to various Gg. Cells were incubated for 24 h with D-[^{14}C] glucosamine (2.5 mCi/ml) in the presence of either GM_3 or GT_{1b} (5–50 mg/ml) and fluorographs prepared as described in Fig. 3.

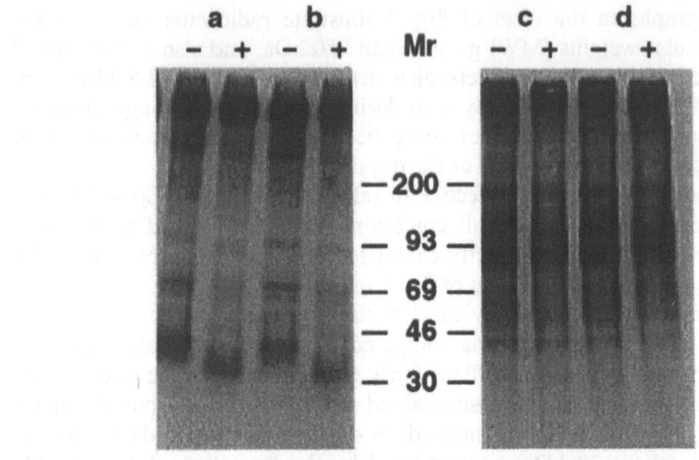

Fig. 5. Fluorograph of D-[^{14}C] glucosamine medium-released (a and b) and cell-associated (c and d) Gp from C6 cells untreated or treated (+) with sialidase. Cells were incubated with either GM_3 (a and c) or NGF (b and d) the isotope and simultaneously with 5 mU of Vibrio cholera neuraminidase for 48 h. Experimental conditions were similar to Fig. 3.

4.2 Release of Extracellular Matrix Proteins

Identification of the [^{14}C]glucosamine labeled Gp released from C6 cells upon stimulation was determined by immunochemical techniques. Unlabeled medium was precipitated, subjected to gel electrophoresis, transferred onto nitrocellulose, and incubated with polyclonal antisera to fibronectin. As shown in Fig. 6, a highly labeled 220 kDa band that reacted with the fibronectin antiserum was present in the culture medium of cells treated with NGF (lane 2), GM_3 (lane 3) and GM_1 (lane 4). Fibronectin release was not detected in control (lane 1) or GD_{1a}-treated cells (lane 5). An almost identical immunoreactive pattern of C6 culture medium proteins was obtained with an IgG preparation that identifies the entire J1 family (31). NGF stimulation of Gp release questioned the release specificity, since C6 glioma cells contain few NGF receptors (23). To examine the role of NGF a polyclonal antiserum directed against NGF was simultaneously added to C6 cells. As shown in Fig. 7, NGF-stimulated release of Gp was inhibited.

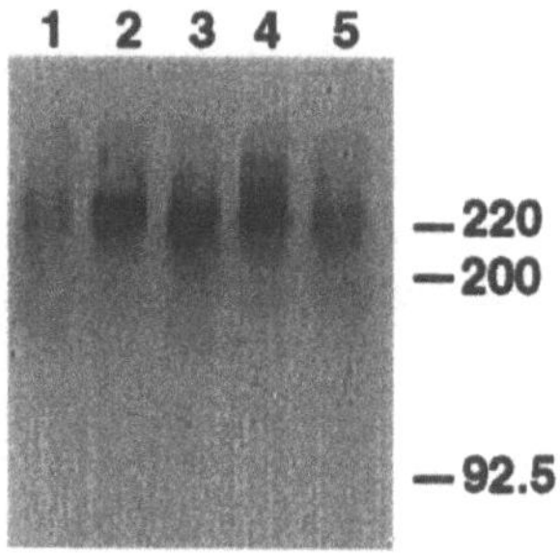

Fig. 6. Western blot of Gp released into the medium using fibronectin antibody. Media collected from control cells (lane 1) or after incubation with either 50 ng/ml NGF (lane 2), 50 mg/ml GM_3 (lane 3), 50 mg/ml GM_1 (lane 4), or 50 mg/ml GD_{1a} (lane 5) for 24 h. The TCA precipitates of the medium were resolved by 7% SDS-PAGE, transferred to nitrocellulose and incubated with a polyclonal antiserum raised against fibronectin (1:200). Immunoreactivity was detected with [^{125}I]protein A. Experiment was done in the absence of added sera.

Thus the effect of NGF appears highly specific on three accounts: (i) anti-NGF inhibits Gp release; (ii) other conventional growth factors (b-FGF, EGF, insulin, hydrocortisone and dibutyryl cAMP) do not cause Gp release, ruling out the possibility that Gp release is induced by impurities in the NGF preparation; and (iii) NGF evokes release of a 220 kDa and other Gp. The released 220 kDa band is comprised of at least of two major discrete ECM components, fibronectin and J1/tenascin (31). Thus, NGF-induced release of ECM Gp by glioma cells should provide new insights into the reciprocal interactions between neurons and glia cells during development. The results in Fig. 7, unexpectedly, also showed that the antiserum to NGF inhibited the GM_3- and GM_1-induced Gp release.

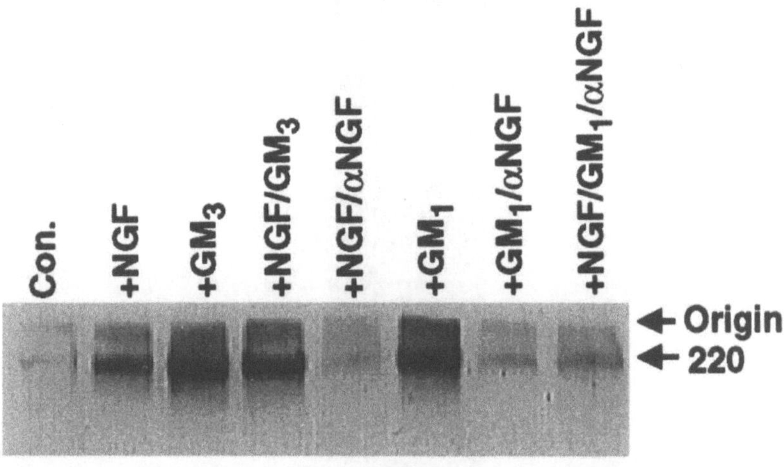

Fig. 7. Effect of anti-NGF serum on release of [^{14}C] glucosamine labeled Gp from C6 cells in the presence of NGF, GM_3 and GM_1 Gg. Experimental conditions were as detailed in Fig. 3, except that designated wells contained anti-NGF sera (1:40 dilution) raised in goat, for the entire period of incubation.

Specifically, the antiserum inhibited the release of the 220 kDa Gp from C6 cells incubated with NGF, GM_1 or GM_3 as observed from the SDS-PAGE profile. GM_3 by itself had no NGF-like activity as assessed by lack of stimulation of neurite outgrowth in PC12 cells (data not shown).

Secretion of Gp by glial cells particularly fibronectin and J1/tenascin are important physiological events during the assembly of the nervous system. Enhanced Gp release by monosialo-Gg appears to be highly specific on several grounds. Firstly, the stimulatory effect of a monosialo residue rules out nonspecific interactions, such as concentration-dependent micellar aggregates, as a cause for release. Secondly, monosialo-Gg induce the release of the similar ECM molecules and radioactively labeled Gp as NGF. The nearly identical released Gp profile in conjunction with the inhibition of release by the anti-NGF sera suggests for the first time, the possibility that Gg-mediated regulation of NGF secretion may take place in glial cells. C6 cells have been shown previously to secrete NGF into the culture medium (13). NGF impurities in monosialo-Gg are improbable and certainly not more than those present in polysialo-Gg preparations. Based on the inhibition studies with the antibody to NGF it would appear that monosialo-Gg may indirectly act on Gp release by stimulating synthesis and/or secretion of NGF or NGF-like molecules, which can in turn regulate synthesis and/or release of fibronectin and J1/tenascin from the cells.

A most plausible explanation to account for this effect is that GM_3 and other monosialo-Gg may evoke an autocrine response to enhance synthesis and release of NGF. This possibility is most likely to exist since Northern blot analysis of the cells indicate an elevation of the NGF mRNA levels after exposing cells to GM_3 (Fig. 8).

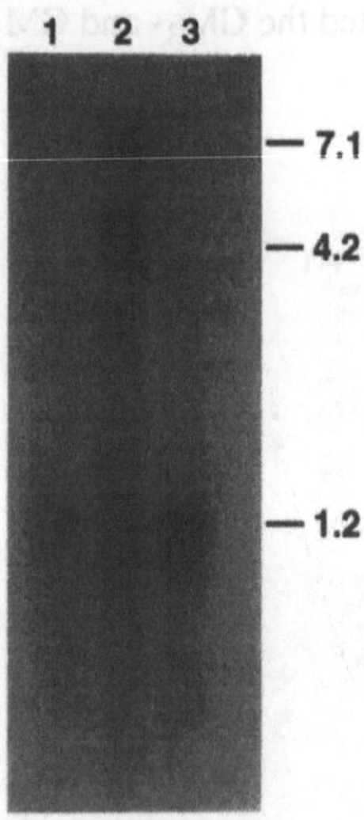

Fig. 8. mRNA levels of NGF in C6 cells treated with GM_3. Northern blot analysis was performed after 24 h exposure of the cells to either 50 mg/ml GM_3 (lane 2) or 50 ng/ml NGF (lane 3). Note the expression of pre-pro NGF (7.1 Kb) and NGF (4.1 Kb) gene products.

At this time the mechanism by which a Gg may exert such an effect on Gp release in C6 cells and the commonality of this phenomenon in other cells remains unknown. It is possible that exogenous monosialo-Gg is taken up by endocytic routes into the cell where it can affect NGF gene expression or its down stream synthetic and secretion processes. Alternatively it could increase the availability of high affinity NGF receptors the presence of which has been most recently noted (10). These alternative possibilities are schematically illustrated in Fig. 9.

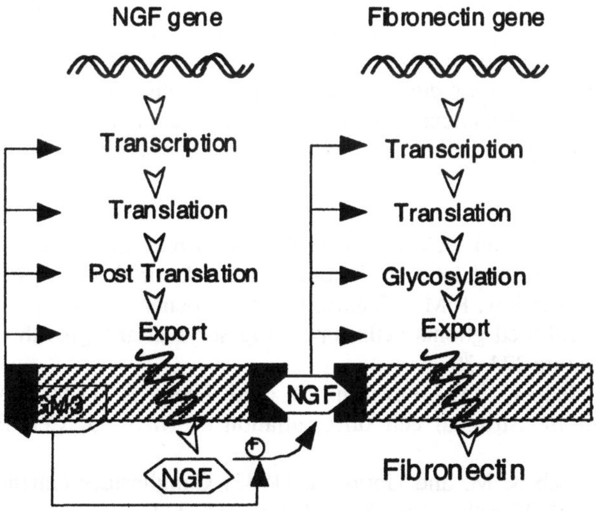

Fig. 9. An hypothetical model for stimulus-coupled Gg activation of NGF and fibronectin gene products secretion. GM_3 can either integrate in the plasma membrane or enter the cell and affect any of the intracellular processes up to the transcriptional level to stimulate production and secretion of NGF. The latter in an autocrine fashion couples indirectly the Gg signal with the release of fibronectin into the extracellular compartment.

According to this scheme the presence of NGF or other related growth factors is necessary to elicit the production and secretion of ECM molecules such as fibronectin or J1-tenascin. Exogenous monosialo Gg acts by either insertion into the plasma membrane or via endocytic internalization to enhance a cascade of events which ultimately couple the NGF release with the ECM Gp secretion. Whether the stimulation of ECM secretion by GM_3 or GM_1 takes place between homologous cells interacting with each other (i.e., glia), whereas termination of the signal may be caused by a negative feedback exerted by cells which contain polysialo Gg (i.e., neurons) remains to be investigated.

5 Summary

D-[^{14}C]glucosamine has been used to study stimulus-coupled secretion of Gp in a variety of neural cells. Exposure of C6 glioma cells to NGF or monosialo Gg enhanced the release of a number of Gp the most prominent of which being 270, 220, and 69 kDa Gp. Several growth factors, hormones, phorbol ester and disialo- and trisialo-Gg did not stimulate secretion. Western blot analysis of the conditioned medium from C6 cells stimulated with NGF or monosialo-Gg, identified one distinct band of approximately 220 kDa for fibronectin and J1/tenascin which comigrated. Antiserum to NGF prevented NGF-stimulated release and also blocked monosialo-Gg-evoked release. These results extend the range of neurotrophic properties attributed to NGF to cells of glial origin and suggest that NGF regulates secretion of ECM proteins. It suggests that monosialo-Gg-stimulated secretion of ECM may be mediated by an NGF or NGF-like molecule also secreted by the C6 glioma cells. The experiments are discussed in the wider context of acquisition of surface molecule and release of growth factors to support cell-cell interaction and differentiation.

6 References

1 .Arenander, A.T. and DeVellis, J. (1980) Glial-released proteins in clonal cultures and their modulation by hydrocortisone. Brain Res. 200, 401–419.

2. Barde, Y.A., Lindsay, R.M., Monard, D. and Thoenen, H. (1978) New factor released by cultured glioma cells supporting survival and growth of sensory neurons. Nature 274, 20.

3. Facci, L., Skaper, S.D., Favaron, M. and Leon, A. (1988) A role of gangliosides in astroglial cell differentiation *in vitro*. J. Cell Biol. 106, 821–828

4. Ferrari, G., Fabris, M. and Gorio, A. (1983) Gangliosides enhance neurite outgrowth in PC12 cells. Dev. Brain Res. 8, 215–222.

5. Grumet, M., Hoffman, S., Crossin, K.L. and Edelman, G.M. (1985) Cytotactin, an extracellular matrix protein of neural and nonneural tissues that mediates neuron-glia interaction. Proc. Natl. Acad. Sci. USA 82, 8075–8079.

6. Hanai, N., Dohi, T., Nores, G.A. and Hakomori, S. (1988) A novel ganglioside, de-N-acetyl-GM_3 ($II^3NeuNH_2LacCer$), acting as a strong promoter for epidermal growth factor receptor kinase and as a stimulator for cell growth. J. Biol. Chem. 263, 6296–6301.

7. Hefti, F., Hartikka, J. and Frick, W. (1985) Gangliosides alter morphology and growth of astrocytes and increase the activity of choline acetyltranferase in cultures of dissociated septal cells. J. Neurosci. 5, 2086–2094.

8. Hoffman, S., Crossin, K.L. and Edelman, G.M. (1988) Molecular forms, binding functions, and developmental expression pattern of cytotactin and cytotactin-binding proteoglycan, an interactive pair of extracellular matrix molecules. J. Cell Biol. 106, 519–532.

9. Kreutter, D., Kim, J.Y.H., Goldenring, J.R., Rasmussen, H., Ukomadu, C., De Lorenzo, R.J. and Yu, R.K. (1987) Regulation of protein kinase C activity by gangliosides. J. Biol. Chem. 262, 1633–1637.

10. Kumar, S., Huber, J., Peña, L.A., Pérez-Polo, J.R., Werrbach-Pérez, K. and DeVellis, J. (1990) mAb 217C recognizes NGFR in C6 glioma cells. Proc. ASN 21, 158.
11. Leon, A., Facci, L., Toffano, G., Sonnino, S. and Tettamanti, G. (1981) Activation of (Na^+,K^+)-ATPase by nanomolar concentrations of GM_1 ganglioside. J. Neurochem. 37, 350–357.
12. Massa, P.R. (1993) Specific suppression of major histocompatibility complex class I and class II genes in astrocytes by brain-enriched gangliosides. J. Exp. Med. 178, 1357–1363.
13. Murphy, R.A., Oger, J., Saide, J.D., Blanchard, M.H., Arnason, B.G.W., Hogan, C., Pantazis, N.J. and Young, M. (1977) Secretion of nerve growth factor by central nervous system glioma cells in culture. J. Cell Biol. 72, 769–773.
14. Nagai, Y. and Tsuji S. (1988) Cell biological significance of gangliosides in neural differentiation and development: Critique and proposals. In R.W. Ledeen, E.L. Hogan, G. Tettamanti, A.J. Yates and R.K. Yu (eds.), New Trends in Ganglioside Research, Vol. 14, Liviana Press, Padova, pp. 329–350.
15. Nakaoka, T., Tsuji, S. and Nagai, Y. (1992) Bimodal regulation of protein phosphorylation by a ganglioside in rat brain membrane. J. Neurosci. Res. 31, 724–730.
16. Rabin, S., Colangelo, A.M. and Mocchetti, I. (1994) GM_1 ganglioside activates the high affinity nerve growth factor receptor trkA. Soc. Neurosci. Abstr. 20, 23.18.
17. Richter-Landsberg, C., Lee, V.M., Salton, S.R.J., Shelanski, M.L. and Greene, L.A. (1984) Release of the NILE and other glycoproteins from cultured PC12 rat pheochromocytoma cells and sympathetic neurons. J. Neurochem. 43, 841–848.
18. Rybak, S., Ginzburg, I. and Yavin, E. (1983) Gangliosides stimulate neurite outgrowth and induce tubulin mRNA accumulation in neural cells. Biochem. Biophys. Res. Commun. 116, 974–980.
19. Schengrund, C.-L. (1995) Evidence that molecules on the surface of one cell can adhere to the oligosaccharide portion of gangliosides on the surface of another cell. Biol. Signals 4, 1–13.
20. Schubert, D. and LaCorbière, M. (1982) The specificity of extracellular glycoprotein complexes in mediating cellular adhesion. J. Neurosci. 2, 82–89.
21. Skaper, S.D., Facci, L., Rudge, J., Katoh-Semba, R., Manthorpe, M. and Varon, S. (1986) Morphological modulation of cultured rat brain astroglial cells: Antagonism by ganglioside GM1. Dev. Brain Res. 25, 21–31.
22. Tsuji, S., Yamashita, T., Matsuda, Y. and Nagai, Y. (1992) A novel glycosignalling system: GQ_{1b}-dependent neuritogenesis of human neuroblastoma cell line, GOTO, is closely associated with GQ_{1b}-dependent ecto-type protein phosphorylation. Neurochem. Int. 21, 549–554.

23. Vinores, S.A. (1983) Increased adhesion response of anaplastic glioma cells to nerve growth factor and the presence of specific receptors. J. Neurosci. Res. 10, 381–395.
24. Westermann, R., Hardung, M., Meyer, D.K., Ehrhard, P., Otten, U. and Unsicker, K. (1988) Neuronotrophic factors released by C6 glioma cells. J. Neurochem. 50, 1747–1758.
25. Yates, A.J., Walters, J.D., Wood, C.L. and Johnson, J.D. (1989) Ganglioside modulation of cyclic AMP-dependent kinase and cyclic nucleotide phosphodiesterase *in vitro*. J. Neurochem. 53, 162–167.
26. Yavin, E. and Yavin, Z. (1974) Attachment and culture of dissociated cells from rat embryo cerebral hemispheres on polylysine-coated surface. J. Cell Biol. 62, 540–546.
27. Yavin, E. and Yavin, Z. (1979) Ganglioside profiles during neural tissue development: Acquisition in the prenatal rat brain and cerebral cell cultures. Dev. Neurosci. 2, 25–37.
28. Yavin, E., Richter-Landsberg, C., Duksin, D. and Yavin, Z. (1984) Tunicamycin blocks neuritogenesis and glucosamine labeling of gangliosides in developing cerebral neuron cultures. Proc. Natl Acad Sci. USA 81, 5638–5642.
29. Yavin, E., Hama, T., Gil, S. and Guroff, G. (1986) NGF and ganglioside stimulate release of glycoproteins in PC12 pheochromocytoma cells. J. Neurochem. 46, 794–803.
30. Yavin, E., Gil, S., Consolazione, A., dal Tosso, R. and Leon, A. (1987) Selective enhancement of tubulin gene expression and increase in oligo(dt) bound RNA in the rat brain after nigrostriatal pathway unilateral lesion and treatment with ganglioside. J. Neurosci. Res. 18, 615–621.
31. Yavin, E., Gabai, A. and Gil, S. (1991) Nerve growth factor mediates monosialoganglioside-induced release of fibronectin and J1/tenascin from C6 glioma cells. J. Neurochem. 56, 105–112.
32. Yavin, Z. and Yavin, E. (1980) Survival and maturation of cerebral neurons on poly-L-lysine surfaces in the absence of serum. Dev. Biol. 75, 454–459.
33. Yim S-H., Yavin E., Hammer J.A. and Quarles, R.H. (1991) Exogenous GM3 ganglioside stimulates process formation and glycoprotein release by cultured bovine oligodendrocytes. J. Neurochem. 57, 2144–2147.
34. Yim, S.H., Farrer, R.G., Hammer, J.A., Yavin, E. and Quarles, R.H. (1994) Differentiation of oligodendrocytes cultured from developing rat brain is enhanced by exogenous GM3 ganglioside. J. Neurosci. Res. 38, 268–281.

Ion Channels and Neurotransmitter Receptors

Responses of Oligodendrocytes to Classical Neurotransmitters: Signaling Pathways

Guillermina Almazan, Amani Khorchid, Hsueh-Ning Liu and Eduardo Molina-Holgado

Department of Pharmacology and Therapeutics, McGill University, 3655 Drummond St., Montreal, Quebec H3G 1Y6, Canada

Keywords: adrenergic receptors, AMPA/kainate glutamatergic receptors, muscarinic receptors, c-fos, second messengers, oligodendroglia

1. Introduction

In the central nervous system (CNS), the time and region-specific pattern of myelination suggests the existence of neuronal signals which regulate oligodendrocyte proliferation and differentiation as well as myelin formation (15,29). Neurotransmitters are a group of molecules of major importance for the CNS. Their main function is to serve as chemical signals of intercellular communication by activation of specific receptors and second messenger systems in postsynaptic cells. However, there is ample pharmacological evidence suggesting that some of these substances regulate trophic biological processes (see 28 and 31 for review).

Oligodendroglial lineage cells possess α_{1A}-adrenergic (11), β-adrenergic (35,47), and muscarinic receptors (12,19) coupled to second messenger systems involving inositol 1,4,5-trisphosphate (IP_3), diacylglycerol (DG), 3',5'-cyclic adenosine monophosphate (cAMP) or Ca^{2+} fluxes. In addition to neurotransmitter receptors, oligodendroglial lineage cells also have GABA- and glutamate-activated channels (3,4,7,22,40).

We have used purified cultures of oligodendrocyte progenitors to characterize the signaling pathways activated by the classical neurotransmitters acetylcholine, noradrenaline and the excitatory amino acid glutamate. To elucidate the potential role(s) of these molecules on oligodendrocyte progenitor development, the expression of the immediate early-transcription factor (IETF) c-fos and proliferation have been assessed under different experimental conditions. The objective of this article is to review some of our recent findings with particular emphasis on the regulation of muscarinic receptors in oligodendrocyte progenitor cultures.

2. Muscarinic acetylcholine receptor activation in oligodendroglial cells

Muscarinic receptors possess seven membrane spanning domains and transduce their biological effects through a variety of associated G-proteins. Type m1

Molecular Signaling and Regulation in Glial Cells
A Key to Remyelination and Functional Repair

receptors (m1,m3 and m5) are positively coupled to phospholipase C (PLC), whereas m2-type receptors (m2 and m4) negatively regulate the formation of cAMP (6). Thus, stimulation of m1-type receptors results in the hydrolysis of phosphatidylinositol 4,5-bisphosphate (PIP_2) leading to the formation of IP_3 and DG. These second messengers cause mobilization of intracellular calcium ($[Ca^{2+}]_i$) and activation of protein kinase C (PKC) which can trigger many cellular responses, including changes in gene expression. Conversely, stimulation of m2-type muscarinic receptors can lead to inhibition of cAMP formation, resulting in decreased activity of the cAMP-dependent protein kinase (PKA).

Muscarinic receptors were first shown to be present in purified myelin by binding studies (24). Functional assays also confirmed that such receptors interact with PLC (12,21,24,36) and the adenylate cyclase system (12,25). Similarly, stimulation of oligodendroglial cell cultures with carbachol increases $[Ca^{2+}]_i$ as shown by Ca^{2+} imaging techniques in single cells (12,19,38) and cell suspensions of bulk-isolated oligodendrocyte progenitors (12). We have shown that production of the different inositol phosphates (IPs) and $[Ca^{2+}]_i$ are both concentration and time-dependent and are preferentially blocked by pirenzepine, a selective M1 antagonist but not by methoctramine, a selective M2 antagonist. Interestingly, the rise in $[Ca^{2+}]_i$ is a consequence of release from intracellular stores as well as influx from the extracellular milieu (12). Polymerase chain reaction carried out with specific m1 and m2 muscarinic receptor primers confirmed that these cells express the two muscarinic receptor subtypes (12) which could in principle mediate the stimulation of PLC and inhibition of adenylate cyclase stimulated by carbachol. These results are in agreement with those reported earlier (25), demonstrating the presence of at least two types of muscarinic binding sites (M1 and M2) in myelin.

2.1 Regulation of carbachol-stimulated phosphoinositide (PI) hydrolysis and $[Ca^{2+}]_i$

Muscarinic receptors mediating PLC activation/$[Ca^{2+}]_i$ are coupled to a pertussis toxin (PTX)-insensitive G protein, since treating oligodendrocyte progenitor cultures with 100 ng/ml PTX prior to the stimulation with carbachol (100 mM) did not reduce IPs formation or the increase in $[Ca^{2+}]_i$ (Figure 1). In support of our observations, the presence of PTX-insensitive Gq/G_{11a} has been reported in myelin (5).

Considerable evidence suggests that several G-protein coupled receptors are regulated by PKC. To examine if activation of PKC affects the muscarinic-stimulated mobilization of $[Ca^{2+}]_i$ or PI hydrolysis, cultures were treated with phorbol-12-myristate 13-acetate (TPA). A 30 min pre-incubation with 1 mM TPA abolished both the carbachol-stimulated PI hydrolysis and the mobilization of $[Ca^{2+}]_i$, suggesting that the receptors are regulated by a PKC-dependent negative feedback mechanism (Figure 1, reference 12). However, the mechanism of regulation is unclear as there are several possible targets for the action of PKC, including the receptor, the putative G protein, and PLC itself.

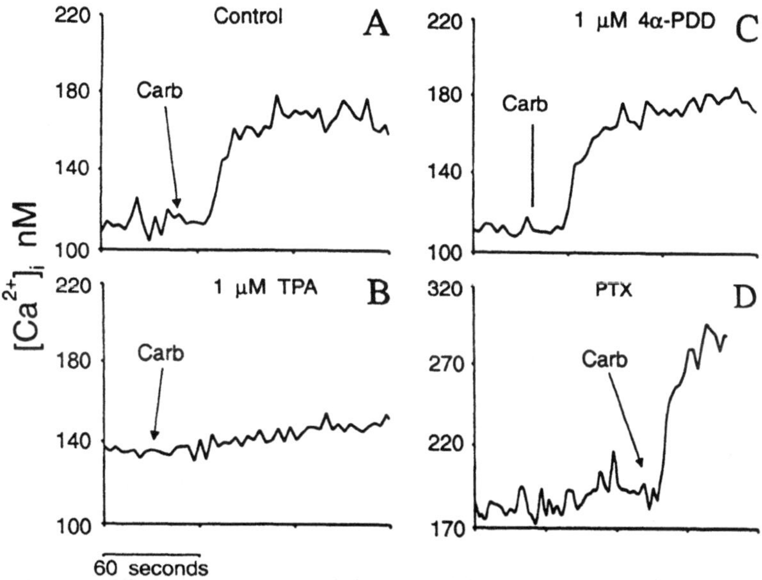

Figure 1. Effect of PKC activation on carbachol-stimulated increases in $[Ca^{2+}]_i$. Cultures were treated with 100 mM carbachol without (A) or with 30 min incubation with 1 mM TPA (B), 4a-PDD, an inactive analog or (D) 100ng/ml PTX. Changes in $[Ca^{2+}]_i$ were measured with the fluorescent dye Fura-2. Data published in reference 12.

G-protein coupled receptors can also be subject to regulation by desensitization or down-regulation following repetitive or prolonged exposure to their own agonist. Desensitization of the b-adrenergic receptor involves a rapid and reversible functional uncoupling of the receptor which is attributed largely to phosphorylation by PKA or by the b-adrenergic receptor kinase, and subsequent sequestration of the receptors away from the cell surface (see for example 41). We were interested to determine whether muscarinic receptors coupled to phosphoinositide hydrolysis are desensitized after a short exposure to the agonist (32). Pre-incubation of the cultures with carbachol (100 mM) produced a rapid and time-dependent decrease of the carbachol-stimulated IPs formation. This functional effect was associated with a concomitant decrease of cell surface receptors, as measured by binding of the hydrophilic ligand [^{3}H]NMS to intact cell monolayers (Figure 2). However, binding with the lipophilic ligand [^{3}H]QNB, which labels the total muscarinic binding sites, was not significantly altered during the first 60 min, suggesting that the receptors are only internalized and not down-regulated. Interestingly, neither PKC nor PKA appears to be involved in this desensitization phenomenon because inhibitors of both kinases do not prevent the decrease in [^{3}H]NMS surface binding or carbachol-stimulated PI hydrolysis associated with agonist pre-treatment. Additional studies will be necessary to further characterize the mechanisms regulating muscarinic receptors in oligodendrocytes, as well as their physiological roles.

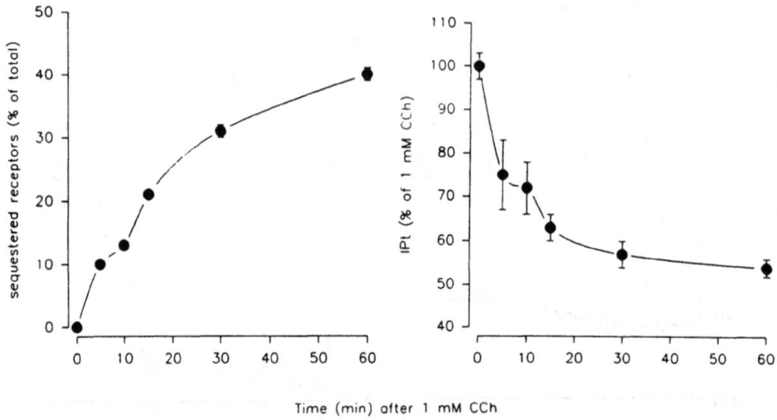

Figure 2. Agonist induced sequestration of muscarinic receptors and desensitization of PI hydrolysis stimulated by carbachol. Cells were pre-incubated for different times with 100 mM carbachol from 0 to 60 min. Sequestered receptors and total receptors were determined by binding with [^{3}H]NMS and [^{3}H]QNB. PI hydrolysis was determined after 10 min challenge to 100 mM carbachol in cells pre-labeled with myo-[^{3}H]inositol.

2.2 Muscarinic receptors activate Mitogen-Activated-Protein Kinase (MAPK) and induce c-fos mRNA in oligodendroglial cells

MAPKs are protein-serine/threonine kinases that are rapidly activated upon stimulation of a variety of cell surface receptors, including G protein coupled receptors (see 9 for review). In collaboration with Jorge Larocca (manuscript in preparation), we have found that stimulation of oligodendroglial cell cultures with carbachol activates a protein kinase which utilizes myelin basic protein as substrate and has a similar size (~40kDa) to MAPKs. It is tempting to speculate that these proteins may be one of the missing links connecting the activation of muscarinic cell surface receptors with changes in gene expression. A group of potential targets may be the IETF with c-fos as the prototype. Neurotransmitter-mediated signals can be translocated to the nucleus via this oncogene, functioning as a "nuclear third messenger" (14). Using Northern blot analysis, we have demonstrated that carbachol causes a time and concentration-dependent increase in c-fos mRNA which is blocked by atropine, a non-selective muscarinic receptor antagonist (1). Regulation of c-fos induction by carbachol involves several molecules, including changes in intracellular Ca^{2+}, PKC and phospholipase A_2 activation as determined with Ca^{2+} chelators and specific inhibitors of these enzymes (13). Thus carbachol-induced changes in oligodendroglial gene expression might be involved in programming long-term changes in development, survival and death. Further elucidation of the molecular events and physiological relevance will be of considerable interest.

3. Adrenergic receptor activation in oligodendroglial cells

Oligodendrocytes respond to noradrenaline (NA) with increases in cAMP (11,30,39) and accumulation of IPs (11,36), effects mediated by stimulation of b and a1-adrenoceptors, respectively. The pharmacological profile of receptors mediating PI hydrolysis suggests the presence of the a_{1A} subtype rather than a_{1B} (11). Thus, WB4101, an a_{1A} selective antagonist, blocks the NE-induced IPs accumulation while chloroethylclonidine, which selectively inactivates a_{1B} receptors, has no effect. The response is also sensitive to the influx of extracellular calcium and over-night pre-treatment with pertussis toxin. In addition, NA-induced PI hydrolysis is negatively regulated by activation of PKC with TPA (Khorchid and Almazan, in preparation). We have recently observed that NA is also capable of inducing c-fos transcription involving both the a_{1A} and b-adrenoceptors. These results are not surprising because the promoter of this proto-oncogene contains multiple regulatory elements, including the serum response element and the calcium and cAMP response element (see 14 for review), all of which are required in concert for the regulation of c-fos expression in transgenic mice (37).

4. Glutamatergic receptors activation in oligodendroglial cells

Evidence for the expression of glutamate receptors in oligodendroglial lineage cells is accumulating. Subpopulations of cultured oligodendrocytes from the spinal cord, cerebellum and optic nerve were reported to depolarize after application of glutamate (3,4,20,34,40), and progenitors release pre-loaded ^{3}H-GABA, taken up by a Na^+-dependent GABA carrier, in response to stimulation by both quisqualate and kainate but not by NMDA (17). Moreover the presence of glutamate receptors in these cells *in vivo* has also been provided by the observations that acutely isolated progenitors respond to glutamate (3), and that cobalt uptake is stimulated by quisqualate in optic nerve preparations (16). Recent studies using molecular and electrophysiological approaches have confirmed the co-expression of multiple glutamate receptor subunits which form ion channels preferentially activated by AMPA or kainate (18,34). The signaling cascade initiated by the binding of glutamate to AMPA/kainate receptors is triggered by depolarization due to Na^+ entry through these ligand-gated receptor channels. Several groups have reported a rise in intracellular calcium (10,19,35,38) which is primarily due to transmembrane Ca^{2+} influx through both Ca^{2+}-permeable glutamate-gated channels and voltage-dependent Ca^{2+} channels.

Treatment	c-fos mRNA (Optical Density)	[^{3}H]thymidine Incorporation (cpm)
Control	170 ± 2	585 ± 54
Glutamate	453 ± 6	446 ± 39
Kainate	502 ± 36	279 ± 31
CNQX	173 ± 60	1007 ± 85
Glutamate + CNQX	206 ± 38	1021 ± 63
Kainate + CNQX	-	782 ± 114
Glutamate + H7	187 ± 10	-
Kainate + H7	181 ± 19	-

TABLE 1. Effects of selective glutamate agonists and antagonists on c-fos induction and [^{3}H]thymidine incorporation in oligodendrocyte progenitors. Cells were exposed to antagonist CNQX (50 mM) or protein kinase C inhibitor H7 (20 mM) prior to glutamate or kainate stimulation. c-Fos mRNA levels were measured by Northern blot analysis followed by densitometry of the autoradiograms. Cultures were exposed to the agonists (100-300 mM) 30 min for c-fos induction and 24 h for [^{3}H]thymidine incorporation. (see for details reference 26)

The expression of voltage-dependent Ca^{2+} channels has been shown to be developmentally regulated in cells of the oligodendroglial lineage (8,23). We have recently found that one of the responses mediated via the rise in intracellular Ca^{2+} is the activation of PLC resulting in IP_3 accumulation (27). Furthermore, we and others have shown that glutamate induces the expression of c-fos and NGFI-A (2,18,26,35). The induction of c-fos was found to be dependent on extracellular calcium influx and the downstream activation of PKC. One important biological consequence of glutamate-stimulation of oligodendrocyte progenitors appears to be the arrest of cellular proliferation at concentrations (1-1000 mm) which are not cytotoxic (26).

The individual cellular responses to glutamate and other neurotransmitters appear to be heterogeneous and dependent on the tissue of origin and maturation stage of oligodendrocytes (18,38). Therefore, the heterogeneity in the Ca^{2+} responses to neuroligands have been suggested to reflect a restricted expression of neurotransmitter receptors in subsets of oligodendrocytes (38). We have also found that the density of muscarinic receptors, determined by [^{3}H]NMS in intact cells, is 3-4 times lower in mature oligodendrocytes than in progenitors (32). Similarly, the carbachol-stimulated PI hydrolysis is considerably lower (12). A possible explanation for our observations is that muscarinic receptors are down-regulated in a selective manner, although changes in coupling and affinity properties of the receptors may also occur.

5. Conclusions

Although most of the experiments discussed in this paper have been conducted in oligodendrocyte cultures at various stages of development, they strongly support the view that classical neurotransmitters play a role(s) in the development and growth of oligodendrocytes. Potentially, neurotransmitters could have wide-ranging effects in oligodendrocytes, including mitosis, cell type specificity, process formation, differentiation and survival.

6. Acknowledgements

Work reported in this article has been supported by grants from the Medical Research Council of Canada and from the John Hopkins Center for Alternatives to Animal Testing, USA. E.M.-H. is supported by a fellowship from the Ministry of Education and Science, Spain. H.-N.L. is supported by a studentship from the Multiple Sclerosis Society of Canada.

7. References

1. Almazan, G. and Cohen, R.I. (1993) Carbachol induces c-fos mRNA expression in oligodendrocytes. Trans. Am. Soc. Neurochem. 24, 176.
2. Almazan, G. and Liu, H.-N. (1994) Glutamate induces c-fos and c-jun gene expression in oligodendrocytes. XIIth Int. Congress of Pharmacology Abs. P18.13.1.
3. Barres, B.A., Koroshetz, W.J., Swartz, K.J., Chun, L.L.Y., and Corey, D.P. (1990) Ion channel expression by white matter glia: the oligodendrocyte glial progenitor cell. Neuron 4, 507-524.
4. Berger, T., Walz, W., Schnitzer, J. and Kettenmann, H. (1992) GABA- and glutamate-activated currents in glial cells of the mouse corpus callosum slice. J. Neurosci. Res. 31, 21-27.
5. Berti-Matera, L.N., Douglas, J.D., Mattera, R. and Goraya, T.Y. (1992) Identification of G protein subtypes in Peripheral Nerve and cultured Schwann cells. J. Neurochem. 59,1729-1735.
6. Birnbaumer, L., Abramowitz, J., and Brown, A. M. (1990) Receptor-effector coupling by G-proteins. Biochim Biophys. Acta 1031, 163-224.
7. Blankenfeld, G. von, Trotter, J. and Kettenmann, H. (1991) Expression and developmental regulation of a $GABA_A$ receptor in cultured murine cells of the oligodendrocyte lineage. Eur. J. Neurosci. 3, 310-316.
8. Blankenfeld, G. von, Verkhratsky, A.N. and Kettenmann, H. (1992) Ca^{2+} channel expression in the oligodendrocyte lineage. Eur. J. Neurosci. 4, 1035-1048.
9. Blumer, K.J. and Johnson G.L. (1994) Diversity in function and regulation of MAP kinase pathways. TIBS 19, 236-240.

10. Borges, K., Ohlemeyer, C., Trotter, J. and Kettenmann, H. (1994) AMPA/Kainate receptor activation in murine oligodendrocyte precursor cells leads to activation of a cation conductance, calcium influx and blockade of delayed rectifying K^+ channels. Neuroscience 63, 135-149.
11. Cohen, R.I., and Almazan, G. (1993) Norepinephrine-stimulated PI hydrolysis in oligodendrocytes is mediated by a_{1A}-adrenoceptors. NeuroReport 4, 1115-1118.
12. Cohen, R.I., and Almazan, G. (1994) Rat oligodendrocytes express muscarinic receptors coupled to phosphoinositide hydrolysis and adenylyl cyclase. Eur. J. Neurosci. 6, 1213-1224.
13. Cohen, R.I., Molina-Holgado, E. and Almazan, G. Regulation of c-fos induction by muscarinic receptors-activation in oligodendrocyte progenitors (submitted for publication).
14. Curran, T. and Morgan, J.I. (1994) Fos: an immediate-early transcription factor in neurons. J. Neurobiol. 26, 403-412.
15. Dutly, F. and Schwab, M.E. (1991) Neurons and astrocytes influence the development of purified O-2A progenitor cells. Glia 4, 559-571.
16. Fulton, B.P., Burne, J.F. and Raff, M.C. (1992) Visualization of O-2A progenitor cells in developing and adult rat optic nerve by quisqualate-stimulated cobalt uptake. J. Neurosci. 12, 4816-4833.
17. Gallo, V., Giovannini, C., Suergiu, R., and Levi, G. (1989) Expression of excitatory amino acid receptors by cerebellar cells of the type-2 astrocyte cell lineage. J. Neurochem 52, 1-9.
18. Gallo, V., Patneau, D.K., Mayer, M.L. and Vaccarino F.M. (1994) Excitatory amino acid receptors in glial progenitor cells: molecular and functional properties. Glia 11, 94-101.
19. Kastritsis, C.H. and McCarthy, K.D. (1993) Oligodendroglial lineage cells express neuroligand receptors. Glia 8, 106-113.
20. Kettenmann, H., Gilbert, P. and Schachner, M. (1984) Depolarization of cultured oligodendrocytes by glutamate and GABA. Neurosci. Lett. 47, 271-276.
21. Khan, D. W., and Morell, P. (1988) Phosphatidic acid and phosphoinositide turnover in myelin and its stimulation by acetylcholine. J. Neurochem. 50, 1542-1550.
22. Kirchhoff, F. and Kettenmann, H. (1992) GABA triggers a $[Ca^{2+}]_i$ increase in murine precursor cells of the oligodendrocyte lineage. Eur. J. Neurosci. 4, 1049-1058.
23. Kirischuk, S., Scherer, J., Moller, T., Verkhratsky, A. and Kettenmann, H. (1995) Subcellular heterogeneity of voltage-gated Ca^{2+} channels in cells of the oligodendrocyte lineage. Glia 13, 1-12.
24. Larocca, J. N., Cervone, A., and Leedeen, R. W. (1987a) Stimulation of phosphoinositide hydrolysis in myelin by muscarinic agonist and potassium. Brain Res. 436, 357-362.

25. Larocca, J.N., Ledeen, R.W., Dvorkin, B., and Makman, M.H. (1987b) Muscarinic receptor binding and muscarinic receptor-mediated inhibition of adenylate cyclase in rat brain myelin. J. Neurosci. **7,** 3869-3876.
26. Liu, H.N. and Almazan, G. (1995) Glutamate induces c-fos proto-oncogene expression and inhibits proliferation in oligodendrocyte progenitors: receptor characterization. Eur. J. Neurosci., in press.
27. Liu, H.N. and Almazan, G. Glutamate stimulates phosphoinositide hydrolysis in oligodendrocyte progenitors (submitted for publication).
28. Lauder, J.M. (1993) Neurotransmitters as growth regulatory signals: role of receptors and second messengers. Trends Neurosci. 16, 233-240.
29. Levine, J. M. (1989) Neuronal influences on glial progenitor cell development. Neuron 3, 103-113.
30. McCarthy, K.D., and de Vellis, J. (1980) Preparation of separate astroglial and oligodendroglial cell cultures from rat cerebral tissue. J. Cell Biology 85, 890-902.
31. Meier, E., Hertz, L. and Schousboe, A. (1991) Neurotransmitters as developmental signals. Neurochem. Int. 19, 1-15.
32. Molina-Holgado, E. and Almazan, G. Regulation of muscarinic receptors in oligodendrocytes (submitted for publication).
33. Notterpek, L.M. and Rome, L.H. (1994) Functional evidence for the role of axolemma in CNS myelination. Neuron 13, 473-485.
34. Patneau, D.K., Wright, P.W., Winters, C., Mayer, M.L. and Gallo, V. (1994) Glial cells of the oligodendrocyte lineage express both kainate- and AMPA-preferring subtypes of glutamate receptor. Neuron 12, 357-371.
35. Pende, M., Holtzclaw, L.A., Curtis, J.L., Russell, J.T. and Gallo, V. (1994) Glutamate regulates intracellular calcium and gene expression in oligodendrocyte progenitors through the activation of DL-a-amino-3-hydroxy-5-methyl-4-isoxazolepropionic acid receptors. Proc. Natl. Acad. Sci. USA 91, 3215-3219.
36. Ritchie, T., Cole, R., Kim, H., de Vellis, J., and Noble, E. P. (1987) Inositol phospholipid hydrolysis in cultured astrocytes and oligodendrocytes. Life Sci., 41, 31-39.
37. Robertson, L.M., Kerppola, T.K., Vendrell, M., Luk, D., Smeyne, R.J., Bocchiaro, C., Morgan J.I. and Curran,T. (1995) Regulation of c-fos expression in transgenic mice requires multiple interdepdendent transcription control elements. Neuron 14, 241-252.
38. Takeda, M., Nelson D.J. and Soliven B. (1995) Calcium signaling in cultured rat oligodendrocytes. Glia 14, 225-236.
39. Vartanian, T., □Sprinkly, T.J., Dawson, G. and Szuchet, S. (1988) Oligodendrocyte substratum adhesioon modulates expression of adenylate cyclase-linked receptors. Proc. Natl. Acad. Sci. USA 85, 939-943.
40. Wyllie, D.J.A., Mathie, A., Symonds, C.J. and Cull-Candy S.G. (1991) Activation of glutamate receptors and glutamate uptake in identified macroglial cells in rat cerebellar cultures. J. Physiology 432, 235-258.

41. Yu, S.S., Lefkowitz R.J. and Hausdorff W.P. (1993) b-adrenergic receptor sequestration. J. Biol. Chem. 268, 337-341.

Ion channel expression and function in astrocytic scars

Harald Sontheimer and Elizabeth Fernandez-Marques

Neurobiology Research Center and Department of Physiology & Biophysics, The University of Alabama at Birmingham, Birmingham AL 35294, USA

Abstract: We have studied electrophysiological properties of glial cells undergoing injury induced proliferation using whole-cell patch-clamp recordings. Mechanical injury to confluent monolayers of spinal cord astrocytes resulted in a transient induction of cell proliferation that lasted for approximately 48h. Astrocytic scars healed by both re-growth into the scar region and by recruitment of newly formed cells. The *in vitro* scar resembled reactive gliosis in many respects including increased expression of GFAP. Cell proliferation was a requirement in the scar healing response and mitogen inhibitors such as cytosine-arabinoside (Ara-C) retarded scar healing.

Cells actively dividing at the scar differed absolutely in their electrophysiology from uninjured astrocytes on the same coverslip but over 300 μm distant from the scar. Proliferating "scarring" cells had resting membrane potentials of ~ -55 mV and expressed large outwardly rectifying K^+ currents but mostly lacked inwardly rectifying K^+ currents. They also expressed prominent voltage-activated Na^+ currents. By contrast non-dividing astrocytes that were not associated with the scar had more negative resting potentials ~ -70 mV, expressed predominantly inwardly rectifying K^+ currents and most cells lacked Na^+ currents. Inhibition of outward K^+ currents by 2 mM 4-aminopyridine (4-AP) inhibited astrocyte proliferation and prevented scar healing suggesting that the up-regulation of 4-AP sensitive K^+ channels is necessary for injury induced proliferation.

Keywords. Gliosis, patch-clamp, video microscopy, time-lapse, ion channel, proliferation

1. Introduction

Unlike most neurons, glial cells can proliferate in the postnatal and even aged brain, where an estimated 2-5% of astrocytes and oligodendrocytes can be mitotically active(1). This pool of proliferating glial cells is thought to participate in the glial response to brain injury by forming glial scars (reactive gliosis), typically associated with traumatic head injury(2), infarct lesions(3), Alzheimer's plaques(4), epileptic seizure foci(5), neurotoxic(6), ischemic(3) and mechanical(2) injury. Proliferation under those conditions may involve formerly postmitotic cells, which are recruited to dedifferentiate and proliferate(7). Such proliferation, considered secondary proliferation, has been shown to be stimulated by growth factors, in particular basic fibroblast growth factor (bFGF) and transforming

Molecular Signaling and Regulation in Glial Cells
A Key to Remyelination and Functional Repair

growth factor beta (TGF-β1)(8-11), and by cytokines such as interleukins IL-1(12) and IL-6(13)or tumor necrosis factors TNF-α and TNF-β(14) released from adjacent cells. Thus glial proliferation is an integral part of normal brain development and wound healing. Yu et al. recently introduced a simple *in vitro* glial injury model in which scarring of astrocytes can be observed that resembles reactive gliosis, the brains response to acute injury, in many features(15,16). Using this *in vitro* glial scar model we have characterized glial cells associated with the newly forming glial scar and identified their properties electrophysiologically. Since most of the glial monolayer remained uninjured, we were able to compare properties of cells associated with the scar to those not associated with it and observed remarkable differences with regards to their membrane properties.

2. Methods

2.1 Cell culture: All studies were obtained on primary cultures of P0 rat spinal cord astrocytes. Cultures were obtained as we have previously described(17-19). In brief, spinal cords between mid-cervical and lower lumbar levels were dissected free and meninges removed. The cords were minced and incubated in an enzyme solution containing Earle's salts, 300 units/ml papain (Worthington), 0.5 mM EDTA and 1.65 mM L-cysteine for 30 minutes at 37°C. The tissue was rotated for 20 min using a Nutator (Clay Adams) placed in a 37° C oven. Tissue pieces were removed from the enzyme solution using a wide bore pipette and rinsed twice in a trituration solution containing 1.5 mg/ml trypsin inhibitor and 1.5 mg/ml Fraction V Bovine serum albumin (Sigma) in complete medium containing Earle's Minimal Essential Medium (GIBCO) and Serum Extender (Collaborative Research). Following the 2 rinses tissue was gently triturated using a silicon-coated Pasteur pipette (Prosil-28, PCR Corporation). The cell suspension was plated onto poly-ornithine/laminin-coated 12 mm circular glass coverslips at a density of $2.5x10^5$/ml.

During time-lapse video microscopy and electrophysiological recordings, cells were maintained in HCO_3 buffered saline (see below).

2.2 Electrophysiology: Techniques for whole-cell voltage-clamp recordings were standard and as previously described(20). Pipettes were made from thin-walled Borosilicat glass (WPI, TW150F-40) and had resistances of 3-5 MOHm after filling with KCl containing solution which contained (in mM): KCl 145, $MgCl_2$ 1, $CaCl_2$ 2, EGTA 10, Hepes (sodium salt) 10, pH adjusted to 7.4 using Tris.

Recordings were made on the stage of an inverted microscope (Nikon Diaphot) equipped with Normarski DIC optics and epifluorescence. Cells were continuously superfused with saline solution at room temperature (22°C) allowing for rapid (<20 sec) exchange of the bath volume. The bath solution contained (in mM): NaCl 122.6, $NaHCO_3$ 25.0, KCl 5.0, $MgSO_4$ 1.2, $CaCl_2$ 1.0, Na_2HPO_4 1.6, NaH_2PO_4 0.4, Glucose 10.5 bubbled with 5% CO_2/95% O_2.

Current recordings were obtained using an Axopatch-1D amplifier (Axon Instruments). Current signals were lowpass-filtered at 3 KHz using an 8 pole

Bessel filter (Frequency Devices), and were digitized on-line at 25 to 100 kHz using a Labmaster TL-125 digitizing board (Axon Instruments) interfaced with an IBM-compatible computer (Gateway 486). The digitizing board also delivered voltage steps to activate voltage-activated currents. Data-acquisition, storage and analysis was done using Pclamp 5.51 (Axon Inst.). For all measurements, capacity compensation and series resistance compensation were used to minimize voltage errors. Recordings were only included from cells that had series resistances (before compensation) of < 10 MOhm and compensation of residual error was typically set to 80%.

To compare currents recorded in cells of different size, age and morphology, current densities were determined by dividing peak currents by whole-cell capacitance. This yielded current density values normalized to membrane area (in pA/pF) which were independent of cell size. Whole cell capacities were read from the series resistance/capacity cancellation circuitry of the amplifier.

2.3 Cell proliferation assays: Incorporation of ^{3}H-thymidine was used as quantitative marker for DNA synthesis(21,22). Cells were incubated with 1 µCi/ml radio-labeled thymidine ([methyl-^{3}H]thymidine) for 120 min (at 37°C). Culture dishes were rinsed once with: ice-cold PBS, 1N perchloric acid, 0.5N perchloric acid, 95% ethanol and solubilized with 0.3N NaOH for 30 minutes at 37° C. An aliquot (50 µl) was used for cell protein determination using the bicinchroninic assay (BCA; Pierce Rockford, IL) as modified by Goldschmidt and Kimelberg(23). The remaining cell suspension was mixed with Ultima Gold, and radioactivity was determined with a scintillation counter. Results were expressed as cpm/µg protein.

2.4. Time-lapse video microscopy: Cells were maintained on the stage of a Nikon Diaphot microscope and observed under phase-contrast illumination using a 20X objective. Cells were maintained in a 35 mm Petri-dish in a Leiden chamber(24)(Medical systems) in which pCO_2 was maintained constant by gassing with 5/95% CO_2/O_2. The stage of the microscope was enclosed by a microincubator (Nikon) that was maintained at 37°C using a circulating air heater (Nikon, NP-2). To reduce evaporation of media and risk of contamination, 1.5 ml sterile mineral oil was added on top of the media in petri-dishes. Under our recording conditions we were able to maintain cells viable and contamination free for at least 1 week, although most recordings only followed cells for 48h. Cells were continuously monitored by a low light CCD camera (Watcom, Edmond Scientific) and images were recorded on a time-lapse video recorder (Panasonic AG-6010s) at 1/60 real time (120h per tape). Each video frame was time stamped on-line to allow reconstruction of time series. Time-stamps were not removed in documentation. To analyze and document, recordings were digitized off-line using an image frame-grabber (Intel video) at either 320*240 or 640*480 pixels.

3. Results

Glial scars could be readily inflicted on confluent monolayers of astrocytes by gently scratching the cell monolayer with a sterile pipette tip (Fig.1) resulting in scars that were ~ 100 µm wide. Using time-lapse video microscopy we determined that within 6-24 hours, the scratch area healed and became confluent with the surrounding cell monolayer. Due to some similarities of this *in vitro* scar to the typical glial injury scar *in vivo*, we simply refer to it as glial scar in this report without claiming that this injury model exactly parallels glial scarring *in vivo*.

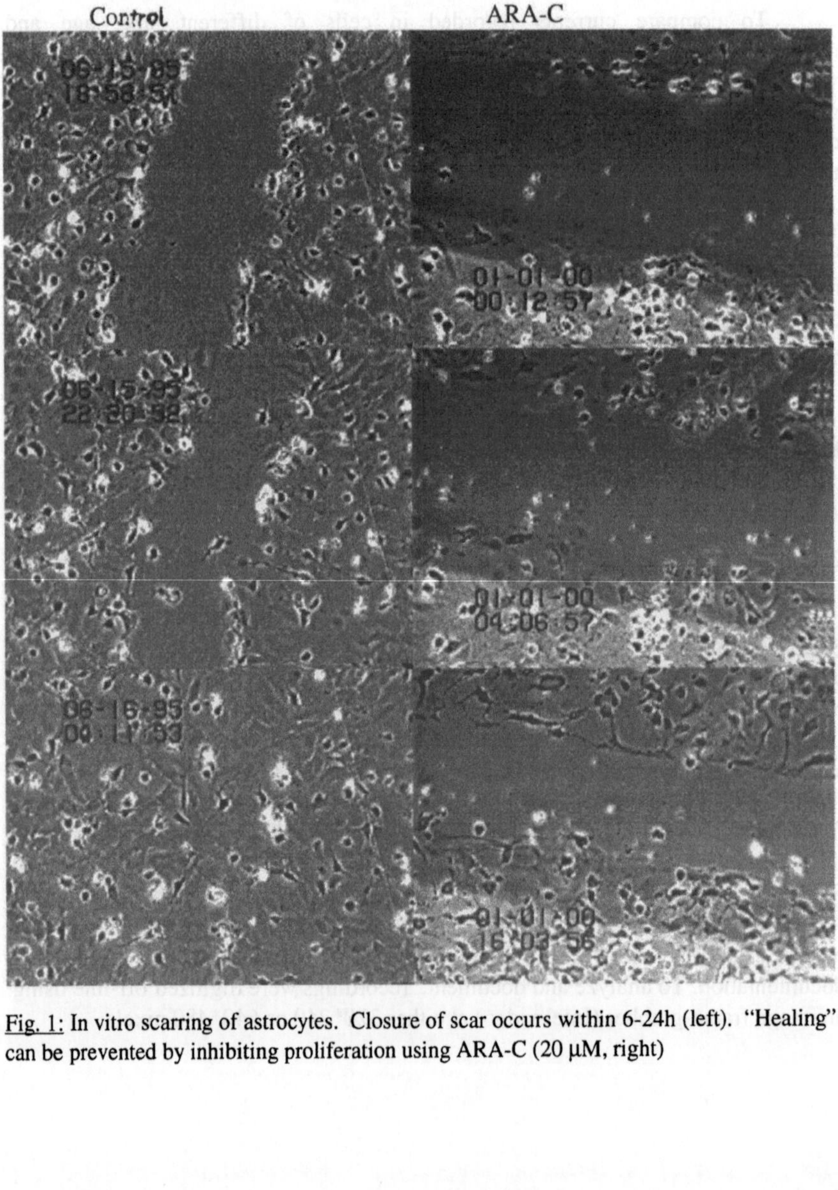

Fig. 1: In vitro scarring of astrocytes. Closure of scar occurs within 6-24h (left). "Healing" can be prevented by inhibiting proliferation using ARA-C (20 µM, right)

We were particularly interested in properties of cells associated with scars including cells type, proliferative state and electrophysiological properties. Cells surrounding the scar showed intense labeling with GFAP antibodies as did the majority of the uninjured cell monolayer (>95%) suggesting that the cells recruited in the scar healing were predominantly astrocytes. The healing of the scar involved both migration of cells as well as the recruitment of newly formed cells, thus required cell proliferation. This conclusion was based on our time-lapse recordings in which under untreated control conditions scar closure occurred at a rate of 5-10 µm/h, whereas in cultures treated with the anti-mitotic agent Ara-C scar closure was severely delayed and always incomplete (Fig. 1). Moreover, in control conditions, we routinely observed cells condensing their cell body, rounding-up and pulling apart giving rise to two sister cells (Fig. 2). After treatment with Ara-C, we did not observe cell division but instead were able to follow the migration of cells into the scar over considerable distances (not shown).

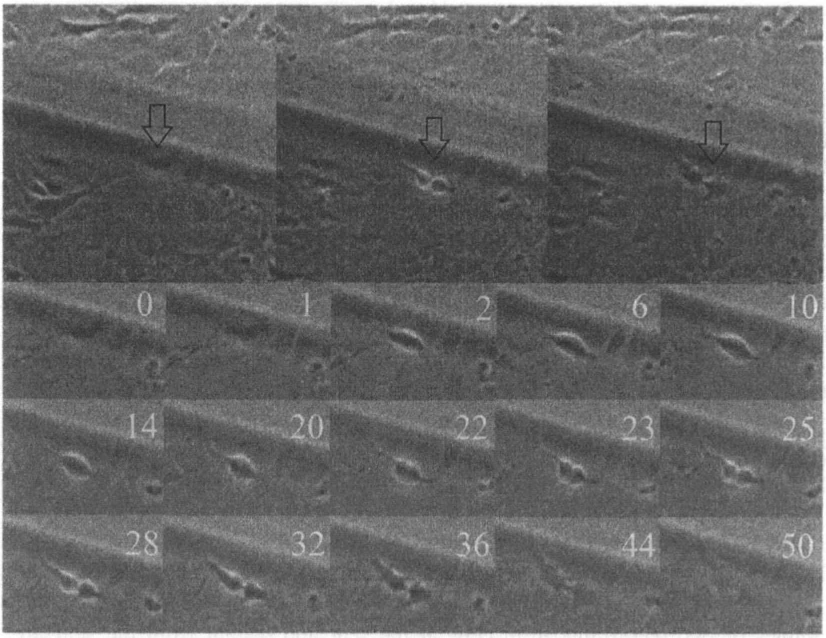

Fig. 2: Actively dividing astrocyte at scar (top row, arrow). Lower 3 rows are a time series (numbers indicated are minutes) of the same cell in close-up.

Using $^{3}H^{+}$-thymidine uptake as a quantitative marker for DNA synthesis, we determined the relative change in scar-induced glial proliferation (Fig. 3). Since scars were inflicted after confluency had been reached (~10-14 days in vitro DIV) proliferation prior to the insult was relatively small and accounted for less than 5% of the proliferation observed 3 days after plating. The increase in proliferation was transient and peaked around 24 hours reaching levels of up to 35% above control (mean 10% S.D. 5.5%). Scratch-induced proliferation was sensitive to the anti-mitotic agent Ara-C (20 μM) which completely inhibited scratch-induced proliferation and inhibited scar healing (Fig. 1).

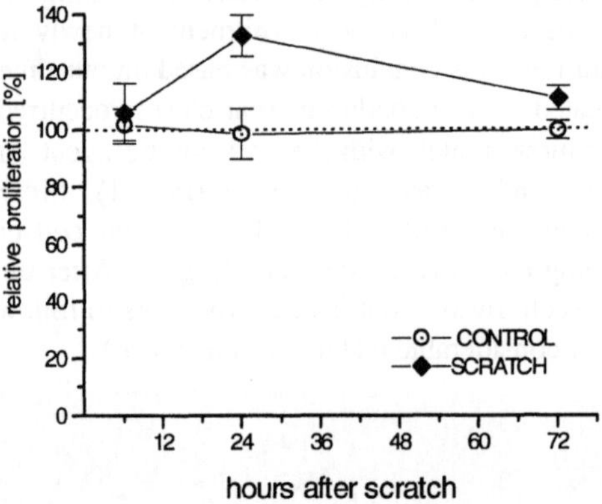

Fig. 3: "Scratch" injury to monolayers of confluent astrocytes causes a transient increase in cell proliferation.

Since we previously demonstrated that normal astrocyte proliferation can be inhibited by exposure of cells to K^+ channel blockers such as 4-AP(25), we were interested in the electrophysiological properties of actively dividing cells at glial scars. We thus scratched confluent monolayers of astrocytes and recorded cells within 6-24 hours while cells were actively proliferating at the scar. Using whole-cell patch-clamp recordings we determined resting membrane potential (determined as entrance potential), whole cell capacitance, input resistance and expression of voltage-activated ion channels. We used recording conditions that did not restrict ionic currents and allowed detection of both inward and outward currents. Moreover, stimulus protocols were used that would detect a variety of voltage-activated currents. Recordings were always also obtained on sister cells on the same coverslip that were at least 300 μm distant from the scratch site. These cells were part of the confluent monolayer and the vast majority of these cells did not divide as judged by bromo-deoxyuridine labeling(26). Representative examples of two recordings from cells at the scar and distant from the scar are shown in Figure 4. Cells at the scar were similar in cell size as judged by their membrane capacitance (16.9±9.4, N=28) as compared to distant control cells (16.6±8.8pF, N=17). However, cells at the scratch had much more positive resting membrane potentials (-49.9±16.5 mV, N=28.) than control cells (-68.6±11.3mV, N=74). Depolarizing voltage steps induced time-and voltage dependent outward currents in both scratch-associated and control cells (Fig.4a,c).

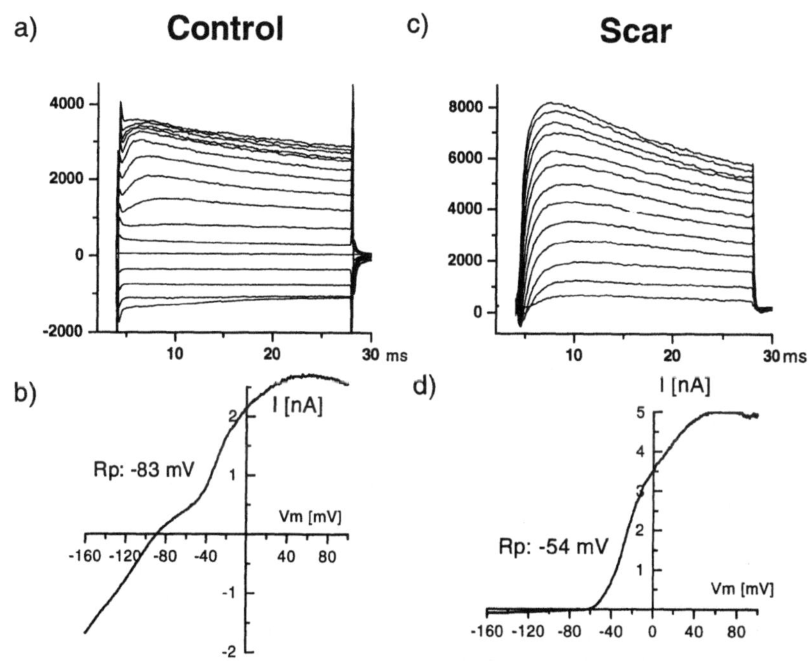

Fig. 4: Representative recordings from a proliferating, scar-associated astrocyte and a non-proliferating control cell 300 μm distant from the scratch.

These currents were mediated by delayed-rectifying K^+ channels and were sensitive to both tetraethylammonium (TEA) and 4-AP (not shown). In scar-associated cells this was the major conductance while in control cells inwardly rectifying K^+ currents (Kir) activated by hyperpolarizing voltage steps predominated. The latter were sensitive to Ba^{2+} and Cs^+ but were insensitive to 4-AP (not shown). The pronounced inward rectification was most visible when ramp-like changes in voltage were applied (Fig. 4 b,d). In control cells these ramps showed large inward currents at potential negative of the resting potential mediated by Kir channels (Fig 4b) which were not observed in scar-associated cells (Fig. 4d). Additionally, most control cells lacked fast, transient inward Na^+ currents, however, 26 out of 27 scar-associated cells expressed pronounced Na^+ currents (Fig. 5a; mean current density 494.4±323.2 pA/pF). These currents were sensitive to Na^+ replacement and reversed at the Na^+ equilibrium potential (Fig. 5b).

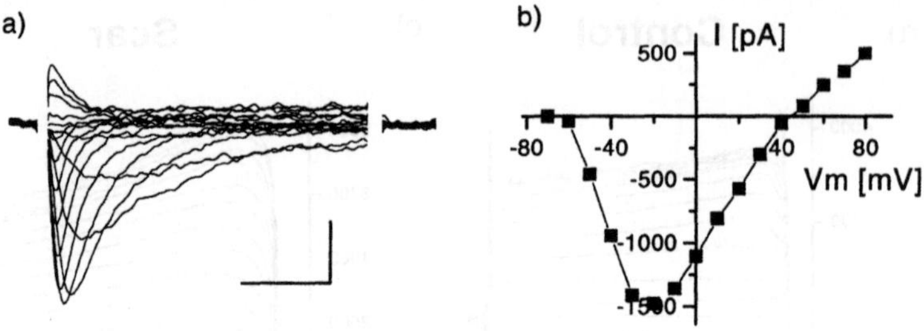

Fig. 5: Astrocytes at the scar express large voltage-activated Na^+ currents. Calibration bar: 500 pA, 2 ms.

Since we previously observed that astrocyte proliferation can be retarded by exposing cells to the K^+ channel blocker 4-AP(25), we seeked to determine whether scar-induced proliferation and scar healing could also be influenced by 4-AP. Indeed, 4-AP inhibited scar-induced proliferation in a dose-dependent manner (Fig. 6) with complete inhibition at concentrations > 500µM, concentrations that are also most efficient in channel block(27). Moreover, in the presence of 4-AP, scars did not heal (Fig. 7) suggesting that the activity of outwardly rectifying 4-AP-sensitive K^+ channels may be a prerequisite for injury induced astrocyte proliferation and thus healing of gliotic scars.

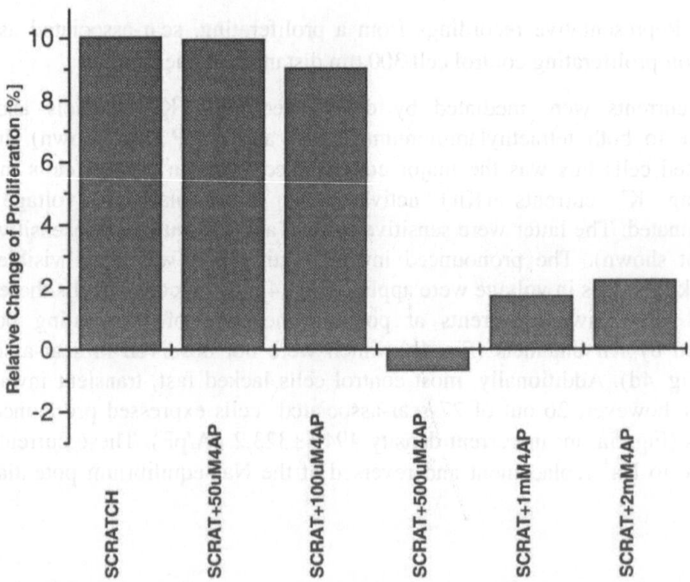

Fig. 6: Scar-induced proliferation can be inhibited in a dose-dependent manner by the K^+ channel blocker 4-AP.

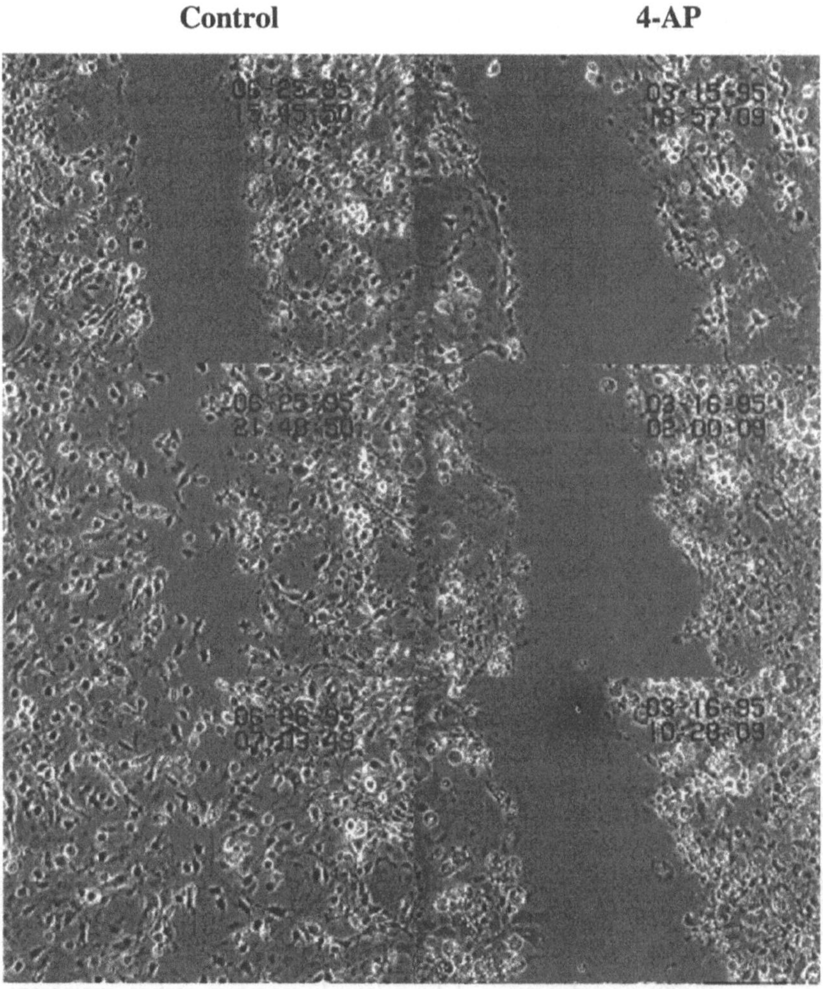

Fig. 7: Scar healing is inhibited by the continuous presence of the K^+ channel blocker 4-AP (2 mM). Middle row after 6h, bottom row after 14 h.

4. Discussion

One of the distinguishing features of glial cells is their ability to proliferate after birth. This is of particular importance in response to brain injury which results in conspicuous gliosis. It is commonly believed that the formation of glial scars involves proliferation of astrocytes near the injury site. One of the hallmarks of reactive gliosis is the up-regulation of the glial intermediate filament recognized by GFAP antibodies(28), however reactive gliosis has recently been observed in mice knock-out mutants in which the GFAP gene is lacking(29). The properties of astrocytes at glial scars and their functional role in brain injury are not well understood.

In this study we show that glial scarring can be induced *in vitro* by scratching a confluent astrocyte monolayer. This scaring results in increased proliferation at the injury site ultimately leading to scar healing. Interestingly, cells recruited in the scar closure differed absolutely in their electrophysiological properties. In particular, they were characterized by a lack of inwardly rectifying K^+ channels that appeared to characterize only non-dividing astrocytes. Since this channel is instrumental in establishing the negative glial resting potential(20) it was not surprising that the lack of Kir in scarring astrocytes resulted in significantly more depolarized membrane potentials in these cells. In these scarring, proliferating cells the main channel expressed were outwardly rectifying K^+ channels and voltage-activated Na^+ channels.

A number of previous studies have suggested that ion channel expression changes as cells proliferate and some of these studies bear striking similarity to our observation here. For example, in lymphocytes, mitogen stimulation likewise resulted in the up-regulation of voltage-activated (outwardly-rectifying) K^+ channels which are transiently expressed at high densities and down-regulated in post-mitotic cells(30). A large conductance K^+ channel is selectively up-regulated in mouse early embryos during M and G1 phases of the cell cycle (31). Studies from a number of other cell types including rat brown fat cells(32), human melanoma cells(33) and breast cancer cells(34), Schwann cells(35), retinal glial cells(36) and astrocytes(25), showed that pharmacological blockade of voltage-activated K^+ channels, including delayed rectifying channels retards cell proliferation and suggest that channel activity may be functionally linked to cell cycle progression. The anti-proliferative effects of K^+ channel blockers (including 4-AP) in the above studies is in line with our observation that 4-AP abolishes scar healing, and supports the notion that channel activity is also required to allow scratch-induced proliferation to occur. Changes in channel complement between proliferating and non-proliferating cells may be very common. For example differentiated, post-mitotic oligodendrocytes(37), astrocytes(20,38) and human leukemia cells(39) are characterized by relatively negative resting membrane potentials mediated by a class of inwardly rectifying K^+ channels(20,38). These channels are absent in their proliferating progenies(37), or their mitogen stimulated offsprings(39).

The factors that control changes in channel expression are in most examples not well understood. It has been speculated that growth factors may play a major role and in some instances it has been demonstrated that growth factor binding acting on the ras/raf signaling cascade can resulting in altered gene expression. For example, stimulation of leukemia cells with macrophage-colony-stimulating factor (CSF-1), which promotes dedifferentiation and proliferation, within minutes reduced the activity of inwardly rectifying K^+ channels(39). In murine fibroblast cell lines, activation of the ras (p21) signaling cascade, which forms an essential element in growth control of these cells, controls the expression of a Ca^{2+}-activated

K^+ channel. This channel can also be induced by epidermal growth factor (EGF) or platelet derived growth factor (PDGF), and channel blockade inhibits proliferation.

Since reactive gliosis results in the release of growth factors including basic fibroblast growth factor (bFGF(40)), interleukin-1 (IL-1(12)) and granulocyte-macrophage colony-stimulating factor GM-CSF(12), it could be speculated that scratch-induced proliferation in our studies may have also resulted in growth factor release from adjacent cells. These in turn would induce astrocyte proliferation and may have induced alterations in channel complement observed. We are presently investigating possible links between growth factor signaling, ion channel expression and astrocyte proliferation.

5. Acknowledgments: This work was supported by NIH grants NS 31234 and NICHD P50HD32901.

6. References

1. Korr, H. (1986) Proliferation and cell cycle parameters of astrocytes. Astrocytes (Fedoroff, S. and Vernadakis, A. eds) pp. 77-127, Academic Press, Orlando, San Diego, New York, Austin.
2. Reier, P. J. (1986) Gliosis following CNS injury: The anatomy of astrocytic scars and their influences on axonal elongation. Astrocytes, Cell Biology and Pathology of Astrocytes (S., F. and Vernadakis, A. eds) pp. 263-324, Academic Press, Orlando.
3. Kraig, R. P.and Jaeger, C. B. (1990) Ionic concomitants of astroglial transformation to reactive species. Stroke 21, III184-7.
4. Murphy, G. M.,Jr., Ellis, W. G., Lee, Y. L., Stultz, K. E., Shrivastava, R., Tinklenberg, J. R.and Eng, L. F. (1992) Astrocytic gliosis in the amygdala in Down's syndrome and Alzheimer's disease. Prog. Brain Res. 94, 475-483.
5. Pollen, D. and Trachtenberg, M. C. (1970) Neuroglia: Gliosis and focal epilepsy. Science 167, 1252-1253.
6. Niquet, J., Ben-Ari, Y.and Represa, A. (1994) Glial reaction after seizure induced hippocampal lesion: Immunohistochemical characterization of proliferating glial cells. Journal Of Neurocytology 23, 641-656.
7. Hatten, M. E., Liem, R. K., Shelanski, M. L.and Mason, C. A. (1991) Astroglia in CNS injury. Glia 4, 233-243.
8. Logan, A., Berry, M., Gonzalez, A. M., Frautschy, S. A., Sporn, M. B.and Baird, A. (1994) Effects of transforming growth factor beta 1 on scar production in the injured central nervous system of the rat. Europ. J. Neurosci. 6, 355-363.
9. Logan, A.and Berry, M. (1993) Transforming growth factor-beta 1 and basic fibroblast growth factor in the injured CNS. [Review]. TIPS 14, 337-342.
10. Logan, A., Frautschy, S. A., Gonzalez, A. M., Sporn, M. B.and Baird, A. (1992) Enhanced expression of transforming growth factor beta 1 in the rat brain after a localized cerebral injury. Brain Res. 587, 216-225.
11. Logan, A., Frautschy, S. A.and Baird, A. (1991) Basic fibroblast growth factor and central nervous system injury. Ann. NY Acad. Sci. 638, 474-476.

12. Giulian, D., Li, J., Li, X., George, J.and Rutecki, P. A. (1994) The impact of microglia-derived cytokines upon gliosis in the CNS. Dev. Neursoci. 16, 128-136.
13. Hariri, R. J., Chang, V. A., Barie, P. S., Wang, R. S., Sharif, S. F.and Ghajar, J. B. G. (1994) Traumatic injury induces interleukin-6 production by human astrocytes. Brain Res. 636, 139-142.
14. Benveniste, E. N. (1992) Inflammatory cytokines within the central nervous system: sources, function, and mechanism of action. Am. J. Physiol., 263, C1-C16.
15. Hou, Y.-J., Yu, A. C. H., Garcia, J. M. R. Z., Aotaki-Keen, A., Lee, Y.-L., Eng, L. F., Hjelmeland, L. J.and Menon, V. K. (1995) Astrogliosis in culture. IV. Effects of basic fibroblast growth factor. J. Neurosci. Res. 40, 359-370.
16. Yu, A. C., Lee, Y. L.and Eng, L. F. (1993) Astrogliosis in culture: I. The model and the effect of antisense oligonucleotides on glial fibrillary acidic protein synthesis. J. Neurosci. Res. 34, 295-303.
17. Sontheimer, H., Ransom, B. R., Cornell-Bell, A. H., Black, J. A.and Waxman, S. G. (1991) Na^+-current expression in rat hippocampal astrocytes in vitro: alterations during development. J. Neurophysiol. 65, 3-19.
18. Black, J. A., Sontheimer, H.and Waxman, S. G. (1993) Spinal cord astrocytes in vitro: Sodium channel immunoreactivity. Glia 7, 272-285.
19. Sontheimer, H., Minturn, J. E., Black, J. A., Ransom, B. R.and Waxman, S. G. (1991) Two types of Na^+-currents in cultured rat optic nerve astrocytes: Changes with time in culture and with age of culture derivation. J. Neurosci. Res. 30, 275-287.
20. Ransom, C. B.and Sontheimer, H. (1995) Biophysical and pharmacological characterization of inwardly rectifying K^+ currents in rat spinal cord astrocytes. J. Neurophysiol. 73, 333-345.
21. Condorelli, D. F., Ingrao, F., Magri, G., Bruno, V., Nicoletti, F.and Avola, R. (1989) Activation of excitatory amino acid receptors reduces thymidine incorporation and cell proliferation rate in primary cultures of astrocytes. Glia 2, 67-69.
22. Avola, R., Condorelli, D. F., Surrentino, S., Turpeenoja, L., Costa, A.and Giuffrida Stella, A. M. (1988) Effect of epidermal growth factor and insulin on DNA, RNA, and cytoskeletal protein labeling in primary rat astroglial cell cultures. J. Neurosci. Res. 19, 230-238.
23. Goldschmidt, R. C.and Kimelberg, H. K. (1989) Protein analysis of mammalian cells in monolayer culture using the bicinchroninic assay. Analyt. Biochem. 177, 41-45.
24. Ince, C., Ypey, D. L., Diesselhoff-Den, D., Viser, J. A. M., De, V.and Van, F. (1983) Micro-CO2-incubator for use on a microscope. J. Physiol. (London) 60, 269-275.
25. Pappas, C. A., Ullrich, N.and Sontheimer, H. (1994) Reduction of glial proliferation by K^+ channel blockers is mediated by changes in pH_i. Neuroreport 6, 193-196.

26. Gratzner, H. G. (1982) Monoclonal antibody to 5-bromo and 5-iododeoxyuridine: a new reagent for detection of DNA replication. Science 218, 474-475.
27. Sontheimer, H., Black, J. A., Ransom, B. R.and Waxman, S. G. (1992) Ion channels in spinal cord astrocytes in vitro: I. Transient expression of high levels of Na^+ and K^+ channels. J. Neurophysiol. 68, 985-1000.
28. Eng, L. F.and Ghirnikar, R. S. (1994) GFAP and astrogliosis. Brain Pathology 4, 229-237.
29. Gomi, H., Yokoyama, T., Fujimoto, K., Ikeda, T., Katoh, A., Itoh, T.and Itohara, S. (1995) Mice devoid of the glial fibrillary acidic protein develop normally and are susceptible to scrapie prions. Neuron 14, 29-41.
30. DeCoursey, T. E., Chandy, G., Gupta, S.and Cahalan, M. D. (1984) Voltage-gated K^+ channels in human T lymphocytes: a role in mitogenesis? Nature 307, 465-468.
31. Day, M. L., Pickering, S. J., Johnson, M. H.and Cook, D. I. (1993) Cell-cycle control of a large-conductance K^+ channel in mouse early embryos. Nature 365, 560-562.
32. Pappone, P. A.and Ortizmiranda, S. I. (1993) Blockers of voltage-gated K^+-channels inhibit proliferation of cultured brown fat cells. Am. J. Physiol. 264, C1014-C1019.
33. Nilius, B.and Wohlrab, W. (1992) Potassium channels and regulation of proliferation of human melanoma cells. Journal of Physiology - London 445, 537-548.
34. Woodfork, K. A., Wonderlin, W. F., Peterson, V. A.and Strobl, J. S. (1995) Inhibition of ATP-sensitive potassium channels causes reversible cell-cycle arrest of human breast cancer cells in tissue culture. J. Cell. Physiol. 162, 163-171.
35. Chiu, S. Y.and Wilson, G. F. (1989) The role of potassium channels in Schwann cell proliferation in Wallerian degeneration of explant rabbit sciatic nerves. J. Physiol. (London) 408, 199-222.
36. Puro, D. G., Roberge, F.and Chan, C. C. (1989) Retinal glial cell proliferation and ion channels: a possible link. Invest. Ophatl. & Vis. Sci. 30, 521-529.
37. Sontheimer, H., Trotter, J., Schachner, M.and Kettenmann, H. (1989) Channel expression correlates with differentiation stage during development of oligodendrocytes from their precursor cells in culture. Neuron 2, 1135-1145.
38. Roy, M.-L.and Sontheimer, H. (1995) Modulation of glial inwardly rectifying K^+ channels by protein kinase A. J. Neurochem. 64, 1576-1584.
39. Wieland, S. J., Chou, R. H.and Gong, Q. H. (1990) Macrophage-colony-stimulating factor (CSF-1) modulates a differentiation-specific inward-rectifying potassium current in human leukemic (HL-60) cells. J. Cell. Physiol. 142, 643-651.
40. Johnson, M. D., Jennings, M. T., Gold, L. I.and Moses, H. L. (1993) Transforming growth factor-beta in neural embryogenesis and neoplasia. Human Pathology 24, 457-462.

Ion Channel Expression in Glial Cells from Trout Central Nervous System

H. Rabe, G. Glassmeier, T.-D. Nguyen and G. Jeserich

AG Zoophysiologie, Universität Osnabrück, 49069 Osnabrück, Germany

Keywords: bony fish, glia cells, patch-clamp technique.

Introduction

Macroglial cells in the vertebrate gray matter are divided into two subtypes, the nonmyelinating astrocytes and the myelinating oligodendrocytes. To astrocytes a variety of functional roles have been ascribed including nutrition of neurons, uptake of amino acid neurotransmitters, and a contribution to extracellular potassium regulation. Furthermore, electrophysiological studies have revealed an astonishing repertoire of voltage-dependent and ligand-activated ion channels in cultured astrocytes as well as in vivo, not unlike those found in neurons (5). Oligodendrocytes, on the other hand, which enwrap axons with an electrically insulating membrane sheath, were originally considered as electrophysiologically inactive elements. Recent patch clamp studies, however, have shown a characteristic pattern of voltage-dependent ionic currents in this cell type, too.

Since the majority of studies in this direction have been performed with mammalian species, very little is known about the ion channel expression of glial cells from lower vertebrates in particular of fish. Studies on glial cells in the CNS of bony fish appear particularly interesting, since in these species neurons and oligodendrocytes spontaneously regenerate after injury.

Astrocytes

For all studies described here dissociated cell cultures of trout brain tissue were used. These were prepared from hatched larvae (stage 32 and 34) or young fish of 3-4 cm body length as previously described (17). In these cultures two morphologically and antigenically distinct cell populations were observed. One of it was intensely immunostained for the astrocyte marker GFAP and exhibited a significant structural diversity: while most of the cells displayed a flattened epitheloid cell shape closely matching type 1 astrocytes of mammalian cell cultures, strongly elongated or spindle shaped cells were frequently observed, too. Immunostainig with the A2B5 antibody, which is a marker for mammalian type 2 astrocytes (23), revealed that trout astrocytes did not express this ganglioside epitope.

Molecular Signaling and Regulation in Glial Cells
A Key to Remyelination and Functional Repair

Electrophysiological recordings were performed using the whole cell variation of the patch clamp technique (12, 14). To avoid electrical coupling cells lying in isolation were used for measurements. Since the flattened epitheliod cells did not voltage-clamp well, most experiments were done on astrocytes with more rounded cell bodies and short processes. In general, there was no clear correlation between individual cell morphology and the pattern of ion channel expression. In about 85% of the cells an average membrane resting potential of -39.5 ± 11.2 mV was measured, while in the remaining population it was -68 ± 3.4 mV on average. Starting from a standard holding potential of -80 mV depolarizing voltage steps ranging from -60 to 10 mV in all cases evoked a large potassium outward current. As judged from its reversal potential and its sensitivity to intracellular CsCl (140 mM) or extracellular 4-AP (1 mM) this current was carried by potassium ions. It activated at potentials positive to -20 mV and reached peak current amplitudes at 120 mV, which varied between 350-4300 pA (**Fig. 1**). According to its voltage dependence and pharmacological features, the potassium current could be separated into two different components: a transient A-type component, which was dependent on the holding potential and which could be completely blocked by 4-AP and a sustained delayed rectifier, which was independent of the holding potential and less sensitive to 4-AP.

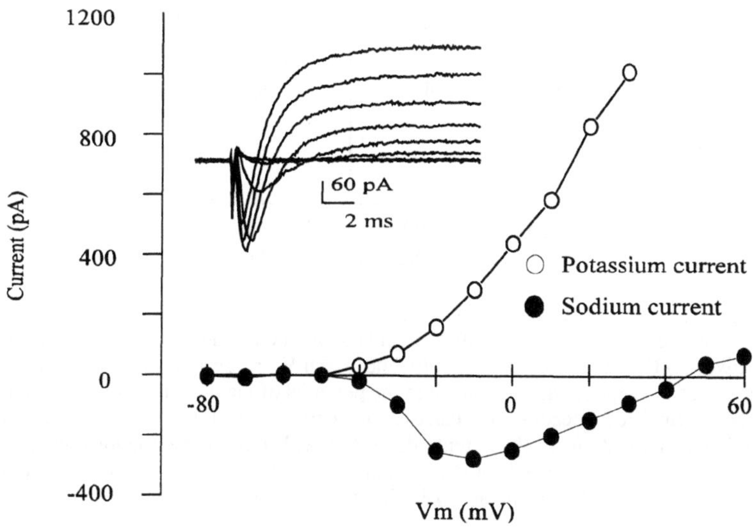

Fig. 1. Current voltage relationship of trout astrocytes sodium inward (•) and potassium outward currents(o). All currents activated at -40 mV. The sodium current peaked at -10 mV and reversed at approximately 45 mV. The inset is an original recording elicited by testpulses ranging from -60 to 10 mV at 10 mV increment.

The steady state inactivation kinetics were tested by using a two pulse protocol: after a prepulse of 500 ms duration, ranging between -140 and -40 mV with 10 mV increment, a testpulse of 60 mV was given to activate the current. The data were fitted to the half maximal inactivation curve using a modified Boltzmann equation $I/Imax = 1 / \{1 + exp[(V - V_h)/K_h]\}$, which yielded a midpoint potential of -87 mV (**Fig. 2**). At potentials positive to -60 mV, the potassium current was completely inactivated. Potassium outwards currents with comparable overall biophysical properties were found in cultured as well as in acutely isolated mammalian cortical astrocytes, too (22, 31). By contrast, rat hippocampal astrocytes exhibited a more depolarized steady state inactivation curve with a midpoint value of -68 mV (34).

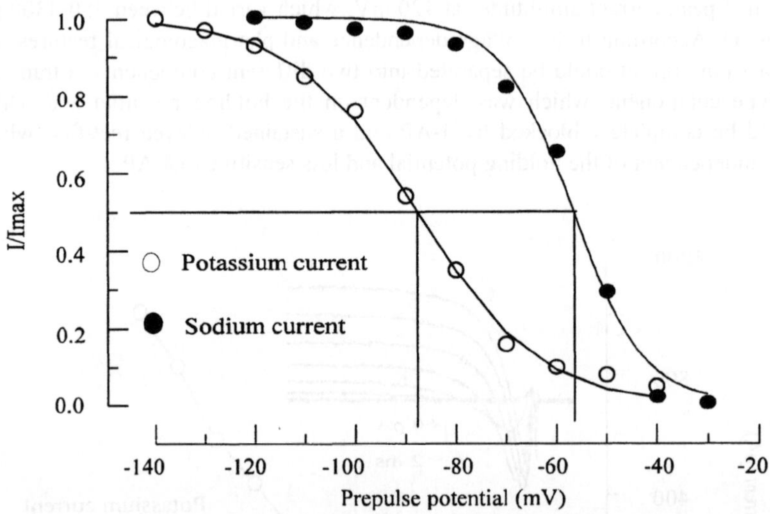

Fig.2. Kinetics of inactivation of sodium (•) and potassium currents (o) measured in cultured trout astrocytes. After prepulses of 500 ms duration, which ranged between -120 and -30 mV or -140 and -40 mV, respectively, currents were activated by stepping the membrane to the potential of maximal current activation. The normalized steady-state inactivation curve was fitted by a modified Boltzmann equation. Both currents decreased with more depolarizing prepulses and were completely inactivated at -40 mV. At -87.6 mV half of the potassium current were inactivated, and at -56.8 mV half of the sodium current.

As mentioned above in a subpopulation of trout astrocytes the resting potential was close to -70 mV. These cells exhibited an inwardly rectifying current, which activated at hyperpolarizing potentials in addition to the potassium outward conductivities (12). As judged from its reversal potential and its complete blockade by 5 mM CsCl or BaCl in the external solution, this current was carried by potassium ions, too. Starting from a holding potential of -60 mV, the cells were hyperpolarized up to -180 mV with 20 mV increments. The conductance activated rapidly and showed a time- and voltage-dependent decay at testpulses negative to -120 mV. In its overall biophysical features the potassium inward rectifier closely resembled those found in mammalian astrocytes (6), albeit a more pronounced voltage-dependent inactivation was established here. Since this current activated at elevated external potassium concentrations as well as at hyperpolarized potentials, it is generally believed to play a role in potassium ion homeostasis (5).

Furthermore, in a subpopulation of trout astrocytes a voltage-gated fast inactivating transient inward current was elicited (12). Isolation of this current was achieved by replacing KCl for CsCl in the pipette solution. Substituting sodium for choline in the bath solution completely abolished this current, identifying it as a sodium current. It became detectable at depolarizing testpulses above -40 mV, peaked at -10 mV and reversed at approximately 45 mV (**Fig 1**). Peak current sizes varied between -160 and -2800 pA. The current density decreased with time in vitro from -114 ± 68 pA/pF after 4 days to -25.8 ± 16.8 pA/pF after 1 week in culture. External application of 1 μM of the sodium channel blocker TTX reduced the current by 50%, whereas at 4 μM TTX it was blocked totally. Hence compared with sodium currents of mammalian astrocytes the TTX sensitivity of trout astrocytes appears rather low (30). The half inactivation kinetics were determined as described above for potassium currents using, however, prepulse potentials ranging from -120 to -10 mV. The inactivation curve exhibited „neuronal like" characteristics with a h_∞ midpoint of -57 mV and a complete current inactivation at a prepulse potential of -40 mV (**Fig. 2**). This markedly differs from the sodium currents of mammalian astrocytes, which typically exhibit h_∞ midpoints of -85 mV. Moreover, mammalian astrocytes typically exhibited a pronounced shift in the half inactivation voltage of the sodium current during in vitro differentiation (3, 28, 29), which was never observed in trout astrocytes. Another distinguishing biophysical feature lies in the decay of the sodium current: in trout astrocytes the current amplitude decreased with increasing testpulse potentials, but reached its peak values faster. The decay currents could be fitted with an exponential function to obtain the τ constant. The τ values attenuated with more depolarized test potentials: at -30 mV and 10 mV τ values of 1.94 ms and 0.7 ms were determined, respectively. For mammalian astrocytes a typical correlation between a fast kinetic behavior and a more depolarized steady state inactivation curve is described (15, 30). Consistent with this, in trout astrocytes fast „neuronal like" decay constants of the sodium current coincided with a more depolarized h_∞ curve.

Ligand gated currents

In addition to the occurrence of voltage-gated ionic currents evidence for the expression of ligand gated ionic conductances was obtained. About 90% of trout astrocytes responded to the external application of the amino acid neurotransmitter glutamate by an inward current in a concentration dependent manner (9). A Michaelis-Menten fit of the dose-response curve revealed a threshold of activation below 10 μM and a current saturation between 1 and 5 mM. A Hill coefficient of 1.1 was calculated from the slope of the regression line of the Hill plot, suggesting the presence of one glutamate binding site per receptor molecule. The ionic selectivity of the glutamate induced current was determined by ion substitution experiments. Whereas chloride and calcium ions seemed not to be involved in the transmitter induced current, substitution of sodium by choline had a strong effect in that the reversal potential shifted from 0 to -60 mV. Besides glutamate, kainate and quisqualate induced a similar inward current, too, while NMDA failed to evoke any detectable response. Altogether these data imply, that trout astrocytes express a non NMDA glutamate receptor type resembling those found in mammalian glial cells. Recently it was demonstrated that astrocytes cultured from rat cortex posses glutamate receptors, while biophysical and pharmacological properties matching these of neuronal non-NMDA receptors, too (1, 7, 26). Glutamate-induced activation of ionic channels was described also for rabbit retinal astrocytes in situ (8) as well as for cultured rat cerebellar type 2 astrocytes (35). In summary the pattern of ion channels appears to be highly conserved between bony fish and mammalian astrocytes, implying a basic functional significance of these channels for glial physiology. The heterogeneity of channel expression observed in the population of astrocytes may well reflect functional diversity among the subtypes of astrocytes (36).

Oligodendrocytes

The second cell type occurring in cultures of dissociated trout brain tissue were oligodendrocytes. These cells were characterized by a multipolar branched morphology and a selective labeling with the monoclonal antibody 6D2, which recognizes an epitope in the carbohydrate portion of the two major myelin glycoproteins of trout CNS, IP1 and IP2 (16, 17).

Patch clamp recordings were similarly done as described above for astrocytes. Oligodendrocytes exclusively exhibited two types of potassium outward currents as concluded from its reversal potentials and the blockade by the extracellular potassium channel blockers TEA and 4-AP (11). Both currents could be separated from each other by using different holding potentials and by their distinctive sensitivities to TEA and 4-AP. The delayed rectifier current (K_D-type) was superimposed by a transient A-type current at a holding potential of -80 mV. The threshold of activation was -40 mV for both components (**Fig. 3**). Furthermore, the transient component was totally blocked by 1 mM 4-AP in the bath solution, leaving only a slightly reduced K_D current. Externally applied TEA on the other hand significantly reduced the total outward current, still leaving an inactivating component. A major difference between the potassium outward currents found in astrocytes and those measured in oligodendrocytes of trout lies

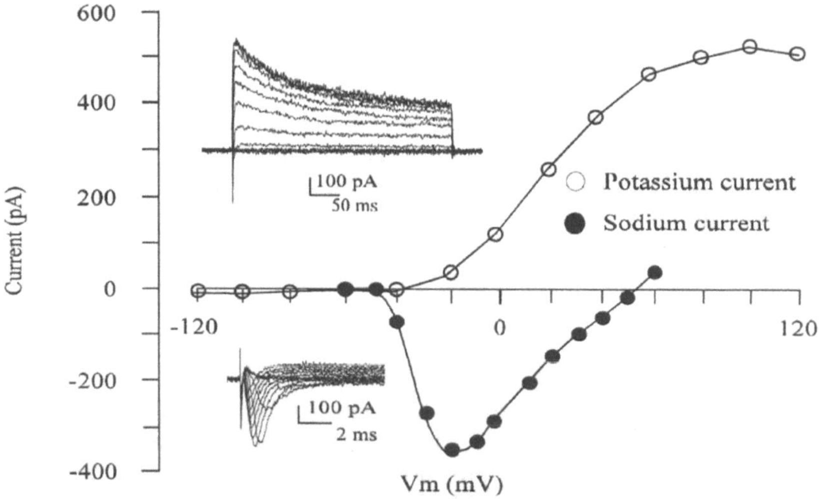

Fig. 3. I/V curves of the sodium and potassium currents measured in trout oligodendrocyte progenitor cells. Cells were held at -80 mV and testpulses up to 120 mV were given. The sodium current peaked at -20 mV and reversed at approximately 55 mV, maximal potassium currents were reached at 100 mV. Insets are original registrations measured in response to the testpulses.

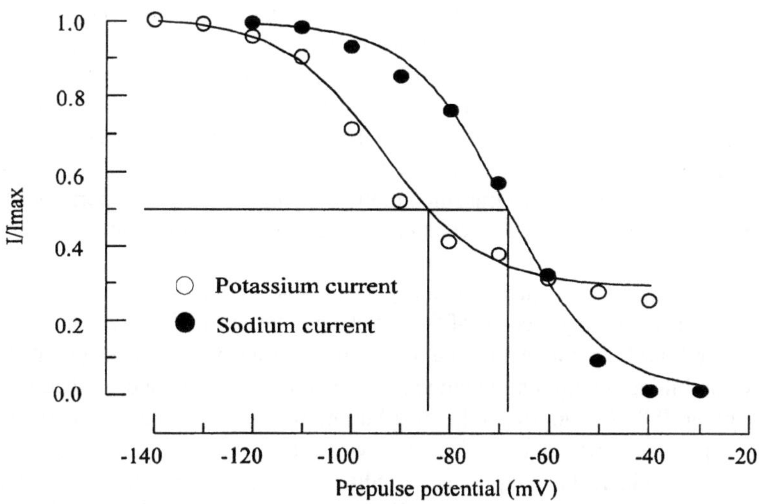

Fig. 4. Steady-state inactivation of sodium current (•) and potassium currents (o) in cultured oligodendrocyte progenitor cells. Currents were evoked as described above. Curves represents fits by a modification of the Boltzmann equation. The sodium current was completely inactivated at a prepulse potential of -40 mV, whereas 30 % of the potassium current was still left at this potential. The midpoint of the h∞ curve was in the case of sodium -88.3 and in the case of potassium -68.6 mV.

in the kinetics of inactivation. While in astrocytes, as mentioned above, the current totally inactivated at prepulse potentials positive to -60 mV, in oligodendrocytes 30% of the current still remained active under these conditions (**Fig. 4**). By contrast the voltages of half inactivation were superimposable in both cases being -88 mV and -83 mV, respectively. Outward potassium currents with similar characteristics regarding voltage dependence of activation, kinetics, and pharmacology were reported also in mammalian oligodendrocytes (5, 21, 24). The inwardly rectifying potassium current elicited at hyperpolarizing testpulses, which is a most typical electrophysiological feature of mammalian oligodendrocytes, was totally absent in cultured trout oligodendrocytes (2, 24, 25), however.

Molecular structure of potassium channels

To allow for a more precise characterization of the ion channel proteins underlying the potassium outward currents of trout glial cells, mRNA extracted from the cultures was used to amplify the potassium channel cDNA fragments in a reverse transcription-PCR approach (RT-PCR), employing degenerated primers chosen from highly conserved regions of voltage-activated potassium channels. A PCR product of 970 bp length was obtained. Nucleotide sequencing of the cloned fragments suggested the presence of two slightly different cDNA sequences. Computer-assisted comparisons (The Wisconsin Package Version. 8, Genetics Computer Group, USA) of the deduced amino acid sequences revealed a highest degree of sequence identity (about 80%) with mammalian Shaker-type potassium channels (33) while Shal-, Shab- and Shaw-like K^+ channels were only between 46% and 50% identical to those of trout glial cells. They were therefore termed TSha1 and TSha2 for **t**rout **Sha**ker-type channels. Hydrophobicity analyses (PC/Gene: the nucleic acid and protein sequences analysis software system, IntelliGenetics Inc. and Genofit SA, Vers. 6.01) of the amino acid sequences revealed a pattern of putative transmembrane domains which was superimposable with that of Drosophila Shaker channel comprising six hydrophobic membrane spanning segments (**Fig. 5**).

The highest degree of sequence identity was found in the S4-segment which is considered as the voltage sensor of the ion channel. Major differences occurred in the extracellular loop connecting the S1 and the S2 transmembrane segment (**Fig. 6**). By combining the immunomagnetic cell separation technique described above (20) with an RT-PCR approach the distribution of potassium channel subtypes in trout astrocytes and oligodendrocytes was comparatively analyzed: while astrocytes exclusively expressed TSha2, oligodendrocytes additionally synthesized TSha1. This cell type specific pattern of ion channel expression may account for the different kinetics of potassium current inactivation measured in the two glial cell types by patch-clamp recording.

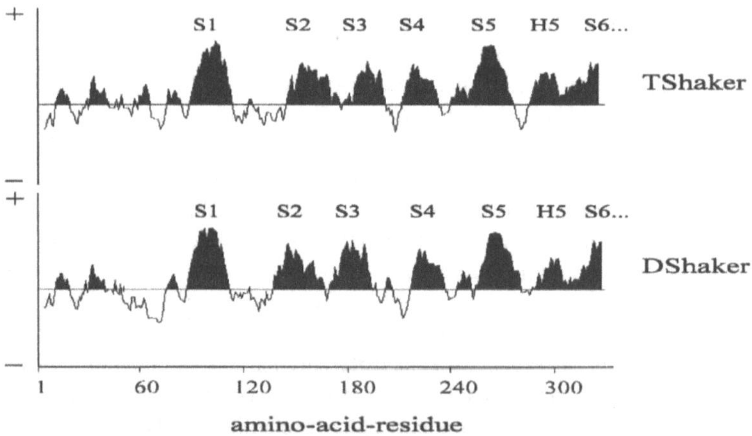

Fig. 5. Hydropathy profiles from voltage-gated potassium channels in cultured glial cells of trout CNS (TShaker) compared with Drosophila (DShaker) K^+-channel amino acid sequence.

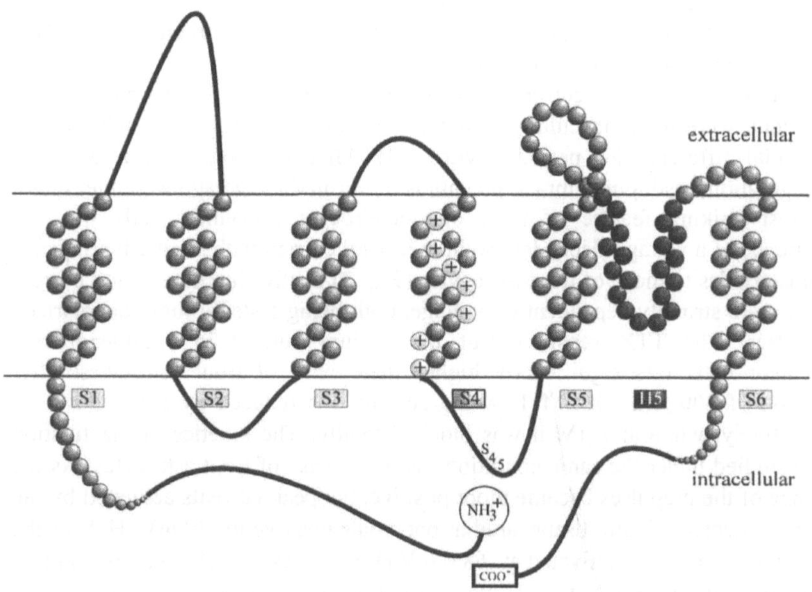

Fig. 6. The two-dimensional topology for a K^+-channel subunit. The α-helical transmembrane segments are numbered, S1-S6. The positively charged S4 segment is thought to act as the voltage sensor and the pore is formed by the loop (H5) between segments S5 and S6. The inactivation particle responsible for rapid inactivation is indicated by a "ball" at the amino terminus.

Oligodendrocyte Precursor Cells

During development of the mammalian brain, cells of the oligodendroglial lineage originate from mitotically active precursor cells, which carry the ganglioside epitope A2B5 on their surface. The initiation of oligodendroglial differentiation is characterized by the expression of galactocerebroside (GC) and myelin specific proteins, such as myelin basic protein (MBP) and proteolipid protein (PLP), in a sequential manner (10). Recent electrophysiological studies have revealed that the antigenic differentiation of oligodendroglial progenitor cells coincides with dramatic changes in ion channel expression (4, 27). Several voltage gated ionic currents are described in mouse cerebrum and rat optic nerve precursor cells, including TTX sensitive sodium currents and three different types of potassium currents. As precursor cells differentiated into oligodendrocytes, the sodium current was lost and replaced by an inwardly rectifying potassium current (2, 32). Interestingly, trout oligodendroglia progenitor cells behaved quite similarly during in vitro differentiation (18, 19, 20). Within 24 h in vitro the $A2B5^+$ cells isolated from larval trout brain typically adopted a bipolar or tripolar morphology. After 3 days the majority of these cells started to express the myelin glycoprotein IP2 and predominantly exhibited a mixed A2B5 and 6D2 positive phenotype. After 6 days in culture most of the $6D2^+$ cells had lost the A2B5 epitope and developed a multipolar shape with numerous fine ramifications. This coincided with the cessation of mitotic activity.

Electrophysiological recordings with the whole-cell mode of the patch-clamp technique were done on cultured trout oligodendrocyte progenitor cells during the initial differentiation period in vitro (13). After attaining a stable recording configuration, resting membrane potentials of -41 mV in average were measured. A most striking feature of trout oligodendrocyte progenitor cells was the occurence of a voltage-dependent sodium inward current with similar biophysical characteristics to those of trout astrocytes (**Fig. 3**). Activation kinetic and current decay were strongly dependent on voltage, both being faster at more depolarized testpulses. The TTX sensitivity of the sodium current in oligodendrocyte precursor cells was significantly higher than these of trout astrocytes: after addition of 100 and 200 nM TTX the current was reduced by 50% and 77%, respectively, while at 1 μM it was blocked totally. The kinetics of inactivation were studied under the same conditions as in the case of trout astrocytes. As the voltage of the prepulses became more positive, the peak currents activated by the testpulse decreased and disappeared at potentials positive to -40 mV. Half of the sodium current was inactivated at -68.6 mV (**Fig. 4**). As the cells acquired a more differentiated phenotype, the sodium current diminished both in amplitude and incidence. The dramatic decrease in current density is shown in figure 5: until the first day in culture the sodium current density was about 60 pA/pF and continuously declined to 10 pA/pF after 5 days in vitro, while after eight days sodium currents were no longer detectable.

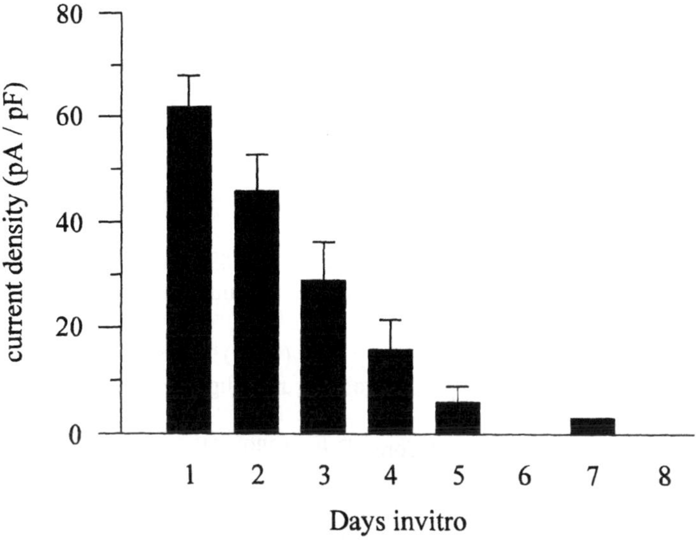

Fig. 7. Sodium current densities of cultured trout oligodendroglia progenitor cells after increasing periods in vitro. Highest densities were recorded during the first 2 days in culture in cells with bipolar morphology. After 6 and 8 days in vitro sodium current were no longer recorded, at the 7th day only two out of eleven cell elicited this current. Therefore no error bar is indicated. Reprint with permission (13, Fig. 4).

In addition to the sodium current trout oligodendrocyte progenitor cells exhibited a potassium outward current, which consisted of two components. In their overall biophysical properties these currents were identical to those of mature trout oligodendrocytes (**Fig. 3 and 4**). As mentioned above, it consisted of a K_D-type delayed rectifier and an inactivating A-type current. Whereas the delayed rectifier was already present in all cells shortly after plating, the inactivating component was observed only after two days in vitro, correlating with the outgrowth of cell processes.

In conclusion, in the present study a striking similarity between the electrophysiological properties of oligodendrocyte progenitor cells from fish brain and those from the mammalian CNS was demonstrated. Furthermore, in both vertebrate species these cells were closely matching in terms of morphology and antigenic phenotype. As a whole the marked phylogenetic conservation of ion channel properties between fish and mammalian oligodenroglial precursor cells advocates for a high significance of ionic currents in glial cells differentiation.

Acknowlegments

The authors would like to thank Professor W. Lueken for ongoing support and Mrs. B. Flenker for expert technical assistance. The studies were supported by the Deutsche Forschungsgemeinschaft (SFB 171/C13). G. Glassmeier and T.-D. Nguyen were supported by the Graduate Collage of Lower Saxony.

References

1. Backus K.H., Kettenmann H., Schachner M. (1989) Pharmacological charac terisation of the glutamtate receptor in cultured trout astrocytes. J. Neurosci. Res. **22**, 274-282.
2. Barres B.A., Chun L.L.Y., and Corey D.P. (1988) Ion channel expression by white matter glia: I. Type 2 astrocytes and oligodendrocytes. Glia **1**, 10-30.
3. Barres B.A., Chun L.L.Y., and Corey D.P. (1989) Glial and neuronal forms of the voltage-dependent sodium channel: characterisation and cell-type distribution. Neuron **2**, 1375-1388.
4. Barres B.A., Koroshetz, W.J., Swartz K.J., Chun L.L.Y., and Corey D.P. (1990) Ion channel expression by white matter glia: The O-2A glial progenitor cell. Neuron **4**, 507-524.
5. Barres B.A., Chun L.L.Y., and Corey D.P. (1990a) Ion channels in vertebrate glia. Annu. Rev. Neurosci. **13**, 441-474.
6. Barres B.A., Koroshetz, W.J., Chun L.L.Y., and Corey D.P. (1990b) Ion channel expression by white matter glia: The type 1 astrocyte. Neuron **5**, 527-544.
7. Blankenfeld G. and Kettenmann H. (1992) Glutamate and GABA receptors in vertebrate glial cells. Mol. Neurobiol. **5**, 31-42.
8. Clark B. and Mobbs P. (1992) Transmitter-operated channels in rabbit retinal astrocytes studied in situ by whole-cell patch clamping. J. Neurosci. **12**, 664-673.
9. Clasen T., Jeserich G., and Krüppel T. (1995) Glutamate-activated ionic currents in cultured astrocytes from trout: Evidence for the occurence of Non-*N*-Methyl-D-Aspartate receptors. J. Neurosci. Res. **40**, 632-640.
10. Gard A.L. and Pfeifer S.E. (1989) Oligodendrocyte progenitors isolated directly from developing telencephalon at a specific phenotypic stage: myelinogenic potential in a defined environment. Development **106**, 119-132.
11. Glassmeier G., Jeserich G., and Krüppel T. (1992) Voltage-dependent potassium currents in cultured trout oligodendrocytes. J. Neurosci. Res. **32**, 301-309.
12. Glassmeier G., Jeserich G., and Krüppel T. (1994) Voltage-dependent sodium and potassium currents in cultured trout astrocytes. Glia **11**, 245-254.

13. Glassmeier G.and Jeserich G. (1995) Changes in ion channel expression during in vitro differentiation of trout oligodendrocyte precursor cells. Glia **15**, 83-93.
14. Hamill O.P., Marty A., Neher E., Sakmann B., and Sigworth F. J. (1981) Improved patch-clamp techniques for high resolution current recording from cell and cell-free membrane patches. Pflugers Arch. **391**, 85-100.
15. Howe J.R. and Ritchie J.M. (1990) Sodium currents in Schwann cells from myelinated and non-myelinated nerves of neonatal and adult rabbits. J. physiol. (London) **425**, 169-210.
16. Jeserich G. and Waehneldt T.V. (1987) Antigenic sites common to major fish myelin glycoproteins (IP) and to major tetrapode PNS myelin glycoprotein P_0 reside in amino acid chain. Neurochem. Res. **12**, 821-825.
17. Jeserich G. and Rauen T. (1990) Cell cultures enriched in oligodendrocytes from the central nervous system of trout in terms of phenotypic expression exhibits parallels with cultured rat Schwann cells. Glia **3**, 65-74.
18. Jeserich G., Müller A., and Jacque C. (1990a) Developmental expression of myelin proteins by oligodendrocytes in the CNS of trout. Dev. Brain Res. **51**, 27-34.
19. Jeserich G. and Stratmann A. (1992) In vitro differentiation of trout oligoden drocytes: Evidance for an A2B5-positve origin. Dev. Brain Res. **67**, 27-35.
20. Jeserich G., Strelau J., and Tönnies R. (1993) An immunomagnetic cell separation technique for isolating oligodendroglial progenitor cells of trout CNS. Neuroprotocols: A Comp. Meth. Neurosci. **2**, 219-224.
21. Kolb H.A. (1990) Potassium channels in excitable and non-excitable cells. Rev. Physiol. Biochem. Pharmacol. **115**, 51-91.
22. Nowak L., Ascher P., and Berwald-Netter Y. (1987) Ionic channels in mouse astrocytes in culture. J. Neurosci. **7**, 101-109.
23. Raff M.C., Abney E., Cohen J., Lindsay R., and Noble M. (1983) Two types of astrocytes in cultures of developing rat white matter: differences in morphology, surface gangliosides and growth characteristics. J. Neurosci. **3**, 1289- 1300.
24. Soliven B., Szuchet S., Arnason B.G.W., and Nelson D.J. (1988) Voltage-gated potassium currents in cultured ovine oligodendrocytes. J. Neurosci. **8**, 2131-2141.
25. Sontheimer H. and Kettenmann H. (1988) Heterogenity of potassium currents in cultured oligodendrocytes. Glia **1**, 415-420.
26. Sontheimer H., Kettenmann H., Backus K.H., and Schachner M. (1988) Glutamate opens Na^+/K^+ channels in cultured astrocytes. Glia **1**, 328-336.
27. Sontheimer H., Trotter J., Schachner M., and Kettenmann H. (1989) Channel expression correlates with differentiation stage during the development of oligodendrocytes from their precursor cells in culture. Neuron **2**, 1135-1145.

28. Sontheimer H., Ransom B.R., Cornell-Bell A.H., Black J.A., and Waxman S.G. (1991) Na^+ current expression in rat hippocampal astrocytes in vitro: Alteration during development. J. Neurophysiol. **65**, 3-19.
29. Sontheimer H., Ransom B.R., and Waxman S.G. (1992a) Different Na^+ currents in P0 and P7-derived hippocampal astrocytes in vitro: Evidance for a switch in Na^+-channel expression in vivo. Brain. Res. **597**, 24-29.
30. Sontheimer H. and Waxman S.G. (1992b) Ion channels in spinal cord astro cytes in vitro. II. Biophysical and pharmacological analysis of two Na^+ current types. J. Neurophysiol. **68**, 1001-1011.
31. Sontheimer H., Black J.A., Ransom B.R., and Waxman S. G. (1992c) Ion channels in spinal cord astrocytes in vitro. 1. Transient expression of high levels of Na^+ and K^+ channels. J. Neurophysiol. **68**, 985-1000.
32. Sontheimer H. and Waxman S.G. (1993) Expression of voltage-activated ion channels by astrocytes and oligodendrocytes in the hippocampal slice. J. Neurophysiol. **70**, 1863-1873.
33. Stühmer W., Ruppersberg, J.P., Schröter K.H., Sakmann B., Stocker M., Giese K.P., Perschke A., Baumann A., and Pongs O. (1989) Molecular basis of functional derversity of voltage-gated potassium channels in mammalian brain. The EMBO Journal vol.**8** no. **11**, 3225-3244.
34. Tse F.W., Fraser D.D., Duffy S., and Mac Vicar B.A. (1992) Voltage acti vated K^+ currents in acutely isolated hippocampal astrocytes. J. Neurosci. **12**, 1781-1788.
35. Usowicz M.M., Gallo V., and Cull-Candy S.G. (1989) Multiple conductance channels in type-2 cerebellar astrocytes activated by excitatory amino acids. Nature **339**, 380-383.
36. Wilkin G.P., Mariott D.R. and Cholewinski A.J. (1990) Astrocyte heterogeneity. Trend Neurosci. **13**, 43-46.

Ion Channel Reorganization During Remyelination

Peter Shrager and Sanja D. Novakovi_

Department of Physiology, University of Rochester Medical Center, Rochester, NY 14642 USA

Keywords: Myelin, Demyelination, Remyelination, Axons, Sodium channel

1. Introduction

Conduction in myelinated axons depends upon both a highly heterogeneous distribution of ionic channels and on passive improvements in cable properties resulting from myelin formation. Under pathological conditions involving demyelination, action potential propagation can be delayed or blocked and recovery requires firstly removal of damaged myelin and secondly association with new glial cells. We have studied these processes extensively in the peripheral nervous system with both electrophysiological and immunocytochemical methods. We show that a complex series of events is initiated by Schwann cell adherence to demyelinated axons and that the recovery of conduction occurs in several stages.

2. Methods

2.1 Demyelination

Xenopus and Lewis rats were anesthetized and the sciatic nerve in one leg surgically exposed. 1 µl of 1% lysolecithin in sterile saline was injected into the nerve [8] via a glass micropipette broken to a tip diameter of 18-20 µm. The wound was sutured and the animal allowed to recover. On the experiment day the animal was sacrificed and the nerve was dissected, desheathed, and dissociated into individual axons with collagenase (3.5 mg/ml).

2.2 Loose Patch Clamp

Pipettes were drawn to a resistance of about 2 MΩ and fire polished. They were sealed to axons with gentle suction until the seal resistance was 6-12 MΩ. The electronic circuit used was based on the design of Stühmer et al. [20], modified to allow automatic correction for variations in the seal resistance [15]. Application of a 40 mV hyperpolarization after each test pulse allowed correction for remaining linear currents. Experiments were run at room temperature .

Molecular Signaling and Regulation in Glial Cells
A Key to Remyelination and Functional Repair

2.3 Optical Recording

Nerves were mounted in a chamber fitted with pairs of Pt wires for stimulation and recording of the compound action potential. They were exposed to the dye RH155 [11], 1-2 mg/ml, for 1.5-2 hrs. The dye solution was then replaced with normal frog Ringer's. A Nikon Diaphot microscope was modified for high intensity illumination [19] at 705 ± 25 nm bandwidth. The image was projected onto a photodiode which served as input to a high gain current to voltage converter. Typically, responses to 64 stimuli were averaged to increase the signal-to-noise ratio. Optical signals are plotted as the fractional change in transmitted light intensity.

2.4 Immunocytochemistry

Axons were fixed in 4% paraformaldehyde and were preincubated in phosphate buffer containing 0.3% Triton X-100 and 10% goat serum for 0.5 hrs. They were then exposed to a rabbit polyclonal antibody raised against an 18-mer peptide forming part of the domain III-IV linker of the vertebrate Na^+ channel linked to keyhole limpet hemocyanin. The antibody was prepared and affinity purified by Dr. S. Rock Levinson. The tissue was rinsed and incubated in biotinylated goat anti-rabbit IgG followed by avidin-FITC [6].

3. Results

3.1 Electrophysiology

In the sciatic nerve myelin is rapidly damaged by lysolecithin injection, and disrupted sites can be seen within 1 day of surgery. Macrophages then appear within the lesion and remove all myelin within an internode, a process that is largely complete after 7 days [6,15]. Schwann cells proliferate and appear adherent to demyelinated axons at 8-9 days post injection. Thin remyelination can be seen at 9-10 days in a limited number of axons. This relatively synchronous temporal progression is advantageous in studies of demyelination and remyelination since it allows an accurate prediction of the state of a large majority of axons within the nerve. There is of course always some heterogeneity in any given preparation, and spatially fibers vary considerably in the length of the demyelinated zone. In addition to light and thin section electron microscopy we have examined this preparation by 3 techniques designed to assess the distribution and role of axonal ion channels.

The loose patch clamp is tolerant of myelin debris, yet sufficiently sensitive to detect about 100 pA of peak Na^+ current in a membrane area of 10-100 μm^2. Na^+ currents could be recorded from myelin-damaged internodes as early as 1 day post-surgery. Fig. 1A illustrates ionic currents recorded from the axolemma of a rat internode at 3 days. Internodal peak Na^+ current densities were virtually constant from 1 to 150 days after injection, suggesting that Na^+ channels are normally present in these regions. This view was strengthened by experiments on *Xenopus* axons that had been severed from

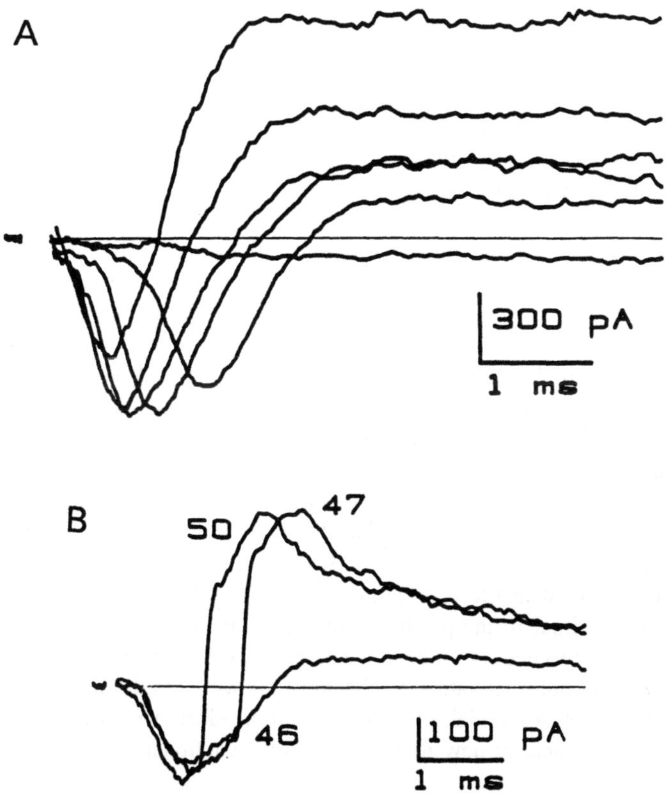

Fig. 1. Loose patch clamp measurement of ionic currents. A, records from a demyelinated internode. This rat axon was 3 days post-injection of lysolecithin. The patch was held 30 mV negative to the resting potential and was depolarized by pulses of 40, 60, 70, 80, 90, and 110 mV. The uncompensated portion of the capacitative transient has been blanked. B, Paranodal currents. Site was a few μm from the end of original myelin. The numbers denote the amplitude of the depolarizing pulse from the resting potential. 13 days post-surgery. Reproduced by permission from [16].

their cell bodies just prior to lysolecithin injection to eliminate the possibility of Na^+ channel synthesis in response to demyelination [14]. Na^+ currents recorded from these axons 6 days later were similar to those from uncut fibers. The density of Na^+ channels in the internode was calculated to be 4% of the nodal density, or about $25/\mu m^2$ in the rat [16]. At paranodal regions inward Na^+ currents were often abruptly interrupted by a transient outward current over a range of test voltages (Fig. 1B). The behavior of this current suggested that it represented a loss of control due to the high density of Na^+ channels in the neighboring nodal zone. These records were seen only at paranodes and thus served as an indicator of the sharp gradient in Na^+ channel density at these sites. Interestingly, this gradient did not dissipate following demyelination, and could be detected throughout the process of remyelination. This suggested that nodal clusters of Na^+ channels do not break up and diffuse laterally into internodes when myelin is disrupted, a process that has been proposed to explain internodal inward currents in demyelinated axons [3]. As will be seen below, immunocytochemical experiments have confirmed and extended this conclusion.

When axons remyelinate new nodes of Ranvier form in regions that previously were internodal. It is of interest to understand the development of these sites. What determines their location? Do they contain a high density of Na^+ channels? Are they functional in conduction? The loose patch clamp afforded some initial clues. Presumptive new nodes could be observed at the light level as gaps between Schwann cell processes. In *Xenopus* fibers peak Na^+ currents at these sites were 2-6 times higher than at random internodal loci indicating that some aggregation had taken place [15]. Further, it could be demonstrated that these Na^+ channels function very early in the repair process. We took advantage of the fact that the patch clamp controls only the extracellular potential. A pipette was sealed to a demyelinated site and the membrane was held at the resting potential. The axon was then stimulated several mm proximal to the patch site by a pair of Pt wires. If a propagating action potential reached the patch, it caused a depolarization of the internal potential which could activate Na^+ channels. It was found that Na^+ channels were functional at internodal sites but only after adherence of Schwann cells nearby. Na^+ channels at new nodes were fully functional.

Adult peripheral myelinated axons may have up to 150 lamellae of highly compact myelin and conduct action potentials with a very high degree of reliability. Measured in *Xenopus*, conduction was typically blocked in acutely demyelinated axons. What level of remyelination is required for the restoration of signal propagation? We wished to make these measurements on single, visually observable fibers and chose an optical technique using a voltage sensitive dye. The sensitivity of this method, developed by Cohen and his colleagues [4,5,7], has been increased to allow recording from single nodes of Ranvier and single demyelinated loci [19]. Fig. 2 shows that a propagating action potential reached the last intact node in an axon, but was blocked at the proximal end of the demyelinated zone. In *Xenopus* the first lamellae of new myelin were seen 12 days post-injection. Successful conduction across a demyelinated zone was found, however, between 8 and 10 days (Fig. 3). This corresponds to a time when Schwann cells adhere and begin to wrap processes around the axon. Thus, just the most minimal extracellular association is sufficient for the restoration of signal propagation. The conduction velocity within the demyelinated zone is, however, very slow, typically 0.5-1 m/s as compared with 15-30 m/s in normal axons of this caliber. As myelin develops over the next few days the conduction velocity rises only slightly but the fraction of axons conducting successfully through the lesion increases dramatically. A computational model has been used to analyze this situation [10,19]. The low density of Na^+ channels in the internode play an essential role is allowing conduction to proceed at these very early stages.

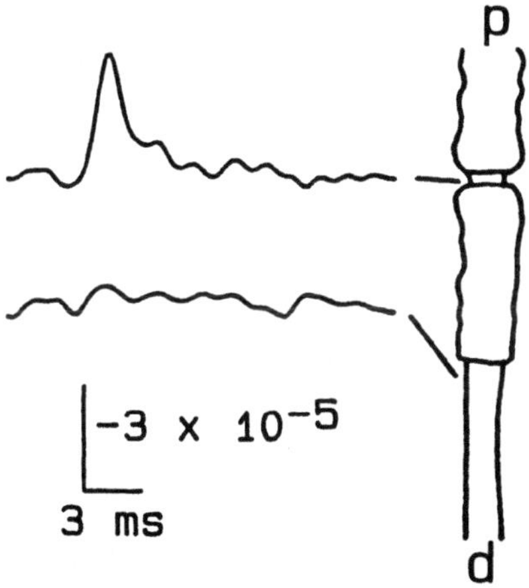

Fig. 2. Optical recording of action potentials shows block at the proximal end of a demyelinated zone. The fiber was stimulated several mm proximal (p) to the recording region. The action potential reached the last intact "driving" node, but was blocked at the first demyelinated heminode. 8 days post-injection. Reproduced by permission from [19].

Optical recording has revealed an interesting alteration of the pattern of action potentials during repetitive stimulation that takes place in demyelinated fibers. In normal axons during a short burst of stimuli spike frequency is limited by the refractory period so that the second or later signals my be absent, but the *first* action potential is always present. In demyelinated axons this is altered in 2 ways. Firstly, as has been observed previously, the refractory period is longer, limiting the maximum frequency of stimulation. However, at stimulus frequencies of 30 – 140 Hz an unusual phenomenon appears. The *initial* action potential in a train may be missing, while subsequent signals are conducted successfully [17]. Fig. 4 illustrates both effects. Optical signals show that integration of signals occurs at the heminode at the proximal end of the demyelinated zone (Fig. 5), as well as at new nodes of Ranvier in axons at later stages of repair. The first spike is reduced in amplitude to a subthreshold response at these sites, but the signal is prolonged sufficiently to change the threshold for subsequent signals. The depolarizing afterpotential involved is analogous to threshold changes involved in superexcitability [2,13]. Pharmacological tests did not reveal any particular ion channel that is responsible for this axonal coding of signals [17]. However, ouabain, a blocker of the Na^+-K^+ ATPase, had small and possibly significant effects on responses. Further, the computational model suggested that the system developed by Barrett and Barrett [1] involving lowered myelin resistance and nodal leakage conductance may be applicable here.

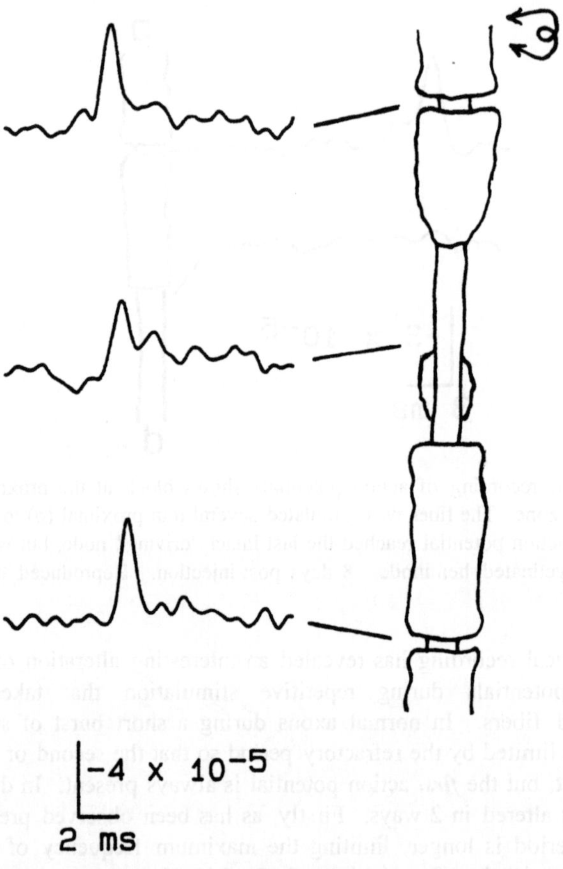

Fig. 3. Restoration of conduction prior to remyelination. An optically recorded action potential successfully propagates through an internode that is simultaneously undergoing myelin removal and adherence of proliferating Schwann cells. The clump of material in the demyelinated zone indicates uncleared myelin debris. The remainder of this region was covered with Schwann cell processes. 10 days post-injection. Reproduced by permission from [18].

3.2 Immunocytochemistry

Exposing normal rat sciatic axons to the Na^+ channel antibody resulted in intense labeling of nodes of Ranvier and much lower, diffuse labeling of myelin (Fig. 6A). If the primary antibody were preabsorbed with the peptide

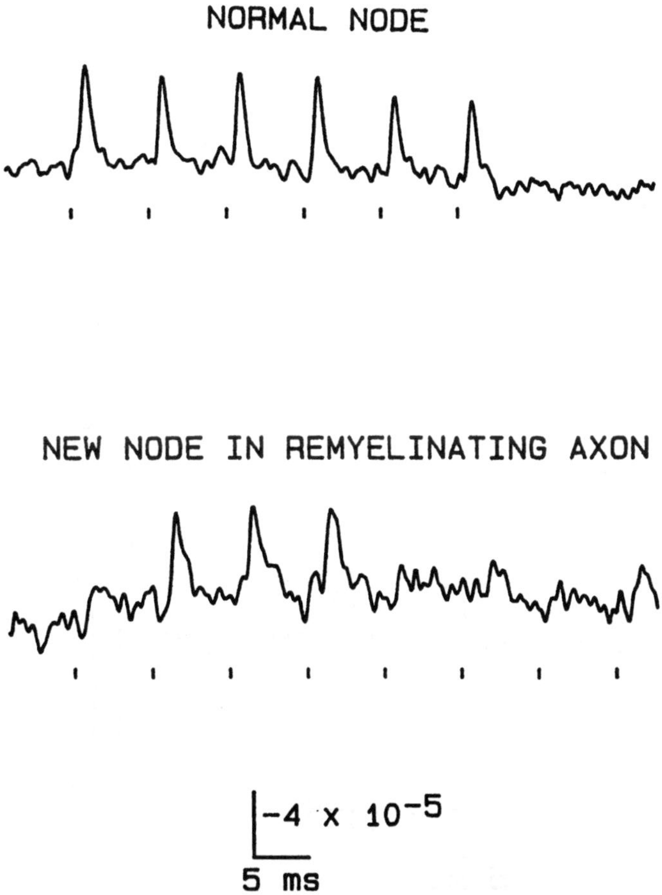

Fig. 4. Responses to bursts of stimuli. Top, a normal node of Ranvier. Bottom, coding of action potentials in a new node of Ranvier in a remyelinating *Xenopus* axon, 22 days post-surgery. The tick marks indicate the times of stimulation, several mm proximal to each site. Reproduced by permission from [17].

antigen the nodal label disappeared, while the diffuse myelin stain remained [6]. The latter was thus deemed non-specific and is not discussed further. Axons examined at 7 days were at the peak of demyelination and adherent Schwann cells were not yet seem. At this stage clusters of Na^+ channels were very rare and were about 1 mm apart. These were likely to be original nodes of Ranvier, and it is noteworthy that here too we found evidence for a lack of lateral diffusion of channels following demyelination. Just a few days later, the situation changed dramatically. As Schwann cells bound to axons and extended processes new clusters of Na^+ channels appeared at their edges (Fig. 6B). As these Schwann cells then grew laterally the clusters appeared to move with them since the distance between 2 clusters associated with the same cell increased over time. Clusters associated with neighboring Schwann cells eventually approached each other and appeared to fuse (Fig. 6C,D), forming a presumptive new node [6]. Over the next few weeks these clusters coalesced into mature nodes, about 270 μm apart (Fig. 6E).

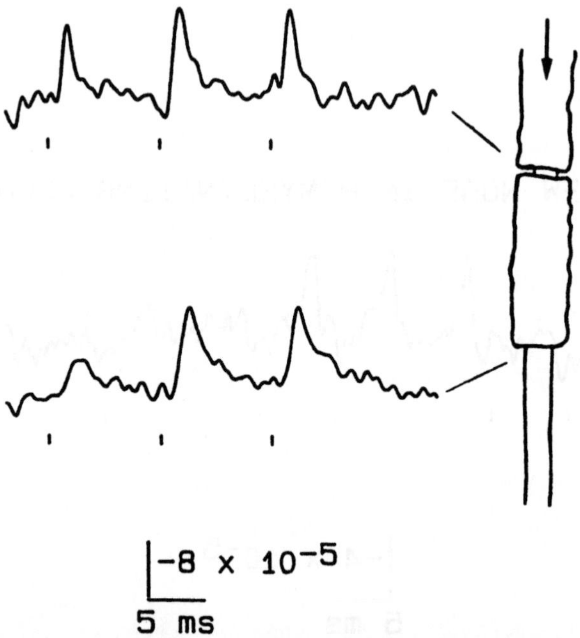

Fig. 5. Integration at the proximal end of a demyelinated internode. The fiber was stimulated several mm proximal (above) the sites shown. Optical signals are illustrated at both the driving node and the beginning of the demyelinated zone. Stimuli were at 10 ms intervals. 12 days post-injection. 26°C. Reproduced by permission from [17].

If Schwann cell proliferation were blocked by inclusion of mitomycin in the lysolecithin injection [9] new clusters of Na^+ channels did not appear. At 7 days post-injection, a time prior to the appearance of new Schwann cells even without mitomycin, Na^+ channel clusters were again seen only at original nodes (both fully and half-demyelinated). However, at 14 days, not only were new clusters absent, but original nodes were virtually gone as well. Thus, not only are proliferating Schwann cells required for the aggregation of channels at new nodes, they are also needed for the preservation of original nodes beyond 7 days as well. What is the origin of the Na^+ channels that cluster in response to Schwann cell adherence? One test has been done in a mouse

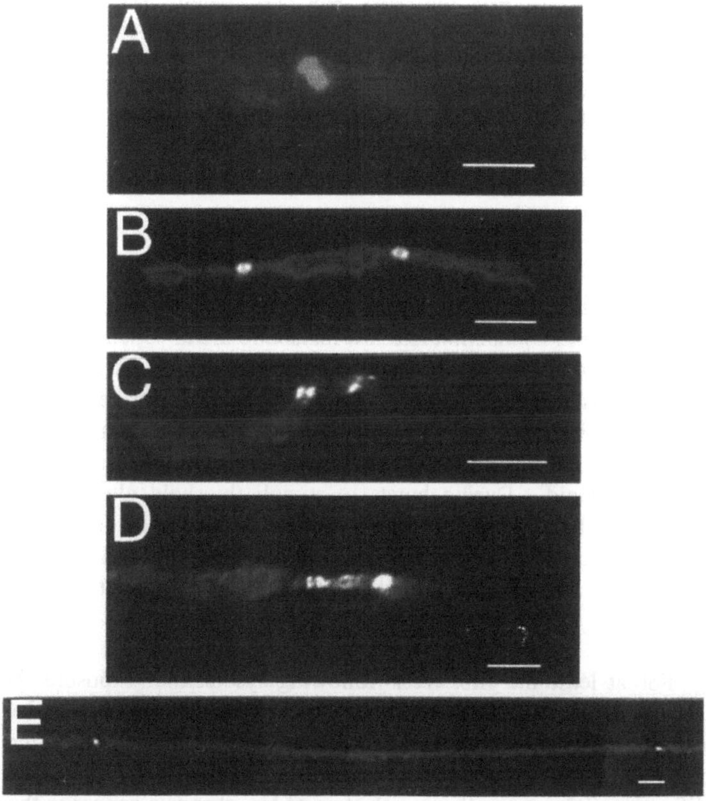

Fig. 6. Immunocytochemistry of Na^+ channels. A, A normal node of Ranvier. B, Na^+ channel clusters at the edges of an adherent Schwann cell, 14 days post-injection. C,D Na^+ channel clusters in close proximity, as though about to fuse, 14 days. E, 2 new nodes of Ranvier at 60 days post-surgery. Scale bars, 25 μm. B and E are reproduced with permission from [6].

strain that is resistant to Wallerian degeneration, C57BL/6/wld (ola) [12]. In experiments analogous to those described above on *Xenopus*, sciatic nerves were injected with lysolecithin and, 3 days later, were transected to disconnect axons from neuronal cell bodies [21]. The 3-day delay was necessary because, despite the resistance of this strain, degeneration in the presence of demyelination was excessive after about 7 days and 8-9 days were needed for Na^+ channel clustering. At the 3-day point, myelin was disrupted, but not yet removed by macrophages, and no new Schwann cells were seen in the lesion area. Na^+ channel clustering proceeded on schedule despite the transection. Aggregates were seen at Schwann cell edges and appeared to fuse as these cells grew. We could not follow the process beyond this point. Thus, the processes of Schwann cell adherence and Na^+ channel clustering, and the movement and apparent fusion of these clusters that may represent the initial stages in new node formation during remyelination do not require continuous communication with the neuronal cell body. Our results also suggest that *de novo* synthesis of channels in response to the demyelination is not required, but with the caveat that the 3-day delay may have allowed a rapid response to lysolecithin action. It appears more likely that the aggregates arise from an existing internodal pool.

4. Discussion

The ion channel distribution in myelinated fibers is highly heterogeneous and is designed for rapid conduction and low metabolic drain. Upon demyelination a low density pool of Na^+ channels in the internodal axolemma is exposed and provides a means whereby action potential propagation can be restored with just the earliest steps in the remyelination process. Conduction through a remyelinating region has a low velocity and is also highly dependent on small changes in threshold. During bursts of stimuli threshold alterations due to long-lasting depolarizing afterpotentials can lead to an unusual coding of action potentials. Thus, while conduction may be restored with minimal Schwann cell association, much pathology may remain until later stages in remyelination.

For at least the first week following lysolecithin exposure the Na^+ channel distribution is relatively unchanged. At nodes of Ranvier densities remain high and channels appear tightly clustered. However, over the next few days reorganization is rapid and extensive. Within 1 day after adherence of proliferating Schwann cells new clusters of Na^+ channels appear in the axon at the edges of glial processes. These Schwann cells then proceed to elaborate myelin and also to grow laterally, and the Na^+ channel aggregates seem to move along with them, staying always close to the ends of processes. As neighboring clusters are caused to approach each other they seem to fuse, forming early new nodes of Ranvier. Schwann cells thus appear to play a major role in determining the organization of axonal ion channels. In support of this hypothesis, if Schwann cell division is blocked by a DNA intercalant aggregation of axonal Na^+ channels is absent. Further, at least the initial stages in new node formation can occur in the absence of continuous signals from the neuronal soma. It appears likely that the internodal Na^+ channels provide a low density pool that can be altered under the influence of adherent Schwann cells. It will be interesting to investigate whether a similar process takes place during normal development, and also to elucidate the molecular mechanism for the glial-axonal communication involved.

5. Acknowledgements

We are grateful to our collaborators: Ellen Brunschweiger, Michael Hines, S. Rock Levinson, Chaim Rubinstein, Elda Tzoumaka, Ian Vabnick, and Jin Wu for their participation in this work. Our research has been supported by grants from the National Institutes of Health (NS17965), the National Multiple Sclerosis Society (RG-1774), and the Lucille P. Markey Charitable Trust Award to the University of Rochester.

6. References

1. Barrett, E.F. and Barrett, J.N. (1982) Intracellular recording from vertebrate myelinated axons: mechanism of the depolarizing afterpotential. Journal of Physiology 323, 117-144.
2. Bostock, H. and Grafe, P. (1985) Activity-dependent excitability changes in normal and demyelinated rat spinal root axons. Journal of Physiology 365, 239-257.
3. Bostock, H. and Sears, T.A. (1978) The internodal axon membrane: electrical excitability and continuous conduction in segmental demyelination. Journal of Physiology 280, 273-301.
4. Cohen, L.B. and Lesher, S. (1986) Optical monitoring of membrane potential: methods of multisite optical measurement. Society of General Physiologists Series 40, 71-99.
5. Cohen, L.B. and Salzberg, B.M. (1978) Optical measurement of membrane potential. Reviews of Physiology Biochemistry & Pharmacology 83, 35-88.
6. Dugandzija-Novakovic, S., Koszowski, A.G., Levinson, S.R. and Shrager, P. (1995) Clustering of Na channels and node of Ranvier formation in remyelinating axons. Journal of Neuroscience 15, 492-502.
7. Grinvald, A. (1985) Real-time optical mapping of neuronal activity: from single growth cones to the intact mammalian brain. Annual Review of Neuroscience 8, 263-305.
8. Hall, S.M. and Gregson, N.A. (1971) The in vivo and ultrastructural effects of injection of lysophosphatidyl choline into myelinated peripheral nerve fibres of the adult mouse. Journal of Cell Science 9, 769-789.
9. Hall, S.M. and Gregson, N.A. (1974) The effects of mitomycin C on remyelination in the peripheral nervous system. Nature 252, 303-305.
10. Hines, M. and Shrager, P. (1991) A computational test of the requirements for conduction in demyelinated axons. Restorative Neurology and Neuroscience 3, 81-93.
11. Lev-Ram, V. and Grinvald, A. (1986) Ca2+- and K+-dependent communication between central nervous system myelinated axons and oligodendrocytes revealed by voltage-sensitive dyes. Proceedings of the National Academy of Sciences U.S.A. 83, 6651-6655.

12. Lunn, E.R., Perry, V.H., Brown, M.C., Rosen, H. and Gordon, S. (1989) Absence of Wallerian degeneration does not hinder regeneration in peripheral nerve. European Journal of Neuroscience 1, 27-33.
13. Raymond, S.A. (1979) Effects of nerve impulses on threshold of frog sciatic nerve fibres. Journal of Physiology 290, 273-303.
14. Rubinstein, C.T. and Shrager, P. (1990) Remyelination of nerve fibers in the transected frog sciatic nerve. Brain Research 524, 303-312.
15. Shrager, P. (1988) Ionic channels and signal conduction in single remyelinating frog nerve fibres. Journal of Physiology 404, 695-712.
16. Shrager, P. (1989) Sodium channels in single demyelinated mammalian axons. Brain Research 483, 149-154.
17. Shrager, P. (1993) Axonal coding of action potentials in demyelinated nerve fibers. Brain Research 619, 278-290.
18. Shrager, P. (1995) Action potential conduction recorded optically in normal, demyelinated and remyelinating axons. In S.G. Waxman, J.D. Kocsis and P.K. Stys (eds.), The Axon, Oxford University Press, Oxford, pp 341-352.
19. Shrager, P. and Rubinstein, C.T. (1990) Optical measurement of conduction in single demyelinated axons. Journal of General Physiology 95, 867-889.
20. Stühmer, W., Roberts, W.M. and Almers, W. (1983) The loose patch clamp. In B. Sakmann and E. Neher (eds.), Single-Channel Recording, Plenum Press, New York, pp 123-132.
21. Tzoumaka, E., Novakovic, S.D., Levinson, S.R. and Shrager, P. (1995) Na+ channel aggregation in remyelinating mouse sciatic axons following transection. Glia In press.

Gap Junction Expression in Myelinating Cells

Rolf Dermietzel

Institute of Anatomy, University of Regensburg, D-93042 Regensburg, Universitystreet 32

Introduction

Gap junctions comprise intercellular channels which allow the direct exchange of small metabolites (glucose, nucleotides, amino acids, etc) as well as the transmission of ions for propagating electrical currents [for reviews see 1, 2]. Accumulating evidence indicates that the biophysical properties of the gap junction channels are variable according to the specific protein complement that forms the channel [3]. The gap junction proteins, collectively called connexins, are oligomerized in hexamers (connexons) which form hemichannels in plasma membranes. When docked to a partner connexon a competent gap junction channel is formed which spans the two adjoining plasma membranes. At present thirteen members of the connexin family have been cloned and their tissue specific expression has been determined [for review see 1]. In the central and peripheral nervous system connexins are differentially expressed with respect to cell specificity and developmental stage [4]. Oligodendrocytes and Schwann cells, the myelin forming cells in the peripheral and central nervous system, have been described to express connexin32 [4, 5], which was first identified and cloned form rodent and human liver cDNAs [6, 7]. The biophysical properties of the gap junction channel formed by connexin32 have been extensively studied. Cell lines, stabely transfected with the connexin32 gene revealed a unitary conductance of the connexin32 channel in the range of 120 pS, showed voltage dependence, and pH sensitivity [8, 9]. From a functional point of view it is of considerable interest that the connexin32 channel is permeable to second messenger including calcium and IP3, and that the strength of coupling is modulated by a PKA dependent phosphorylation [10].

The recent discovery that a specific form of the Charcot-Marie-Tooth syndrome (CMTX1, see below) correlates with mutations in the connexin32 gene [5], has re-issued the search for functional significance of gap junctions in peripheral and central myelin [11].

Here I report on the identification of connexin32 in the myelinating cells of the peripheral and central nervous system peripheral. A brief account of its expression under experimental and developing conditions is enclosed.

1. Connexin32 is a gap junction forming protein in myelinating cells

First evidence that myelinating cells in the central nervous system express connexin32 came from *in situ* studies of immunocytochemically labelled cryostat

Molecular Signaling and Regulation in Glial Cells
A Key to Remyelination and Functional Repair

sections. Using a panel of different antibodies directed to three gap junction proteins, we were able to colocalize connexin32 with cells carrying MBP epitopes [4]. This initial finding was further extended to cultured oligodendrocytes. In contrast to cultured astrocytes [11, 13] adult bovine oligodendrocytes showed a low strenght of coupling as evidenced by the gj which was in the range of 3.9±0.9 nS. This is three times less than in cultured astrocytes. The unitary conductance of gap junction channels in cultured oligodendrocytes shows an accumulation of single channel events in the range of 100 pS. This is in agreement with the channel size expected for connexin32. However, further single channel events clustered around 30-40 pS indicating the presence of a second channel which has not been identified so far. The existence of a second connexin gene expressed in oligodendrocytes is also suggested by the CMTX1 phenotype and by transgenic connexin32 knock out mice (Nelles et al. in preparation). The lack of central symptoms in both cases renders a compensatory expression of an additional connexin gene most likely. Alternatively, a dominant negative effect of the connexin32 mutations could, for some unknown reasons preferably affect myelinating Schwann cells while being compensated in oligodendrocytes. Screening of extracted mRNA form adult bovine oligodendrocytes with a panel of six different connexin cDNAs, however, failed to detect one of the more prominent connexins in oligodendrocytes cultures. It seems reliable that a less prominent or unknown connexin is expressed in oligodendrocytes which might serve for compensation.

2 Topological differences in connexin32 expression between Schwann cells and oligodendrocytes

Within the central nervous system oligodendrocytes are responsible for myelin formation, and they provide the biological machinery for maintenance and constant trophic support of the myelin sheath [14]. Schwann cells constitute the cellular analogues in the peripheral nervous system. With respect to the functional relevance of gap junctional coupling in myelinating cells I consider it beneficial to recapitulate some of the essential differences in the cytoarchitecture between central and peripheral myelin which primarily derive from their differences in the myelinating process.

While in the peripheral nervous system a one to one ratio exists in the number of Schwann cells and myelinated internodes the ratio of oligodendrocytes to the number of internodes is much smaller in the CNS. There are 20 to 40 times as many internodes as oligodendrocytes in the optic nerves [15], and the number is even more variable in the spinal cord where the ratio varies between 18 and 60 [16]. In addition, from reconstructions of serial electron microscopical sections it is known that more than one individual oligodendrocyte contributes to the formation of a single internode [17].Thus we have to envisage a complex arrangement of oligodendrocytes within a single internodal length of central myelin as compared to the stereotype organization of the myelin sheath in peripheral nerves. A final aspect which should be addressed is the presence of cytoplasmic relics in the mature sheath which originally derive from the myelinating cells during the process of compaction. In the peripheral sheath these processes are restricted to the inner cytoplasmic collar, the paranodal or lateral processes, and the Schmidt-Lantermann incisures. Central myelin lacks Schmidt-Lantermann incisures and the cytoplasmic relics are confined to the inner, outer and lateral (paranodal) cytoplasmic processes. Moreover, while in the peripheral myelin sheath the nucleus of the Schwann cell is always enclosed in the outer cytoplasmic collar the soma of oligodendrocytes is distant from the myelin sheaths connnected via attenuated cytoplasmic processes with the outer cytoplasmic portions of the myelin sheath [18]. This attenuated pathways put some physical restrictions on the speed and efficiency of signal trafficking within the cytoplasmic compartment of a single oligodendrocyte. Considering the obligatory metabolic support of the myelin sheath it is suggestive to separate conceptually between „long-living“ metabolites of the myelin shcath, i. c. the constitutive membrane proteins (proteo-lipid protein, myelin-basic protein etc) with half lives of more than hundred days and „short-living“ metabolites including energy supply. The internodal compartment of the myelin sheath is also unique in its bipartite composition. An excessive part of bradytrophic „material“ is laid down in form of compact myelin, in contrast to the multiple cytoplasmic incisures which require a different schedule of energy and metabolite support. Especially the paranodal cytoplasmic loops are considered to be highly active partners of the axon, serving cooperative functions for the propagation of axon potentials [19] and their molecular requirements seem to be essentially different from the compact myelin sheath.

It is therefore not surprising when channels of the gap junction type have been described exclusively between the various cytoplasmic portions of the central and peripheral myelin sheath [20-22]. However, the original observation of gap junctions in myelin, has gained little attention until recently, when the connexin32 mutations had been discovered [5]. It was initially surprising that the phenotype resulting from apparent loss of function mutation of connexin32 is limited to peripheral nerves, because this connexin is a major component of gap junctions of diverse tissues including liver, exocrine pancreas and epithelium of the small intestine [23, 24].

There are several functional roles that connexin32 might play in Schwann cell biology which could be affected by functional changes of the connexin32 gene resulting from the mutations. First, connexin32 hemichannels in paranodal membranes might operate *per se,* without assembly into complete double membrane gap junction structures. Such behavior has been observed for connexin46 expressed in oocytes [25], for an unidentified connexin in retinal horizontal cells [26] and has been proposed for connexin43 hemichannels in macrophages stimulated with ATP [27]. Second, reflexive gap junction structures could provide necessary nutrient exchange between the Schwann perinuclear region and the distant cytoplasmic processes, which otherwise would be separated from the rest of the cytoplasm by a quite tortuous pathway. Third, if transnodal gap junctions occur (as observed in regenerating chick nerves by Tetzlaff [21], they could provide a pathway for signalling from one internode to another under either normal or injury-evoked conditions. Such a function might be important in response to injury including both de- and regenerating events.

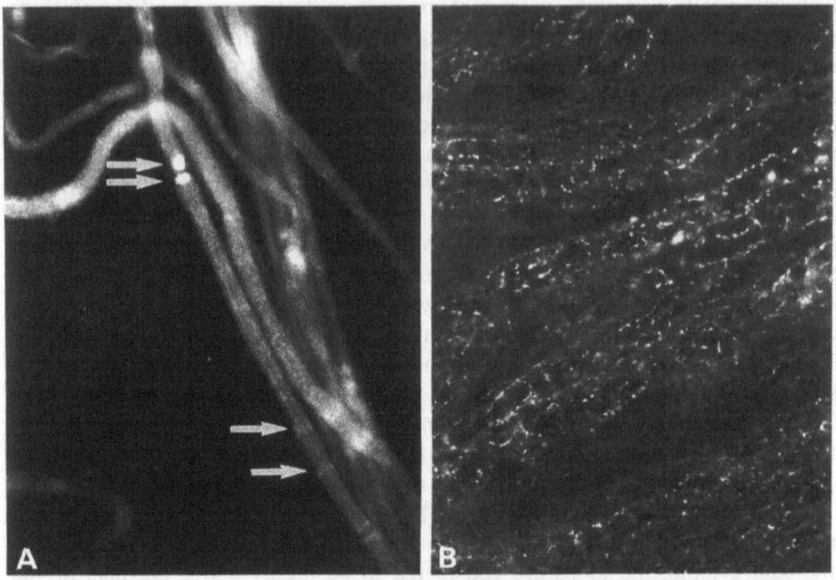

Fig. 1 Immunocytochemical detection of connexin32 in peripheral nerve (A., teased fibers of sciatic nerve) and central myelin (B., white matter of spinal cord). Note the differences in localization. Arrows in A indicate immunolabeling at the node of Ranvier and Schmidt-Lantermann incisures. Mag. x 1300

Gap junctions in non-regenerating peripheral nerves, however, have been reported to be extremely rare, in contrast to the situation in central myelin where junctional assemblies have been reported between the outer cytoplasmic loops and the soma of quiescent oligodendrocytes as well as between paranodal processes [20-22]. Connexin32 immunoreactivity in the peripheral nerve was reported to be confined to Schmidt-Lantermann incisures and the paranodal portions of the node of Ranvier [5]. However, the low incidence of gap junction assemblies at these sites leaves open alternative interpretations concerning the structural identity of the immunolabel. Interestingly, the immunolabelling with connexin32 antibodies in central myelin reveals an entirely different pattern, with extensive immunoreactive material being present in the internodal region [11] (Fig. 1A, B). In addition, its expression is highly variable with regional specificities. The highest amount is found in white matter of the spinal cord with a decrease in order (brain stem>cerebellar white matter>corpus callosum) (Fig. 2A, B).

Studies on the time course of connexin32 expression during myelinogenesis in the CNS indicates that the onset of connexin32 expression coincides with the beginning of myelination, around P5 to P7 in rat spinal cord. A characteristic shift of connexin32 localization is also prevalent during the myelination process showing a preferred cytoplasmic localization at early stages of myelination (P5-P15), and a redistribution to the peripheral processes, i. e. in the internodal cytoplasmic extensions, when myelination has succeeded. Apparently, the final assembly and organization of gap junctions is achieved in the mature myelin sheath.

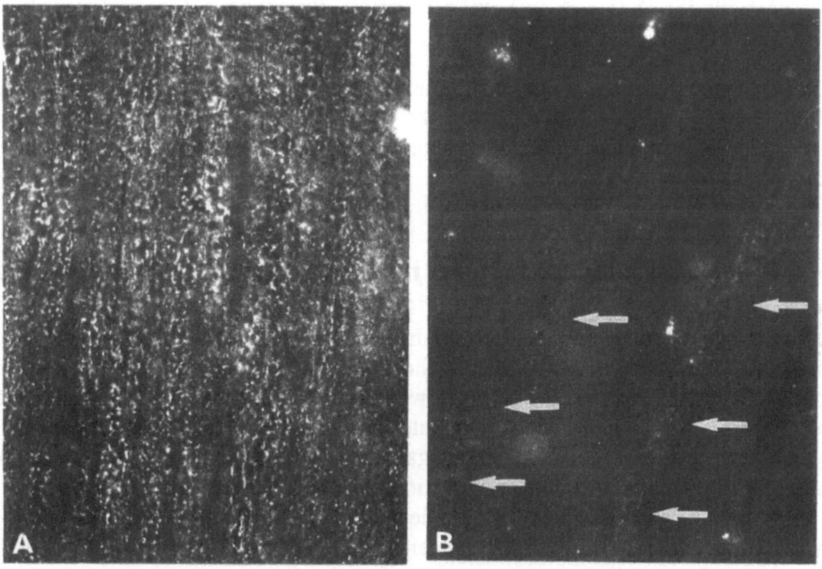

Fig. 2 Differences of connexin32 immunoreactivity in central myelin. A) spinal cord white matter, B) corpus callosum. see text. Mag. x 1300

Studies on peripheral nerves under de- and regenerating conditions [Chandross et al. in preparation] also revealed a strong correlation of connexin32 expression with myelinating events. Initial downregulation of connexin32 expression after nerve crush (day 1 to day 7) is followed by upregulation during the remyelinating phase (day 12 to day 21). Thus, the expression pattern of connexin32 in myelin-forming cells under regenerative conditions is apparently governed by the same genetic program as is the case for the more classical myelin proteins [28].

As already discussed, the functional significance of gap junction channels within the myelin sheath remains an open issue. The favorite idea is that they form reflexive junctions between peripheral processes of the same cell. They can thus be considered to be autocellular. The question then arises how the myelinating cells (Schwann cell as well as oligodendrocyte) benefit from additional shortcuts since large cytoplasmic continuities, albeit extended and tortuos, do already exist. One reasonable explanation is that gap junction channels provide regulated pathways between peripheral cytoplasmic portions of the myelin sheath which, because of their gating properties and selectivity, are responsive to changes of physiological conditions, i. e. transjunctional potentials, pH, etc. Furthermore, the extreme variability of connexin32 expression in central myelin indicates that local factors, either in form of structural specificities of the cytoarchitecture of the myelin sheath or of the local physiological environment, requires differences in the degree of „reflexivness".

The interpretation that most of the gap junctions in central myelin sheath represent the reflexive type was corroborated by microinjection of Lucifer Yellow or Neurobiotin® in oligodendrocytes of slices from spinal cord and corpus callosum (H. Kettenmann, personal communication). No dye transfer was achieved after injection of one of the tracers into electrophysiologically identified oligodendrocytes. Speculative though the idea that the complement of myelinic gap junctions provide responsive shortcuts for autocellular regulation is a compelling hypothesis.

3 A functional concept for gap junction coupling in myelinating cells

A bidirectional route can be conceptualized in terms of signal transfer and metabolite trafficking which involves the gated gap junction pathway: The first route provides an centrifugal pathway where the soma serves as the source and metabolites (signals) are delivered into the periphery (i. e. the peripheral cytoplasmic processes). This centrifugal pathway can be expected to carry preferentially metabolites and gap junctions would provide effective shortcuts for equilibration and buffering of gradients within the complex network of overlapping and interdigitating oligodendrocytic processes (Fig. 3.). Evidently, this form of coupling is not required for the Schwann cell and extensive internodal gap junctions are therefore missing.

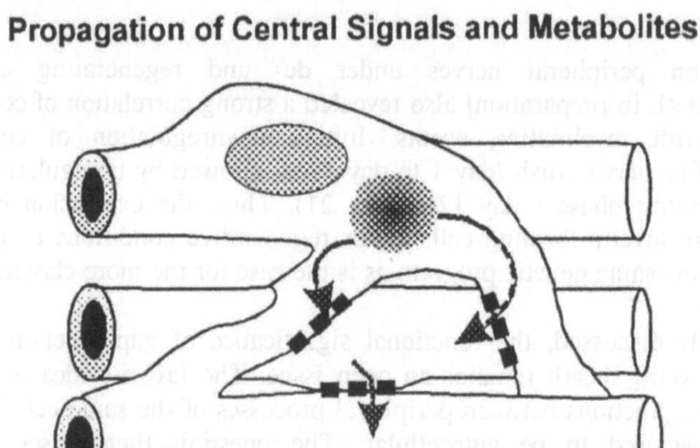

Fig. 3. Schematic drawing of centrifugal pathways for signal and metabolite transport in autocellulary coupled oligodendrocytes.

The second pathway can be considered to allow the transmission of information from the periphery to the soma and ultimately to the nucleus for transcriptional control. This route would involve primarily second messenger exchanges, and if one considers differences in local responsiveness of the plasma membrane of oligodendrocytes due to a polarized distribution of receptor domains or ion channels, gap junctions would allow coordination of local responses (Fig. 4). Dissipation or local restriction of signals in the complex periphery of oligodendrocytes would thus depend on the actual status of the gap junction channels.

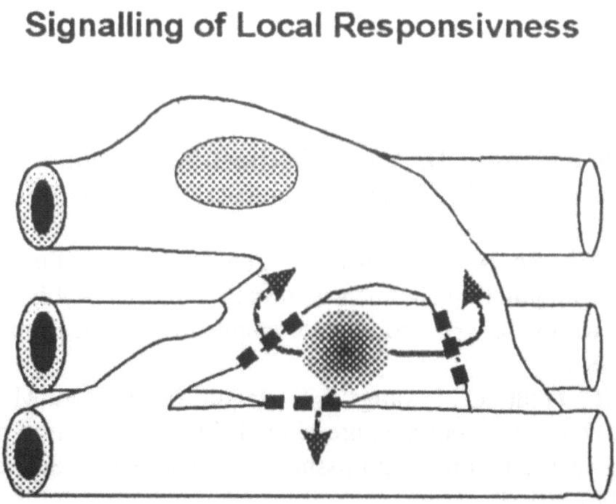

Fig. 4. Schematic drawing of centripetal pathways in autocellular coupled oligodendrocytes.

In this context the question of heterotypic coupling, i. e. between oligodendrocytes and astrocytes seems of considerable interest [29, 30]. In this respect the anatomical situation of oligodendrocytes is significantly different from that of the Schwann cell. We have recently suggested that pairing of oligodendrocytes with astrocytes would request heterotypic hemichannel formations [13]. Considerning that the major gap junction protein in astrocytes is connexin43 a heterotypic configuration which pairs connexin32 (oligodendrocytic) and connexin43 (astrocytic) hemichannels seems feasible. Alternatively, the second yet unidentified oligodendrocytic connexin may serve as the partner-hemichannel in this configuration. Nevertheless, the channel properties in this gap junction set should be regarded to be rectifying, an assumption which has gained support by the demonstration of unidirectional dye transfer in oligodendrocytic/astrocytic pairing in the retina [31]. Since a promiscuous signal transfer form oligodendrocytes into the astroglial environment via gap junctions seems highly unlikely this form of junction can be considered to be locally restricted and polarized in nature.

Although essentially speculative at present this concept allows to explain the differences in the distribution of gap junction channels between oligodendrocytes and Schwann cells. Furthermore, it provides direction in which further studies on the functional significance of gap junctions in myelinating cells should be perused.

Acknowledgements

This work was supported by a grant from the Deutsche Forschungsgemeinschaft (Schwerpunkt Glia) to R. D.

References

1. Dermietzel, R. and Spray, D.C., Gap Junctions in the brain: where, what type, how many and why, Trends in Neurosci., 16 (1993) 186-192.
2. Beyer, E.C., Gap Junctions, Int. Rev. Cytol., 137 (1993) 1-36.
3. Spray, D.C., Moreno, A. P., Eghbali, B., Chanson, M. and Fishman, G. I. Gating of gap junction channels as revealed in cells stably transfected with wild type and mutant connexin cDNAs, Biophys. J., 62 (1992) 48-50.
4. Dermietzel, R., Traub, O., Hwang, T. K., Beyer, E., Bennett, M.V.L. and Spray, D. C. Differential expression of three gap junction proteins in developing and mature brain tissue, Proc. Natl. Acad. Sci. USA, 86 (1989) 10148-10152.
5. Bergoffen, J., Scherer, S.S., Wang, S., Scott, M. and Bone, L.J, Paul, D.L., Chen, K., Lensch, M.W., Chance, P.F. and Fischbeck, K.H., Connexin mutations in X-linked Charcot-Marie-Tooth disease, Science, 262 (1993). 2039-2042.
6. Kumar, N.M. and Gilula, N.B., Cloning and characterization of human and rat liver cDNAs coding for a gap junction protein, J. Cell Biol., 103 (1986) 767-776.
7. Paul, D. C., Molecular cloning of cDNA for rat liver gap junction protein. J. Cell Biol., (1986) 103, 123-134.
8. Moreno, A.P., Eghbali, B. and Spray, D.C., Connexin 32 gap junction channels in stably transfected cells: Equilibrium and kinetic properties, Biophys. J., 60 (1991) 1267-1277.
9. Spray, D.C., Ginsberg, R.D., Morales, E.A., Gatmaitan, Z. and Arias, I.M., Electro-physiological properties of gap junctions between dissociated pairs of rat hepatocytes, J. Cell Biol. 103 (1986) 135-144.
10. Saez, J.C., Connor, J.A. and Spray, D.C., Hepatocytes gap junction are permeable to the second messengers inositol 1, 4, 5 phosphate and calcium ions, Proc. Natl. Acad. Sci., USA, 86 (1989) 2708-2712.
11. Spray, D.C. and Dermietzel, R., X-linked dominant Charcot-Marie-Tooth disease and other potential gap-junction diseases of the nervous system, Trends Neurosci., 18 (1995) 265-262.

12. Giaume, C., Fromaget, C., Aoumari, A.E., Cordier, J., Glowinski, J. and Gros, D., Gap junctions in cultured astrocytes: Single-channel currents and characterization of channel forming protein, Neuron, 6 (1991) 133-143.
13. Dermietzel, R., Hertzberg, E.L., Kessler, J.A. and Spray, D.C., Gap junctions between cultured astrocytes: Immunocytochemical, molecular, and electrophysiological characterization, J. Neurosci., 11 (1991) 1421-1432.
14. Bunge, R.P., Glial cells and the central myelin sheath, Physiol. Rev., 48 (1968) 197-251.
15. Peters, A. and Vaughn, J.E., Morphology and the development of the myelin sheath, In: Myelination, Davidson, A. N. and Peters, A., (eds.), 1970, Charles C. Thomas: Springfield, Ill. 3-79.
16. Matthews, M.A. and Duncan, D., A quantitative study of the morphological changes accompanying the initiation of myelin production in the peripheral nercous system, J. Comp. Neurol., 142 (1971) 1-22.
17. Knobler, R.L., Stempak, J.G. and Laurencin, M., Nonuniformity of the oligodendriglial ensheathment of axons during myelination in the developing rat central nervous system. J. Ultrastruct. Res., 55 (1976) 417-432.
18. Bunge, R.P., The development of myelin and myelin-related cells, Trends Neurosci., 4 (1981) 175-177.
19. Neumcke, B. and Staempfli, R., Sodium currents and sodium-current fluctuations in rat myelinated nerve fibres, J. Phsiol., 329 (1982) 163-184.
20. Dermietzel, R., Schunke, D. and Leibstein, A., The oligodendrocytic junctional complex, Cell Tiss. Res., 193 (1978) 61-72.
21. Tetzlaff, W., Tight junction contact events and temporary gap junctions in the sciatic nerve fibres of the chicken during Wallerian degeneration and subsequent regeneration, J Neurocytol., 11 (1982) 839-58.
22. Schnapp, B. and Mugnaini, E., Membrane architecture of myelinated fibers as seen by freeze-fracture, In: Physiology and Pathobiology of axons, Waxman, S.G. Waxman (ed.) 1978, Raven Press: New York. 83-123.
23. Hertzberg, E.L. and Skibbens, V., A protein homologous to the 27,000 dalton liver gap junction protein is present in a wide variety of species and tissue, Cell, 59 (1984) 61-69.
24. Dermietzel, R., Leibstein, A., Frixen, U., Janssen-Timmen, U., Traub, O. and Willecke, K. Gap junctions in several tissues share antigenic determinants with liver gap junctions, Embo J., 3 (1984) 2261-2270.
25. Paul, D.L., Ebihara, L., Takemoto, L.J., Sumson, K.J. and Goodenough, D.A., Connexin46, a novel lens gap junction protein, induces voltage- gated currents in nonjunctional plasma membrane of Xenopus oocytes, J. Cell Biol., 115 (1991) 1077-1989.

26. DeVries, S.H. and Schwartz, E.A., Modulation of an electrical synapse betwen solitary pairs of catfish horizontal cells by dopamine and second messengers, J. Physiol., (Lond.), 414 (1989) 351-375.
27. Beyer, E.C. and Steinberg, T.H., Evidence that the gap junction protein connexin-43 is the ATP- induced pore of mouse macrophages, J. Biol. Chem., 266 (1991) 7971-7974.
28. Gupta, S.K., Poduslo, J.F. and Mezei, C., Temporal changes in Po and MBP gene expression after crush injury of the adult peripheral nerve, Mol. Brain Res., 4 (1988)133-141.
29. Massa, P.T. and Mugnaini, E., Cell-cell interactions and characteristic plasma membrane features of cultured rat glial cells, Neurosci., 14 (1985) 695-709 Neuroscience. 1985.
30. Kettenmann, H. and Ransom, B.R., Electrical coupling between astrocytes and between oligodendrocytes studied in mammalian cell cultures, Glia, 1 (1988) 64 - 73.
31. Robinson, S.R., Hampson, E. C., Munro, M. N. and Vaney, D. I. Unidirectional coupling of gap junctions between neuroglia, Science, 262 (1993) 1072-1074.

Part 2 Genes and Proteins of Myelin

Structure, Regulation and Function

CNS Myelination: What can we learn from phylogeny?

Robert M. Gould and Warren D. Spivack

Laboratory of Membrane Biology, New York State Institute for Basic Research in Developmental Disabilities,Staten Island, New York 10314-6399

Abstract. The morphology of embryonic and adult spiny dogfish and clearnose skate spinal cords were compared with rat. Myelination occurred in a nearly identical fashion in all three specimens. For example, oligodendrocytes appear first in the ventricle funiculus (vf), coordinate with the development of large (one micron diameter) axons. Since developing dogfish and skate neural cells express the same cell-type-specific antigens, immunofluorescence was used to study their spatial and temporal appearance. In all cases, the order of appearance was: radial glia (GFAP$^+$), neuronal perikarya (P$^-$ (non-phosphorylated) neurofilament), axons (P$^+$ neurofilament) and oligodendrocytes (O4$^+$, R-mAb$^+$, O1$^+$, MBP$^+$, P$_0$/PLP$^+$).The morphological and biochemical similarities demonstrate that many features of myelination are conserved in these divergent lineages. Thus, differences between them, which include: lack of tight junctions (chondrichthye myelin), replacement of the P$_0$ isoforms in fish with PLP in mammals, lack of exon 2-containing MBP isoforms, myelin-associated glycoprotein (MAG), myelin oligodendrocyte protein (MOG), CNPase and MBP mRNA translocation (chondrichthye), are all derived features; differences do not alter myelin function. Preliminary evidence suggests that myelinating oligodendrocytes are related to glial cells that ensheath, but do not myelinate, large caliber axons in lamprey spinal cord. MBP-like molecules are detected in lamprey spinal cord by RT-PCR (reverse transcriptase-polymerase chain reaction). Furthermore, polyclonal antibodies to mammalian and shark MBP recognize antigens expressed by lamprey spinal cord glial cells. We conclude that the understanding of human myelination will be enriched through comparative studies of vertebrates that lack (lamprey and hagfish) and display (non-mammalian vertebrates) myelination.

Key words. Chondrichthyes, clearnose skate, electron microscopy, evolution, immunocytochemistry, lamprey, myelination, myelin basic protein, spinal cord, spiny dogfish

Introduction. There is a dichotomy in the evolution of the vertebrate nervous system. Jawless fishes have no myelin whereas, all gnathostomes have full myelination. The dominance of myelin in the CNS and PNS of chondrichthyes, teleosts, amphibians, reptiles, birds and mammals suggests that myelination originated in ancestors that

Molecular Signaling and Regulation in Glial Cells
A Key to Remyelination and Functional Repair

lived over 450 million years ago, before vertebrate classes evolved as distinct lineages. To test the hypothesis that all vertebrate myelination is homologous, we compared neural development and myelination in chondrichthyes (spiny dogfish, *Squalus acanthias*, and clearnose skate, *Raja eglanteria*) with mammals (rat) using morphology, immunocytochemistry and molecular biology. The high degree of similarity between these animals, coupled with published data (following sections) provide solid support for this hypothesis.

1.1 Morphology of vertebrate CNS myelination. Comparative anatomists have accumulated large amounts of data consistent with the view that all vertebrate nervous systems are constructed from a common plan (1,13). This view has been strengthened by recent pathway tracing, neurotransmitter localization and ultrastructural studies. Examples that illustrate the high degree of similarity of CNS myelin among cartilaginous (10,21) and teleost fishes (8,21), amphibians (19,21), reptiles (28), birds (15) with mammals (18) are referenced.

1.2 X-ray diffraction studies of CNS and PNS myelinfrom members of each vertebrate class ((11) and references therein), complement light and electron microscopy (see 1.1). Together they show that myelin sheaths in all vertebrate are spirally-wrapped membranes with dimensions correspondent with axon size. They display characteristic major dense and intraperiod (IPL) lines, where cytoplasmic and extracellular leaflets of the myelinating cells plasma membrane compact. Differences in IPL periodicity among vertebrates are consistent with protein differences, namely fish CNS contains P_0 and has a wider IPL than PLP-containing tetrapod CNS myelin.

1.3 Biochemical studies, originally based on Western blots (reviewed in (278)) demonstrated that all vertebrate myelin sheaths contain the same proteins, namely two-four alternatively-spliced MBP isoforms and either P_0 or PLP (tetrapod CNS), but not both. Molecular cloning studies verify these findings. MBP isoforms from mammals, chick (30) and chondrichthyes (16,20,22), PLP/DM-20 (DM-20 is the only alternatively-spliced variant) from mammals, teleost fishes (following chapter) and chondrichthyes (12), and P_0 from mammals, teleost fishes (23,24) and shark (20) have highly similar amino acid sequences, especially at the purported functional domains. Likewise, myelin lipid profiles from non-mammalian vertebrates share features, including high amounts of galactocerebroside and sulfatide (both with normal and hydroxy-fatty acids), high amounts of cholesterol, and low amounts of phospholipid, though high portions of ethanolamine plasmalogen, (summarized in (11))characteristic of mammalian myelin.

2 Materials and Methods. Spiny dogfish, *Squalus acanthias*, and little skates, *Raja erinacea*, were obtained at the Marine Biological Laboratory (MBL), Woods Hole, MA. Clearnose skates, *Raja eglanteria*, were obtained from Dr. Carl Luer, Mote

Marine Laboratory, Sarasota, FL. Pregnant Wistar rats were obtained from Charles River Laboratory, Wilmington, DE. All studies received prior approval from the Institutions' (NYS Inst. for Basic Research and MBL) Animal Welfare Committees.

2.1 Preparation of Tissues and Microscopic Examination
Small dogfish and skate embryos were chilled in ice cold water . After embryos became anesthetized, the dorsal surfaces of brains and spinal cords were exposed and the embryos immersed in fixative, 4% glutaraldehyde (light and electron microscopy) or 4% paraformaldehyde (immunocytochemistry and *in situ* hybridization) in 0.1 M cacodylate buffer, pH 7.3, 12% sucrose (1000 mOsm). After fixation (16-24 h at 4°C), the tissues were dissected and transferred to 0.1 M cacodylate buffer, pH 7.3, containing 18% sucrose (1000 mOsm). Thin segments of adult dogfish spinal cord were fixed by immersion. Large (9, 13 and 20-25 cm long) dogfish embryos were perfusion fixed. Rat embryos were anesthetized with halothane vapor, dorsal surfaces of brains and rostral spinal cords exposed, and the embryos immersed in fixative, 4% glutaraldehyde or 4% paraformaldehyde in 0.1 M cacodylate buffer, pH7.3, for 16-24 h at 4°C. Postnatal rats were anesthetized and perfusion fixed. Afterwards, brains and spinal cords were transferred to 0.1 M cacodylate buffer, pH 7.3. Glutaraldehyde fixed material was embedded in Spurr or Epon resin. Sections were cut at 0.5 microns for light microscopy and 80 to 90 nm for electron microscopy. Light microscope sections were stained with toluidine blue and ultrathin sections were stained with uranyl acetate and lead citrate. For immunocytochemistry, fixed, buffer-washed tissues were either embedded in paraplast and sectioned at 6 microns or cryoprotected in 30% buffered sucrose and embedded in OCT compound (Miles Laboratory, Naperville, IL) and sectioned at 10-20 microns.

2.2 Immunocytochemistry Paraffin sections were dewaxed, equilibrated with PBS, blocked with 1% BSA in PBS for 1 h, and incubated overnight at 4°C with primary antibody (see Results) in block solution, They were washed with PBS, incubated for 1 h with secondary antibody (fluorescent or biotinylated, Jackson ImmunoResearch Laboratories or for mouse IgG3, Southern Biotechnology) with 0.1 ug/ml DAPI (4,6-diamidino-2-phenylindole, Sigma) . After biotinylated secondary antibody incubation and wash, sections were incubated with streptavidin-FITC or TR conjugates for 1 h (Jackson ImmunoResearch Labs). After they were washed, sections were mounted with DABCO (diazabicyclo[2.2.2]octane) antifade (9). For frozen sections, the procedure of Hardy and Reynolds (6) was followed. Briefly, cryosections were incubated with sodium borohydride (0.5 mg/ml in PBS, 2 x 5 min) and washed thoroughly. With glycolipid antibodies sections were incubated overnight at 4°C without permeabilization. For antibodies to myelin proteins, neurofilament proteins or glial fibrillary acidic protein (GFAP), sections were permeabilized with methanol or acid alcohol (GFAP) for 10 min at -20°C, washed with PBS and incubated with primary antibody. Subsequent visualization steps were carried out exactly as described for paraffin sections.

2.3 *In situ* Hybridization The procedure of Breitschopf et al. (3) was followed. Briefly, paraffin sections, from adult spiny dogfish and rat spinal cords, were dewaxed, treated with 1 N HCl, acetic anhydride, and proteinase K. They were incubated overnight at 55°C with digoxygenin-labeled cRNA probes. Sense and antisense cRNAs were prepared by reverse transcription of linearized plasmids containing sequences (200-500 bp long) of rat and dogfish MBP, mouse PLP and dogfish P_0. After posthybridization washing, bound cRNAs were visualized with the anti-digoxygenin-alkaline phosphatase antibody-NBT/BCIP detection method.

2.4 Reverse Transcriptase-Polymerase Chain Reaction (RT-PCR).
Tissues, frozen directly after dissection and stored at -70°C, were extracted for total RNA and PCR products generated, subcloned and sequenced as described previously (22).

3 Results.

3.1 Adult CNS myelin in spiny dogfish and rat are highly similar in organization and structural detail. Fortuitously, adult dogfish and rat spinal cords are nearly identical in cross-section (Figs. 1a and b). They are similarly organized, with white matter tracts surrounding the gray matter. Large motoneurons are located in distinctive ventral horns (vh). The largest myelinated fibers are in the vf. Dogfish fibers are larger than rat fibers (also Figs. 1c and d); presumably the larger diameter is for increased conduction in the much longer and colder dogfish spinal cord. Oligdoendrocytes (*) are seen scattered amid the myelinated fibers, which segregate into clusters of similar sized fibers. With the electron microscope (not shown), comparisons between dogfish and rat oligodendrocytes in terms of sizes and shapes of soma and nuclei, disposition of heterochromatin, cytoplasmic density and organelle distribution show numerous similarity. Comparisons of myelin sheath structure; distinct inner and outer tongue assemblies, abutment of neighboring sheaths and paranodal specializations confirm similarity of design. So far, only a higher content of Schmidt-Lanterman incisures and a larger intraperiod line (related to presence of P_0 instead of PLP) distinguish adult dogfish and rat CNS myelin sheaths.

3.2 Early myelination in dogfish and rat spinal cords are highly similar
Myelination begins in both rat and dogfish spinal cords with the appearance of oligodendrocytes in the vf (not shown). The medial longitudinal fascicles of both species begins to become myelinated at the same time as the vf. At early development, more oligodendrocytes are in rostral versus caudal vf sections. Many early oligodendrocytes in the vf line the pial surface. Soon after the cells appear, thin

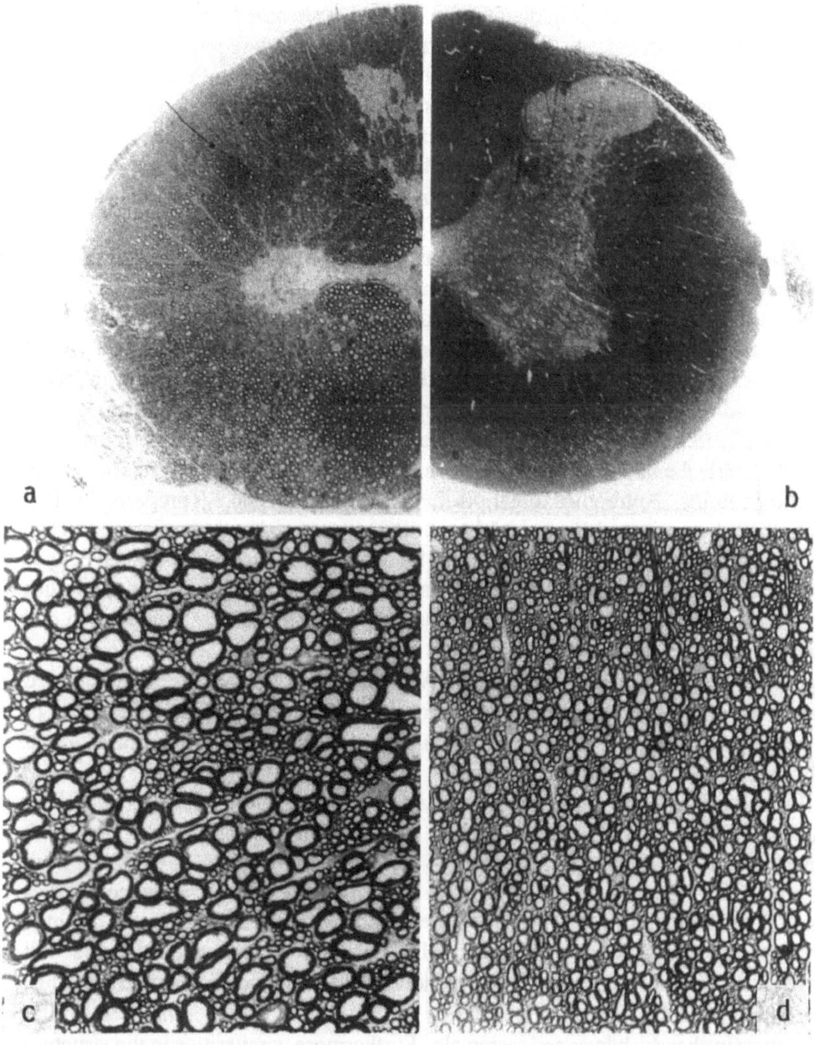

Fig 1: Light micrographs of adult spiny dogfish (a,c) and rat (b,d) spinal cords. Figures a and b are printed at 15X, whereas c and d are printed at 140X. Symbols are defined in the text

myelin sheaths can be seen. With the electron microscope (not shown), the largest (approximately one micron in diameter) axons are separated from smaller axons by oligodendrocyte processes. First loose and then compact wrappings appear. With time most axons in the spinal cord become myelinated. In both dogfish and rat, myelination occurs on a tract by tract basis and within each tract, larger fibers are myelinated well before small fibers. Differences between dogfish and rat are the size distribution of myelinated fibers and the time over which myelination occurs. Rat spinal cords are myelinated postnatally; dogfish spinal cord myelination is prenatal.

3.3 The same neural cell-specific markers are developmentally expressed in a like manner in dogfish, skate and rat. Antibodies to oligodendrocytes (O1, O4, R-mAb, MBP, PLP and P_0), neurons (phosphorylated and non-phosphorylated neurofilament proteins) and radial glia/astrocytes (GFAP) were used to monitor differentiation of neural cells. In 9 cm long dogfish embryos, 5-6 week skate embryos and P1 and P3 rat pups, O1, O4 and R-mAb each exclusively labeled the actively myelinating vf. When two ages were used, a larger area was labeled in the older animal. Nearly the same regions were labeled by polyclonal antibodies raised against myelin proteins. Antibodies to gel purified dogfish MBP and P_0 (prepared by Drs. Richard Rubenstein and Pankaj Mehta at IBR) were used with dogfish and skate. Mammalian MBP antibodies (provided by David Colman, Mount Sinai, Joyce Benjamins, Wayne State and Art McMorris, Wistar Foundation) and PLP antibodies (provided by Pam Knapp, Univ. of Kentucky) were used with rat. In dogfish and skate spinal cord and nearby roots, there was no indication that MBP or P_0 were expressed prior to the onset of myelination. Whether or not the mRNAs are expressed prior to myelination, as has been shown for rat DM-20 (7,25,26), rat and chicken P_0 in peripheral tissues (2,29), and for MBP-like molecules (14,17), require *in situ* hybridization and RT-PCR analysis.

Like oligodendrocyte antigens, neuron and astrocyte antigens appeared on similar time schedule in chondrichthyes and rat. GFAP-labeled radial glial cell processes were the earliest differentiation marker to appear. P^- neurofilament (Rm044, from Virginia Lee, University of Pennsylvania) appeared next in motoneuron cell bodies after GFAP but before phosphorylated fneurofilament forms(NF61, from Dr. Inge Grundke-Iqbal, IBR and LCM 3.7, from Dr. Lee) appeared in axons in the vf. All these antigens appeared before oligodendrocyte/myelin-specific markers. Thus, development of the spinal cord in terms of cell differentiation follows a similar timecourse in chondrichthyes and mammals. Furthermore, similarities in the structure of glycolipids expressed by oligodendrocytes, neurofilaments expressed in neurons and and axons and GFAP expressed in radial glial cells and astrocytes demonstrate coordinate molecular conservation.

3.4 Myelin basic proteins from all vertebrates are highly homologous. High conservation in the amino acid sequences of exon I and III (pre-GOLLI designation)

Table I

Partial Amino Acid Sequences of Myelin Basic Proteins

EXON I	
Sea Lamprey	MAS-STLDHARH--GS-PRHRD---SGLLDSLGRFFGGDR-HVPR-RGFGK
Spiny Dogfish	MASATTSDHAKQAGGAHSRQRD---SGLLDQLGQLFGQE--GSRKVPEKGK
Goldfish	MATASTTDHARH--GS-PRHRD---SGLLDSLGRFFGGDRSRSP--AGFGK
Marine Toad	MASASTTDHARH--GS-PRHRD---SGLLDSLGRFFGGDRHRSP--AGFGK
Garter Snake	MATASTSDHARH--GS-PR-RYREPS-LLDSLGRFFGGDK-HYPR-R--GK
Chicken	MASASTTDHARH--GS-PRHRD---SGLLDSLGRFFGGDR-HVPR-RGFGK
Rat	LATASTMDHARH--GFLPRHRD---TGILDSIGRFFSGDR-GAPK-RGSGK
EXON III	
Sea Lamprey	DIH-AA-R------ASHVGSIPQ--RSQHGRPGDDNPVVHFFKNM
Spiny Dogfish	E---PATRSVLMAPTTHKAH--QGAR----RQTDDSPVVHFFKNM
Goldfish	DIH-AA-R------ASHVGTIPQ--RSQHGRPGDDNPVVHFFKNM
Marine Toad	DIH-AA-R------ASHVGSIPQ--RSQHGRPGDDNPVVHFFKNM
Garter Snake	DYHH-ASR-V-----SHVSSLPH-PRSYHGRYVDDNPVVHFFKNM
Chicken	DIH-AA-R------------IPQ--RSQHGRPGDDNPVVHFFKNI
Rat	DSH---TR------TTHYGSLPQ--KSQ--RTQDENPVVHFFKNI

Sequences (both directions) were obtained from PCR products (several clones each) subcloned into PCR-TMII vector (Invitrogen). The most highly conserved sequence, PVVHFFK (underlined in Table), here and in all MBPs sequenced, lies within the reverse primer. Although the high conservation may be a primer artifact and needs confirmation, the same sequence has been obtained in 14 chondrichthyes, cloned with a downstream primer. This sequence at the C terminus of exon III, corresponds to a recognized phospholipid binding sequence motif (G,A,P)$X_{4\text{-}5}$(K,R)

Table II

Percent Identity of MBP Protein Sequences

	Exon	Spiny Dogfish	Goldfish	Marine Toad	Garter Snake	Chick	Rat
S. Lamprey	I	46	81	86	72	72	54
	III	39	97	100	68	80	58
Dogfish	I		50	50	46	37	31
	III		39	39	38	44	38
Goldfish	I			95	70	65	51
	III			97	68	80	56
Toad	I				68	68	52
	III				68	80	58
Snake	I					57	44
	III					55	51
Chick	I						73
	III						52

allowed us to design primers that amplified cDNAs from MBP mRNAs in every vertebrate class (examples are given in Table I) in protein kinase C (5) and synaptotagmin (4). This possibility fits the function of MBP, which would be expected to bind phospholipids in the cytoplasmic leaflet during compaction. Other conserved features in exons I and III are the five aligned basic residues (H, K, R) and three aligned acidic residues (D,E) in each. Other notable features include the lack of cysteine and the 4-5 aligned potential phosphorylation sites (S,T). Further information can be obtained from amino acid identity comparisons (Table II). Surprisingly, the sea lamprey sequence is very similar to goldfish and toad. The spiny dogfish and rat sequences are somewhat more different from the other sequences than any of the other sequences are from each other. These comparisons do not take into account the numerous gaps required to align these sequences. Interestingly the dogfish sequence and other chondrichthye MBP sequences (not shown) contain several stretches of amino acids that are not present in species from other classes.

3.5 MBP-like molecules are present in glial cells in the sea lamprey spinal cord Because of the sequence similarities between lamprey MBP and other MBPs, polyclonal antibodies to mammalian and dogfish MBP were tested with frozen sections of lamprey spinal cord. Of eight polyclonal antibodies tested, six produced a rather specific pattern in which label surrounded each of the large caliber axons (not shown). This is precisely where glial cells that ensheath these giant axons are located and suggest that these glial cells, like gnathostome oligodendrocytes, express MBP-like molecules. Efforts to label paraffin sections with these antibodies have failed as well as Western blots to determine the size of the MBP-like molecules. Future immunocytochemical and Western blot studies will be with antibodies directed to specific lamprey MBP sequences and to obtain evidence as to whether lamprey glial cells also express P_0.

4 Discussion To the best of our knowledge, myelination in all adult gnathostomes examined is highly similar. Examination of micrographs in numerous manuscripts on non-mammalian vertebrates show that the structural and ultrastructural arrangements of myelin sheaths are remarkably conserved.This result suggests that all common features existed in a single "primitive" ancestor or that evolutionary pressures caused substantial parallel development. Clearly two cells myelinate all gnathostome nervous systems, oligodendrocytes myelinate the CNS and Schwann cells myelinate the PNS. Most recognized differences in mammalian CNS and PNS myelin are present in non-mammalian vertebrates. To validate the impact of comparative studies, we felt it was essential to make a careful comparative study of the structural and ultrastructural features of myelinated fibers in a non-mammalian specimen with a mammal. The properties of spiny dogfish spinal cord (and spinal root, not discussed) myelin and myelination based are for all intensive purposes identical to those of rat and other mammals (18). So too, are developmental parameters assessed both with morphological criteria and immunocytochemistry. These results pave the way for studies that assess the impact of known (see Abstract) and yet to be discovered differences between members of different vertebrate classes.

5 Acknowledgments We would like to thank New York State for supporting these studies. Ed Gilland and Bob Baker (NYU Medical Center) introduced us to the spiny dogfish and helped us with handling, dissection and anatomy. Kasia Hammar (Marine Biological Laboratory) helped with preparation of tissues and initial electron microscopy. Numerous colleagues, though especially Victor Friedrich, Jr. (Mount Sinai Medical Center), provided us with antibodies and protocols for immunoflouroescence. Richard Rubenstein and Pankaj Mehta (IBR) prepared polyclonal antibodies to MBP and P_0 and Dick Weed and Dan Klitnick (IBR) helped with light microscopy and photography needed for these studies. Dr. Eleazar Cohen, NIH, pointed out phospholipid binding motifs and analyzed myelin protein sequences for potential motifs.

6 References.

1. Ariëns Kappers, C. U., G. C. Huber, and E. C. Crosby. (1936) The Comparative Anatomy of the Nervous System of Vertebrates Including Man. MacMillan, New York.

2. Bhattacharyya, A., E. Frank, N. Ratner, and R. Brackenbury. (1991) P_0 is an early marker of the Schwann cell lineage in chickens. Neuron 7:831-844.

3. Breitschopf, H., G. Suchanek, R. M. Gould, D. R. Colman, and H. Lassmann. (1992) In situ hybridization with digoxigenin-labeled probes: sensitive and reliable detection method applied to myelinating rat brain. Acta Neuropathol. (Berl.) 84:581-587.

4. Chapman, E. R. and R. Jahn. (1994) Calcium-dependent interaction of the cytoplasmic region of synaptotagmin with membranes. J. Biol. Chem. 269:5735-5741.

5. Clark, J. D., L.-L. Lin, R. W. Kriz, C. S. Ramesha, L. A. Sultzman, A. Lin,Y., N. Milona, and J. L. Knopf. (1991) A novel arachidonic acid-selective cytosolic PLA_2 contains a Ca^{2+}-dependent translocation domain with homology to PKC and GAP. Cell 65:1043-1051.

6. Hardy, R. and R. Reynolds. (1991) Proliferation and differentiation potential of rat forebrain oligodendroglial progenitors both *in vitro* and *in vivo*. Development 111:1061-1080.

7. Ikenaka, K., T. Kagawa, and K. Mikoshiba. (1992) Selective expression of DM-20, an alternatively spliced myelin proteolipid protein gene product, in developing nervous system and in nonglial cells. J. Neurochem.. 58:2248-2253.

8. Jeserich, G. and T. V. Waehneldt. (1986) Bony fish myelin: evidence for common major structural glycoproteins in central and peripheral myelin of trout. J. Neurochem. 46:525-533.

9. Johnson, G. D., R. S. Davidson, K. C. McNamee, G. Russell, D. Goodwin, and E. J. Holborow. (1982) Fading of immunofluorescence during microscopy: a study of the phenomenon and its remedy. J. Immunol. Methods 55:231-242.

10. Kemali, M., E. Sada, A. Miralto, and G. Zummo. (1983) Central myelin in the shark *Scyllium stellare* (Elasmobranchii, Selachii). Zeit. Mikrosk. Anat. Forsch. ,Leipzig 97:3-14.

11. Kirschner, D.A. and A.E. Blaurock. (1992) Organization, phylogenetic variation, and dynamic transitions of myelin. In R.E. Martenson, (ed) Myelin: Biology and Chemistry. CRC Press, Boca Raton, FL.. 3-78.

12. Kitagawa, K., M. P. Sinoway, C. Yang, R. M. Gould, and D. R. Colman. (1993) A proteolipid protein gene family: Expression in sharks and rays and possible evolution from an ancestral gene encoding a pore-forming polypeptide. Neuron 11:433-448.

13. Kuhlenberg, H. (1975) The Central Nervous System of Vertebrates Vol. 4: Spinal Cord and Deutoerencephalon. S. Karger, Basel..

14. Mathisen, P. M., S. Pease, J. Garvey, L. Hood, and C. Readhead. (1993) Identification of an embryonic isoform of myelin basic protein that is expressed widely in the mouse embryo. Proc. Natl. Acad. Sci. USA 90:10125-10129.

15. Matthieu, J.-M., F. X. Omlin, H. Ginalski-Winkelmann, and B. J. Cooper. (1984) Myelination in the CNS of mld mutant mice: comparison between composition and structure. Dev. Brain Res. 13:149-158.

16. Milne, T. J., A. R. Atkins, J. A. Warren, W. P. Auton, and R. Smith. (1990) Shark myelin basic protein: amino acid sequence, secondary structure, and self-association. J. Neurochem. 55:950-955.

17. Nakajima, K., K. Ikenaka, T. Kagawa, J. Aruga, J. Nakao, K. Nakahira, C. Shiota, S. U. Kim, and K. Mikoshiba. (1993) Novel isoforms of mouse myelin basic protein predominantly expressed in embryonic stage. J. Neurochem. 60:1554-1563.

18. Peters, A., S. L. Palay, and H. deF. Webster. (1991) The fine structure of the nervous system: Neurons and their supporting cells. Oxford University Press, New York.

19. Playford, D. E. and S. A. Dunlop. (1993) A biphasic sequence of myelination in the developing optic nerve of the frog. J. Comp. Neurol. 333:83-93.

20. Saavedra, R. A., L. Fors, R. H. Aebersold, B. Arden, S. Horvath, J. Sanders, and L. Hood. (1989) The myelin proteins of the shark brain are similar to the myelin proteins of the mammalian peripheral nervous system. J. Mol. Evol. 29:149-156.

21. Shinowara, N. L., W. B. Beutel, and J. P. Revel. (1980). Comparative analysis of junctions in the myelin sheath of central and peripheral axons of fish, amphibians and mammals: a freeze-fracture study using complementary replicas. J. Neurocytol. 9:15-38.

22. Spivack, W. D., N. Zhong, S. Salerno, R. A. Saavedra, and R. M. Gould. (1993) Molecular cloning of the myelin basic proteins in the shark, *Squalus acanthias*, and the ray, *Raja erinacia*. J. Neurosci. Res. 35:577-584.

23. Stratmann, A. and G. Jeserich. (1995) Molecular cloning and tissue expression of a cDNA encoding IP1 - a P0-like glycoprotein of trout CNS myelin. J. Neurochem. 64:2427-2436.

24. Takei, K., K. Kitamura, K. Banno, and K. Uyemura. (1993) Major glycoproteins in carp CNS myelin: Homology to P0 protein with HNK-1/L2 carbohydrate epitope. Neurochem.Int. 23:239-248.

25. Timsit, S., S. Martinez, B. Allinquant, F. Peyron, L. Puelles, and B. Zalc. (1995) Oligodendrocytes originate in a restricted zone of the embryonic ventral neural tube defined by DM-20 mRNA expression. J. Neurosci. 15:1012-1024.

26. Timsit, S. G., L. Bally-Cuif, D. R. Colman, and B. Zalc. (1992) DM-20 mRNA is expressed during the embryonic development of the nervous system of the mouse. J. Neurochem. 58:1172-1175.

27. Waehneldt, T. V. (1990) Phylogeny of myelin proteins. Ann. NY Acad. Sci. 605:15-28.

28. Yanes,C., Monzon-Major,M., De Barry,J., Gombos,G. (1992) Myelin and myelinization in the telencephalon and mesencephalon of the lizard, *Gallotia galloti* as revealed by immunohistochemical localization of myelin basic protein. Anat. Embryol. (Berl.) 185: 475-87.

29. Zhang, S.-M., R. Marsh, N. Ratner, and R. Brackenbury. (1995) Myelin glycoprotein P_0 is expressed at early stages of chicken and rat embryogenesis. J. Neurosci. Res. 40:241-250.

30. Zopf, D., V. Sonntag, H. Betz, and E. D. Gundelfinger. (1989). Developmental accumulation and heterogeneity of myelin basic protein transcripts in the chick visual system. Glia 2:241-249.

The Proteolipid Protein Gene Family in the Central Nervous Systems of Fish

Yoshida, M.[1], Kitagawa, K.[2], and Colman, D.R.[1]

[1]The Brookdale Center for Molecular Biology, The Mount Sinai School of Medicine, One Gustave Levy Place, New York, New York, 10029

[2]First Department of Internal Medicine,,Osaka University School of Medicine, Fukushima 1 chome, Fukushimak, Osaka 533, Japan

1 Introduction

The proteolipid proteins (PLP and DM-20) of the mammalian central nervous system are the major integral membrane proteins of the myelin sheath. These proteins have four hydrophobic domains that are embedded in the bilayer (Popot et al., 1991; Weimbs and Stoffel 1992). They are not glycosylated, and they function in adhesive interactions that generate the intraperiod line. The finding that only a very few variations in PLP/DM-20 sequence are normally tolerated, taken together with the knowledge that many single amino acid substitutions yield dysmyelinating phenotypes, has led to the notion that in order to function correctly in myelination, the mammalian PLP/DM-20 have remained essentially invariant throughout evolution. This notion was challenged by immunocytochemical (see Waehneldt, 1990 for review) and new molecular biological findings (Kitagawa et al., 1993; Yan et al., 1993) which showed that in cartilaginous fish (Kitagawa et al., 1993) and mouse (Yan et al., 1993), proteolipid-like proteins whose sequences were quite divergent from the mammalian PLP/DM-20 could be detected. These findings led to the inescapable conclusion that the proteolipid family originated quite early in vertbrate evolution - at least 400 million year before present (400 MYR BP)

Specifically, there seem to be at least six discrete proteolipid/DM-20 related polypeptides, derived from at least four genes, expressed in certain vertebrate (mouse, human, shark, and lungfish). Not all PLP/DMs are expressed in each species, but certain evolutionary patterns already seem clear. We've termed this family the **PLP/DM family** because of the close resemblance of each of these polypeptides to the DM-20 isoform that is a splice product of the proteolipid gene in tetrapods. In terms of sequence relatedness, each of the polypeptides resembles the one we term DMα, and this one is the closest in terms of primary sequence to that of the DM-20 polypeptide itself. All of the DM polypeptides discovered to date with the exception of PLP itself, lack the specific peptide segment that is uniquely found in the tetrapodal proteolipid protein. This, taken together with the

Molecular Signaling and Regulation in Glial Cells
A Key to Remyelination and Functional Repair

fact that DMα, DMβ, and DMγ are all products of different genes suggest that the primodial DM polypeptide may have originated in organisms even prior to myelination. Gene duplication events then gave rise to the variety of DM genes that we see now in contemporary vertebrates. It may even be envisioned that the primordial DM might have been able to form homo-oligomers with itself similar to the way the α subunit of the nicotinic acetylcholine receptor is capable of forming functional channels in this way. In support of the idea that these newly described proteolipids may have ionophoretic channel or pore forming properties, the localization of one of them, DMβ, also termed M6a (Lagenaur et al. 1992) is found on the apical surfaces of proximal convoluted tubules in kidney and in the choroid plexus, two epithelial derivatives that have extensive ion transporting capabilities. In addition M6a is found in growth cones where ion transport may be extremely important in determining growth rates and direction of the developing axon.

These findings called into question two major concepts about proteolipid function. The first, discussed above, was that the proteolipids must remain invariant in order to fulfill their function. The second, less obvious idea, was the generally accepted idea that proteolipid proteins and protein zero (Po) the major glycoprotein in the peripheral nervous system, did not naturally co-exist in the same membrane.

Since we initially examined (Kitagawa et al., 1993) only two species of relatively ancient cartilaginous fish (*Squalus acanthias* and *Torpedo californica*) for the presence of the new proteolipids, we thought it worthwhile to expand our studies to include representatives of the bony fish (teleosts), and so perhaps identify new members of the proteolipid family, and explore their patterns and extent of expression. Teleosts are in fact the largest and most successful vertebrates, with 23,500 living species, as compared with 4000 mammalian species. We found a surprisingly wide diversity of proteolipid protein forms in teleosts, indicating a much higher degree of sequence tolerance in these proteins than was previously suspected.

2 Method

2.1 Polymerase chain reactions:

Total brain RNAs were isolated by RNA zol B (BIOTECX LAB) and reverse-transcribed for PCR. For reverse transcription, total RNA was primed by oligo (dT) (Pharmacia), and incubated at 40°C for 60 min with MuMLV reverse transcriptase (Boerhinger-Mannheim) and then incubated at 99°C for 5 min for inactivation of reverse transcriptase. Resulting first strand cDNA was used for PCR amplification. PCR amplification was performed for 40 cycles according to the following schedule: 94°C for 1min, 42°C for 2min, and 72°C for 3min. An additional aliquot of Taq polymerase (Promega) was added after 20 cycles. As the positive control experiment, the RT-PCR reactions using total RNA from mouse brain were performed by same schedule. The primers used for PCR were three degenerate oligonucleotides. A 5' primer was made according to the amino acid sequence found in the 1st hydrophobic domain of rat PLP/DM-20, and two 3' primers were designed to match the sequence of the 3rd and 4th hydrophobic domains (see Fig. 1).

The 5' primer had the following sequence:

plp 1: TT(T/C)TT(T/C)GGIGTIGCI(T/C)TITT(T/C)TG(T/C)GGITG(T/C)GG

The 3' primers were:

plp 3: TG(A/G)CAIGTIGTCCAIGT(A/G)TT(A/G)AA(A/G)TAIAT(A/G)TA

plp 4: GC(A/G)AA(A/G)TT(A/G)TAIGTIGCIGCIATCAT(A/G)AA

An aliquot of each PCR reaction was analyzed by 1.2 % agarose gel electrophoresis and stained with ethidium bromide. Resulting PCR products were gel-purified and ligated in the cloning plasmid, PGEM 7Z (Promega). We obtained several positive clones by means of blue-white colony selections. DNA sequence analysis was performed on plasmid DNAs using the ABI computerized automated DNA sequencing.

2.2 Computer analysis:

All analyses for hydropathy plots, database searching and multiple sequence alignments were performed with the Genetic Computer Group (GCG) suite of programs (Devereaux et al. 1984). Hydropathy plots were drawn using the Pepplot program, based on the average hydrophobicity of a residue-specific Kyte and Doolittle hydropathy index over a window of nine residues. Two amino acid sequences were compared using the BestFit program. The "sequence-relatedness" diagram was derived with the PileUp program.

3 Results and Discussion

We carried out RT-PCR using three degenerate primers, plp1, plp3 and plp4, which correspond to part of the hydrophobic domains I, III and IV of the rat PLP sequence (Fig. 1). We obtained amplified DNA fragments from cDNA of mouse, African lungfish (*Protopterus dolloi)*, African bichir (*Polypterus palmas)*, and zebrafish (*Brachiodanio rerio*) brains. Fig. 2 shows the agarose gel of both mouse and *Protopterus* PCR products. Two bands of 370 bp and 480 bp were obtained from the mouse brain template cDNA by amplification with the plpl/plp3 pair (Fig. 2b) and two bands of 610 bp and 710 bp with the plpl/plp4 pair (Fig. 2c). These molecular sizes corresponded to the expected sizes of mouse DM-20 and PLP primed by each primer set, respectively. On the other hand, only one band of 370 bp was detected from *Protopterus* brain cDNA template by amplification with the plpl/plp3 pair (Fig. 2d) and also one band of 610 bp with the plpl/plp4 pair (Fig. 2e). This molecular size corresponded to mouse DM-20 amplified by both primer pairs. The band corresponding to mouse PLP was not obtained by the amplification of either the plpl/plp3 or the plpl/plp4 pair. We obtained similar results with *Polypterus* and *Brachiodanio* PCR (data not shown).

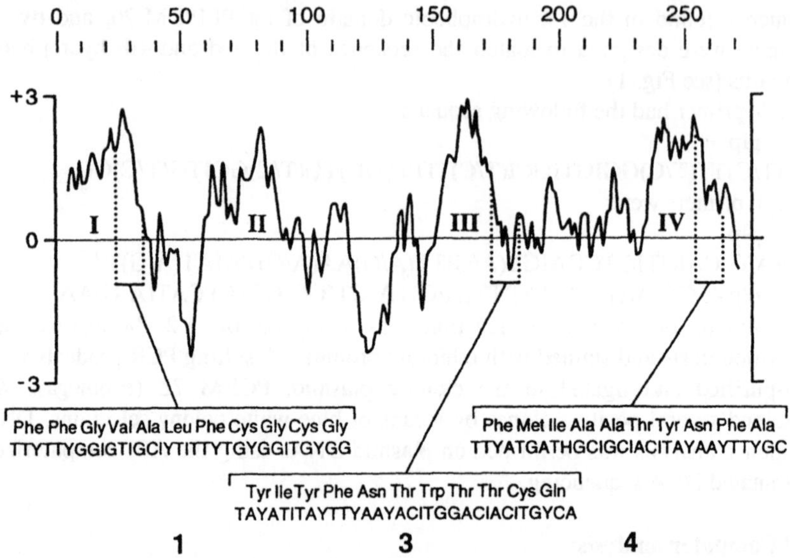

Fig. 1 Hydrophobicity plot of rat PLP and primer construction. The thick lines in the plot indicate those peptide regions within rat PLP from which degenerate oligonucleotide primers were designed.

Using these PCR products, we sequenced several clones from *Protopterus*, *Polypterus* and *Brachiodanio*. The PCR cloning revealed partial sequences of several DM-20 like proteins that exist in fish CNS. To date, it has been observed that one DM-20 related molecule exists in *Protopterus* CNS, two in *Polypterus*, and two in *Brachiodanio* CNS. The amino acid sequences of these newly cloned DM-20 like proteins were compared with *Squalus* DMα, DMβ, and DMγ (Kitagawa et al. 1993) using the BestFit program in GCG and designated "DMα", "DMβ", or "DMγ", according to the similarity with those of *Squalus*. *Protopterus* and *Polypterus* "DMα" showed 60.0% and 61.7% identity to *Squalus* DMα, respectively. *Brachiodanio* "DMβ" showed 89.5% identity to *Squalus* DMβ. *Polypterus* and *Brachiodanio* "DMγ" showed 64.9% and 76.3% identity to *Squalus* DMγ, respectively. Although these molecules have some homology each other, there are still significant differences among these amino acid sequences.

Furthermore, we compared these DM sequences with mouse DM20, M6a and M6b and then, produced a sequence-relatedness diagram to illustrate the relationships between these proteins (Fig. 3). Clearly the DM-20 like protein fall into three groups, a "neuronal" form (mouse M6a, *Braciodanio* DMβ, and

Squalus DM*β*) and a "white matter" form (mouse M6b, *Polypterus* DM*γ*, *Brachiodanio* DM*γ*, and *Squalus* DM*γ*) and the "DM20/PLP" closer relatives (mouse DM20, *Protopterus* DM*α*, *Polypterus* DM*α*, and *Squalus* DM*α*).

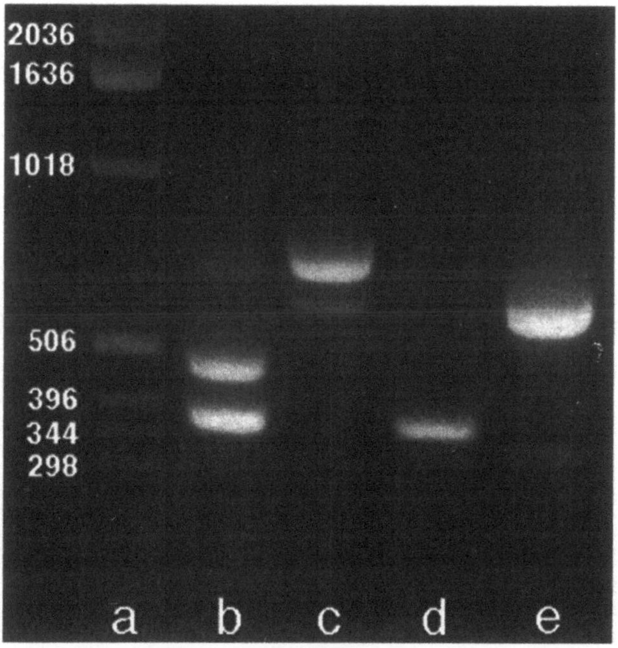

Fig. 2 RT-PCR of mouse and *Protopterus* brains.
We carried out RT-PCR reactions using three degenerate primers, plp1, plp3 and plp4, which correspond to the hydrophobic domains I, III and IV of the rat PLP sequences. PCR products from the mouse brain template cDNA, amplified with the plp1/plp3 pair showed two bands of 370 and 480 bp (lane **b**) and products with plp1/plp4 pair showed two bands of 610 and 710 bp (lane **c**). These molecular sizes corresponded to the expected sizes of mouse DM-20 and PLP primed by each primers, respectively. PCR product from *Protopterus* brain cDNA template, primed by plp1/plp3 pair showed only a band of 370 bp (lane **d**) and product by plp1/plp4 pair showed a band of 610 bp (lane **e**). lane **a**: molecular standard.

There are several studies which identified and characterized cDNA clones encoding PLP and DM-20 from several species (Milner et al. 1985, Dautigny et al. 1985, Gardinier et al. 1986, Schaich et al. 1986, Kronquist et al. 1987). These studies indicated that the evolutionary conservation of PLP/DM-20 sequences is remarkable. By contrast, our present study revealed that the divergence of the proteolipid protein gene family members in lower vertebrates must be quite large, and that there is a wide variety of DM-20 like protein in fish.

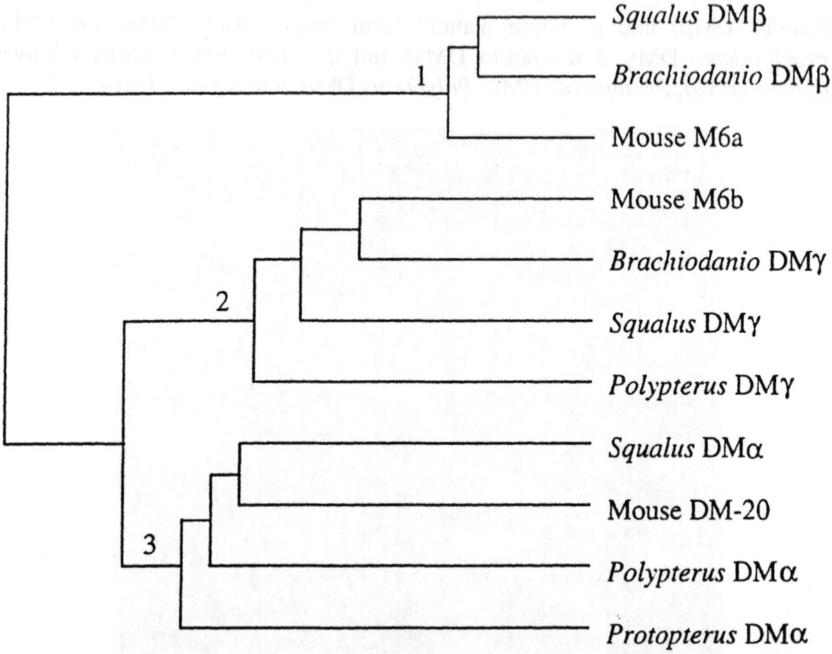

Fig. 3 Dendrogram showing the sequence relatedness amongst proteolipid protein gene family members.
1. neuronal form
2. white matter form
3. white matter "DM-20/PLP" closer relatives

4 Acknowledgements

This work was supported by grants to DRC from the National Institutes of Health (NINDS NS 20147) and the National Multiple Sclerosis Society (NMSS RG2636-A-3). This is manuscript number 208 from The Brookdale Center for Molecular Biology.

5 References

Dautigny A., Mattei M.-G., Morello D., Alliel P.M., Pham-Dinh D., Amar L., Arnaud D., Simon D., Mattei J.-F., Guenet J.-L., Jolles P., and Avner P. (1986) The structural gene coding for myelin-associated proteolipid protein is mutated in jimpy mice. Nature 321, 867-869

Devereaux J., Haeberli P., and Smithies O. (1984) A comprehensive set of sequence analysis programs for the VAX. Nucl. Acids Res. 12, 387-395.

Gardinier M.V., Macklin W.B., Diniak A.J., and Deininger P.L. (1986) Characterization of myelin proteolipid mRNAs in normal and jimpy mice. Mol. Cell. Biol. 6, 3755-3762

Kitagawa K., Sinoway M.P., Yang C., Gould R.M., and Colman D.R. (1993) A proteolipid protein gene family: expression in sharks and rays and possible evolution from an ancestral gene encoding a pore-forming polypeptide. Neuron 11, 433-448

Kronquist K.E., Crandall B.F., Macklin W.B., and Campagnoni A.T. (1987) Expression of myelin proteins in the developing human spinal cord: cloning and sequencing of human proteolipid protein cDNA. J. Neurosci. Res. 18, 395-401

Lagenaur C., Kunemund V., Fischer G., Fushiki S., and Schachner M. (1992) Monoclonal M6 antibody interferes with neurite extension of cultured neurons. J Neurobiol. 23, 71-88

Milner R.J., Lai C., Nave K.-A., Lenoir D., Ogata J., and Sutcliffe J.G.(1985) Nucleotide sequence of two mRNAs for rat brain myelin proteolipid protein. Cell 42, 931-939

Popot J.L., Dinh D.P., and Dautigny A. (1991) Major myelin proteolipid: the 4-a-helix topology. J. Membr. Biol. 120, 233-246

Schaich, M. , Rudzinsky, R.M., and Stoffel, W. (1986) Cloned proteolipid protein and myelin basic protein cDNA. Transcription of the two genes during myelination. Biol. Chem. Hoppe-Seyler 367, 825-834.

Waehneldt T.V. (1990) Phylogeny of myelin proteins. Ann. NY Acad. Sci. 605, 15-28

Weimbs T., and Stoffel W. (1992) Proteolipid protein (PLP) of CNS myelin: positions of free, disulfide-bonded and fatty acid thioester-linked cysteine residues and implications for the membrane topology of PLP. Biochemistry 31, 12289-12296

Yan Y., Lagenaur C., and Narayanan V. (1993) Molecular cloning of M6: identification of a PLP/DM20 gene family. Neuron 11, 423-431

Molecular Structure of Myelin Genes in the CNS of Trout

G. Jeserich, J. Strelau, A. Stratmann

University of Osnabrück, Dept. Animal Physiology, Barbarastr. 11, D-49086 Osnabrück

1 Introduction

Myelin is an extension of the glial plasma membrane which is wrapped around axons in a typical multilamellar fashion to generate an insulating sheath against transmembrane ionic currents. Membrane adhesion of the adjacent lamellae is of unique strength and it is largely conveyed by the concerted interaction of a limited number of major protein constituents, which are specific to this membrane and not found elsewhere. In the CNS of all higher vertebrate classes the pattern of myelin proteins is remarkably constant, comprising a pair of hydrophobic four transmembrane domain proteins (PLP, DM-20) which are critically involved in the adhesion of the outer leaflets as well as a series of basic proteins (MBP) which accomplish the tight apposition of the cytoplasmic membrane faces (13). In the PNS myelin PLP is replaced by glycoprotein Po, which is a member of the IgG superfamiliy of adhesion molecules (25) and unlike PLP has only a single transmembrane segment (14). By contrast phylogenetically older vertebrate classes such as fishes show a quite divergent combination of major structural myelin proteins: as revealed by western blot analysis (23) as well as cDNA sequencing (17) in the CNS of cartilaginous fishes (sharks and rays) PLP is replaced by a Po-like glycoprotein, whilst the myelin basic protein appears closely related in primary structure to its mammalian counterpart (17). In the CNS myelin of trout, a teleost or bony fish species, four major protein constituents were separated by SDS-PAGE ranging from 13kD over 23kD, 26kD and 36kD apparent molecular weight, respectively (5). The fastet migrating component was identified by immunoblot analysis as a basic protein exhibiting an isoelectric point close to 10 (6). The two glycoproteins of intermediate size (termed IP1, IP2) both do not cross-react with anti-PLP antibodies but show a strong immunolabeling with antibodies against Po instead (8). The unglycosylated 36kD peptide (termed 36K) is not immunostained with antibodies against any of the mammalian myelin proteins and hence seems to be a novel myelin constituent for which a molecular equivalent has not yet been identified in other vertebrate classes. Immunostaining with an antiserum raised against trout 36K shows a strong and selective labeling of myelinated fiber tracts in all regions of trout brain as well as their supporting glia, emphasizing its myelin-specific localization (7). Furthermore the antiserum proved instrumental to characterize the myelin-forming glia of fish CNS morphologically: in primary cultures as well as in the intact tissue these cells exhibited a multipolar ramified structure closely matching

Molecular Signaling and Regulation in Glial Cells
A Key to Remyelination and Functional Repair

oligodendrocytes in the mammalian brain. Altogether this shows that in fish CNS Po-like molecules are indeed expressed by oligodendrocytes, though in mammalian species this glycoprotein is restricted to Schwann cells in the PNS. Another remarkable feature that fish oligodendrocytes share with mammalian Schwann cells is their ability to remyelinate axons after injury, contrasting with mammalian oligodendrocytes which have only a poor capacity for regeneration. Since it can be expected that the mechanisms controlling myelin gene expression are critically involved in this phenomenon a detailed structural analysis of the genes coding for myelin proteins in the CNS of fish was of priority interest.

2 Structure of myelin proteins in trout CNS

2.1 Structure of IP1/IP2-glycoproteins

A molecular characterization of the two intermediate glycoproteins (IP1, IP2) of trout CNS myelin which both have a high binding affinity for anti-Po antibodies was performed in a reverse transcription-PCR approach. For this degenerate primers based on the known aminoterminal sequence of trout IP1 (19) as well as a phylogentically conserved region around the glycosylation site of rat and shark Po were chosen (14, 17). The nucleotide sequence of the amplified 300 bp fragment exhibited a significant similarity with the corresponding region of glycoprotein Po. This fragment was digoxgenin-labeled and used for screening a trout brain cDNA library constructed in λ-ZAPII by plaque filter hybridization. A single positive clone having an insert size of 2.03 kb was isolated and sequenced. It exhibited a unique open reading frame of 606 bp predicting a protein of 202 amino acids as well as 1.311 bp of 3'- and 72 bp of 5'-untranslated sequence. The ATG-codon was followed by a cleavable signal sequence (22) of 22 amino acids to yield a mature peptide of 180 amino acids which started with an aminoterminal isoleucine residue exactly as expected from the biochemical sequencing data (19). The calculated molecular weight of the predicted peptide is 20, 300 dalton which agrees well with the value previously determined for the deglycosylated form of IP1 by SDS-PAGE, while it appears too small for IP2 (6). Hydropathy analysis (11) identified a single transmembrane segment separating the molecule into a large N-terminal extracellular domain of 124 amino acids and a short intracellular region of 30 residues. A computer assisted secondary structure prediction of the extracellular domain using the Chou-Fasman algorithm (2) suggested an immunoglobulin-like conformation consisting of seven antiparallel ß-sheets with alternating ß-turns (Fig. 1) similar as it occurs in rat and shark Po (17, 24). A detailed amino acid sequence comparison of trout IP1 with those of rat and shark Po revealed a significant structural similarity, which was strongest in the extracellular domain (60%) and much lower in the cytoplasmic portion (22%). The extracellular domain contains a single glycosylation acceptor site which in trout IP1 is located at an identical position (Asn^{93}) as in the Po isoforms of rat and shark. The same is true for the two cystein residues (Cys^{21} and Cys^{98}) which are required for stabilizing an immunoglobulin fold structure by disulfide bonding and hence are essential for the adhesive properties of the protein (27). Closer inspection of the extracellular portion additionally revealed three longer stretches of high phylogenetic conservation: two of them falling into the third and fifth β-strand, respectively, are quite remarkable, since they consist of a series of hydrophobic residues framed by positive and negative charges. Such an arrangement appears particularly suitable to provide homophilic intermolecular adhesiveness, since it allows hydrophobic interactions to act in concert with electrostatic attraction. Future site-directed mutagenesis experiments will have to verify the physiological relevance of this structural motif.

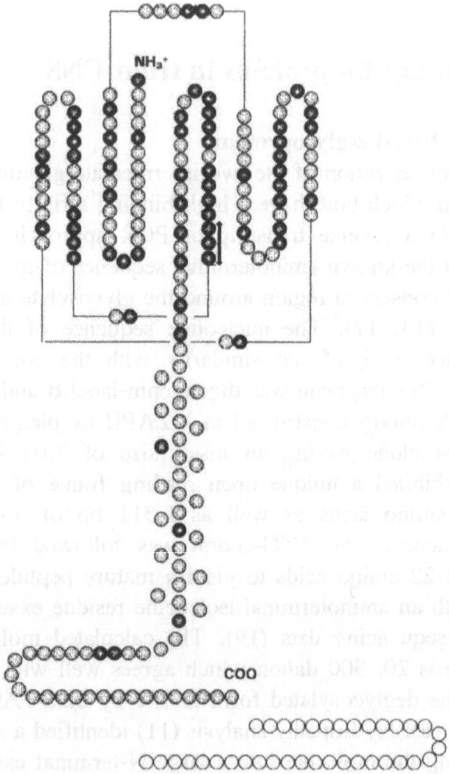

Figure 1: Hypothetical model illustrating the molecular structure of trout IP1. The secondary structure of the extracellular domain was adapted from the model of rat Po as suggested by Wells et al. (1993) taking into account the β-sheet pattern predicted for trout IP1 by Chou-Fasman analysis. The antiparallel β-sheets are oriented vertically. The chain of empty circles at the carboxyterminus indicates the deletion in the cytoplasmic domain of IP1 as compared with those of rat and shark Po. The arrow indicates the potential N-glycosylation site, darker symbols represent residues conserved among the trout, rat and shark sequence. Reproduced with permission from Stratmann and Jeserich (1995).

The intracellular region of IP1 is by 38 amino acid shorter than those of the Po isoforms of rat and shark (Fig. 1), and except for a high proportion of basic amino acid residues does not show any appreciable homology with the primary structure of Po (14). Aiming to identify also the cDNA encoding the larger isoform (IP2) the RACE-technique (4) was applied. Using IP1 specific oligonucleotides in combination with an oligo $(dT)_{18}$ primer a second cDNA sequence was obtained which up to the TGA-stop codon was identical with those of IP1. In this sequence, however, the termination codon was changed into GGC for glycine, which extended the reading frame by another 57 bp thus elongating the protein at its carboxyterminus by 19 amino acids. Unfortunately the sequence prematurely ended in an A-rich region within the coding region. As a whole it is concluded that the two IP isoforms of trout CNS myelin are generated by alternative splicing of a common pre mRNA. The observation that the two IP isoforms differ by length of their carboxyterminal cytoplasmic domain is well consistent with previous 2D electrophoresis investigations (6) demonstrating that IP2 is significantly more basic than IP1 due to a longer cationic cytoplasmic tail. For rat Po it was recently shown that the length of the cytoplasmic region critically influences the adhesive properties of the extracellular domain, probably by supporting clustering of the molecules in the plane of the membrane via molecular interactions with the cytoskeleton (26). It is interesting to note in this context that expression of the two IP-isoforms in trout brain is developmentally regulated, with the larger splice variant (IP2) occurring well before the shorter one (IP1) during oligodendrocyte differentiation (5, 9). A similar correlation is found in the mammalian CNS for the two isoforms of MAG, which differ by length of their intracellular domain, too (18). It is well conceivable that IP2 in fish CNS conveys the initial membrane contact during the early phase of myelination, while IP1 later on helps to further stabilize and compact the myelin sheath. Due to its shorter cytoplasmic tail IP1 is probably more easily integrated into the growing stacks of myelin than IP2.

2.2 Structure of the IP1 gene

To analyse the structural organization of the IP gene a trout genomic library constructed in λ-DASH-II was screened under high stringency conditions using the digoxigenin-labeled 300 bp PCR-fragment described above as a probe. One positive clone was isolated containing an insert of 18 kb which was digested with restriction endonucleases and analyzed by Southern blot hybridization. Nucleic acid sequencing revealed the presence of three exons separated by two introns of 1800 and 747 bp length, respectively (Fig. 2). The exon-intron boundary sequences exactly matched the GT-AG rule (1). Exon 1 encodes the 5'-untranslated region as well as the first 15 residues of the signal peptide. Exon 2 spans the remainder of the signal peptide as well as the initial 49 residues of the extracellular domain. Exon 3 covers another 31 residues of the extracellular domain but then it prematurely terminates at a Sau3A restriction site, the enzyme used for size fractionation of the genomic DNA. Hence from this clone only a partial exon-mapping could be performed. Nevertheless the data obtained thus far point to a striking phylogenetic conservation of myelin gene structures between mammals and fish (Fig. 2) since the location of introns identified in trout IP1 is exactly matching those occurring in the rat Po gene (15). Furthermore in both

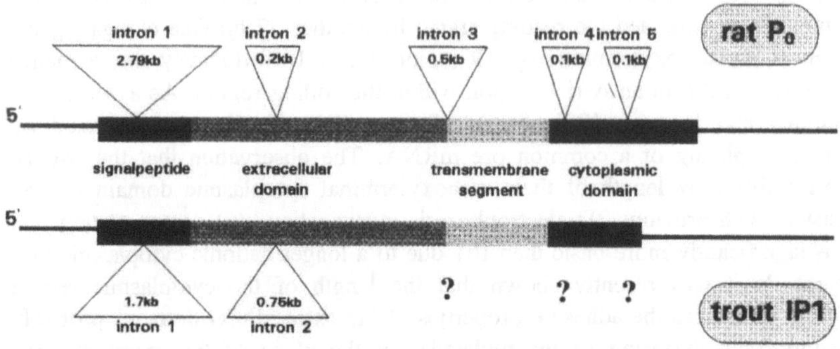

Figure 2: Structural organization of the trout IP gene compared with those of rat Po. In both cases the seggregation of exons correlates well with the functional domains of the protein.

cases the junctions between exon 2 and exon 3 fell within codons. The conserved position of intron 2, splitting the extracellular immunoglobulin domain into two half-domains is particularly noteworthy, since it supports the notion that the variable region of immunoglobulins might have evolved from two separate ancestral genes (15).

The transcription initiation site was determined at 124 bp upstream the first ATG codon by primer extension analysis. Interestingly enough the 3'-flanking sequence immediately following the transcription start of IP was nearly identical through nine nucleotides with those of the mammalian PLP gene while compared with the Po gene no such similarity was found. This nonanucleotide is part of an identical repeat in exon 1. The 5'-flanking sequence of exon 1 in the trout IP gene shows a TATA-box homologue (TATAAAA) at nt -25 to -33 bp upstream the transcription initiation site. The TATA-box is preceeded by a brain specific SP1-like binding site ending at a distance of 37 nt. A similar motif has been identified in the proximal promoter region of mammalian myelin genes PLP, MBP and MAG as well. Furthermore an inverse CAAT-Box is located at position -106 from the transription start. At nucleotide -258 an AP1-binding site is present, suggesting that the jun-fos protooncogene complex may play a role in the regulation of IP gene expression. Furthermore a cyclic AMP binding element (CRE) was found at nt -484, which is consistent with the observation that IP gene expression in cell culture can be accelerated by dibutyryl-cAMP treatment. Furthermore a variety of structural motifs shared with the promotor regions of the mammalian MBP and PLP gene could be identified. The functional significance of these promotor elements remains to be experimentally analysed in the near future.

2.3 Tissue expression of IP1 mRNA

To examine tissue expression of the IP1 gene digoxigenin labeled riboprobes were prepared from the cDNA clone by in vitro transcription and used for Northern blot analysis with poly $(A)^+$ RNA extracted from mature trout brain and liver. With the antisense riboprobe a single transcript of 2.0 kb was detected in brain tissue, while poly $(A)^+$ RNA from trout liver did not give a specific

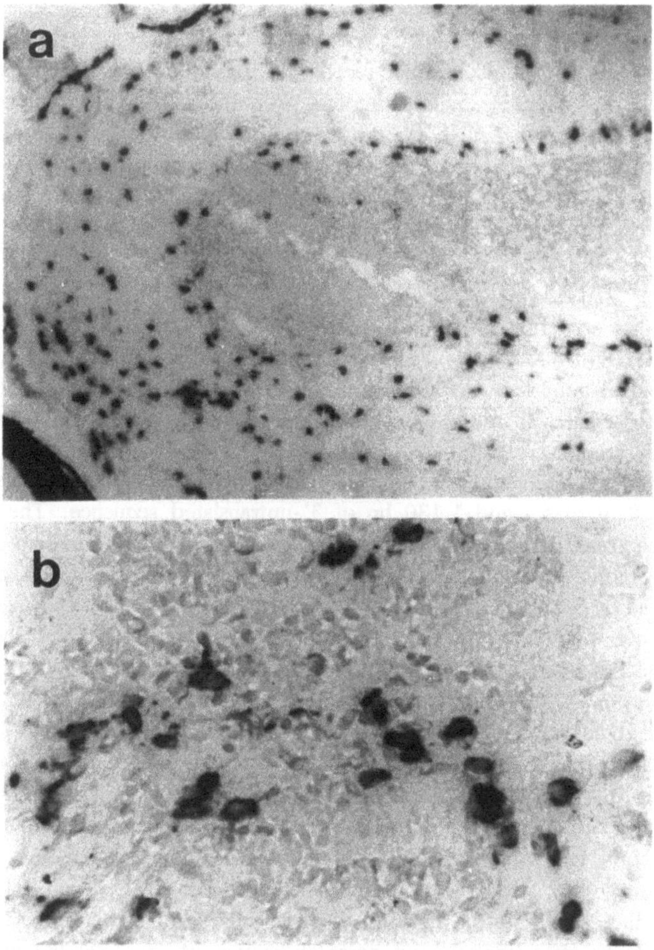

Figure 3: In situ hybridization localization of the IP1 transcript on transverse sections of trout optic tectum. Labeling was generated with a DIG-labeled antisense RNA probe. In (a) the overall distribution of oligodendrocytes in the nerve fiber layers of the optic tectum is shown at low magnification. (b) Perinuclear staining of oligodendrocyte cell bodies at higher magnification.

reaction. The occurence of only a single mRNA species for both IP1 and IP2 is easily explained by the limitations of resolution. When hybridizations were conducted with the sense riboprobe instead the signal was reduced to background levels, confirming the specificity of the probes. Localization of the IP1 transcript in trout brain tissue was performed by in situ hybridization on frozen tissue sections. In the spinal cord, midbrain, cerebellum and optic tectum labeling was confined to oligodendroglial cell bodies associated with myelinated fiber tracts (Fig.3), while unmyelinated brain regions showed no reaction. As expected Schwann cells in peripheral nerve fibres were intensely stained, too, whereas any non-neuronal tissues such as liver, muscle, gills, skin and cartilage were entirely negative.

Although glycoprotein Po is generally considered as a major myelin membrane adhesion molecule (3), recent cell transfection studies suggest that it additionally has neurite-growth promoting properties (20). The occurence of two Po-like proteins in fish CNS fiber tracts, which readily regenerate after injury tempts to speculate that these glycoproteins could be involved in supporting regenerative neuritic outgrowth in fish. Future cell transfection experiments using an expression plasmid containing the IP1 cDNA of trout may help to test this hypothesis.

3.1 Primary structure of the 36K protein

To gain insight into the molecular structure of the novel 36K myelin protein constituent of bony fish CNS a trout brain expression library constructed in λgt 11 was screened with a polyclonal anti-36K antiserum, which on western blots selectively recognizes 36K (7). A positive clone with a largest insert size of 2.1 kb was isolated and sequenced. It exhibited an open reading frame of 951 bp as well as 18 nt of 5'- and 1,130 bp of 3'-untranslated sequence. The predicted protein consisted of 317 amino acids, corresponding to 35,448 daltons molecular weight, which compares well with the apparent molecular size of 36K previously determined by SDS-PAGE (5). No evidence for a signal peptide cleavage site was found at the aminoterminus (22). Hydrophobicity analysis (11) did not reveal a clear transmembrane segment, although the predicted protein generally displayed a high proportion of nonpolar residues. Furthermore no consensus sequence for N-linked glycosylation was present, which is in line with previous biochemical data, demonstrating that 36K is not glycosylated. All taken together it is concluded that 36K is not a transmembrane protein but is associated with the cytoplasmic surface of the glial plasmalemma as a peripheral membrane protein. This is consistent with previous immunohistochemical investigations (9) done with freshly dissociated oligodendroglial cells of trout CNS, which did not show a cell surface staining after direct incubation with polyclonal anti-36K antibodies (Fig.4). Only after fixation and permeabilization of the cells the intracellular epitopes of 36K became visible.

A search for homologous proteins in the database revealed no significant structural similarity with any of the myelin proteins sequenced so far, confirming that 36K is a novel structural entity of myelin. A computer-assisted secondary structure prediction according to Chou-Fasman (1978) revealed a strikingly regular $\beta\alpha\beta$-folding pattern closely resembling the Rossmann-fold typical to the

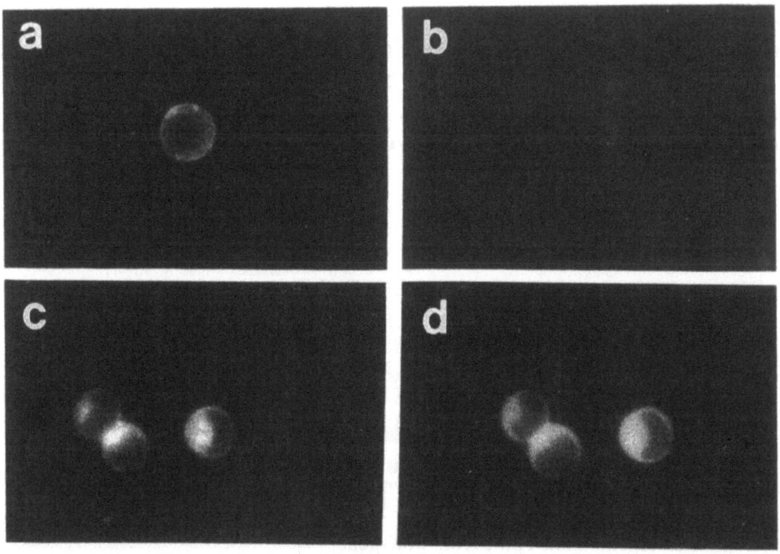

Figure 4: Immunohistochemical loacalization of IP2 (a, c) and 36K (b, d) in freshly dissociated glial cells of trout CNS. In unfixed cells (a, b) IP2 shows a distinct membrane labeling, whereas no 36K staining is visible under these conditions. After fixation and permeabilization both 36K and IP2 exhibit a characteristic intracellular labeling. Reprinted with permission from Jeserich et al (1990).

cofactor-binding domain of dinucleotide-depending enzymes, in particular of NADP-linked oxidoreductases (16). Though the overall sequence identity within this protein family is quite moderate, all of its members share this characteristic folding pattern of seven parallel β-sheets with alternating α-helices. Sequence alignment of 36K with the NADP-binding aminoterminal region of a variety of quite different members of the single domain reductase/epimerase/oxidoreductase superfamily (RED, 12), such as β-hydroxisteroid-dehydrogenase from pig liver, protochlorophyllid-reductase from green plants and human carbonylreductase revealed that almost all of the relevant structural motifs common to this protein superfamiliy occur also in the 36K sequence. Most importantly between the C-terminal end of the first the β-strand and the ensuing α-helix in all cases a hydrophobic glycin-rich fingerprint motif is found which is critically involved in NADP(H)-binding (12, 16). Altogether this suggest that 36K is a novel myelin enzyme, the substrate of which still has to be identified, however.

3.2 Structural organization of the 36K gene

Using the digoxigenin-labeled 36K cDNA as a probe a 19kb DNA fragment which encompassed the entire coding region including its 5'-flanking sequence was isolated from a trout genomic library and characterized by restriction mapping and southernblot analysis. Nucleic acid sequencing revealed that the 36K gene is divided into five exons by four relatively short introns of about 100 bp length each (Fig. 5). The location of introns closely corresponds to the boundaries of protein domains: exon 1and 2 each encode a single β-strand/α-helix unit, while exon 3 additionally includes part of β-strand 4. Intron 2 splits the two mononucleotide half-domains of the predicted NADP(H)-binding region. Exon 4 is largest and covers most of the remaining coding region, while exon 5 spans the carboxyterminal end together with the 3'-untranslated sequence.

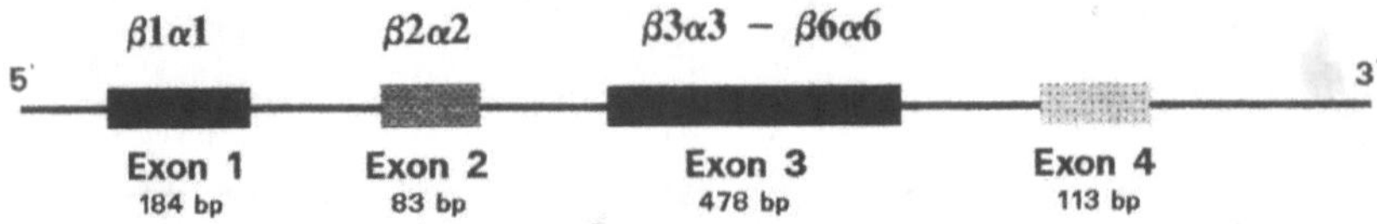

Fig 5: Structural organization of the trout 36K gene

The transcription initiation site was determined by primer extension analysis. A modified TATA-box was identified between nt -24 and -31 upstream the transcription start. Besides a variety of general eucaryotic cis-regulatory elements such as CAAT-boxes a number of structural elements shared with the proximal promotor region of mammalian myelin genes, in particular MPB were identified: at position -317 e.g. a complete consensus sequence for the NF1-binding motif of the JCV-enhancer was found. Furthermore three homologous sequences partially matching the MyT1-binding site identified by Kim and Hudson (1992) in the PLP gene promotor occured. Further studies will be necessary to evaluate the functional activity of these potential cis-regulatory motifs and its corresponding trans-acting factors during development and regeneration of myelinated fiber tracts in the fish CNS.

3.3 Tissue distribution of the 36K transcript

Northern blot analysis of poly (a)$^+$ RNA extracted from mature trout brain using digoxigenin-labeld riboprobes revealed a single mRNA species of about 2.0 kb size confirming that we had obtained a full length cDNA. In situ hybridization with 36K antisense riboprobes showed a selective labeling of oligodendrocyte cell bodies in all myelinated brain regions of trout. In general the tissue distribution of the 36K transcript appeared virtually identical with those of the IP1 mRNA. A major difference, however, was the entire absence of the 36K transcript in Schwann cells of peripheral myelinated fiber tracts, excatly as expected from previous electrophoretic as well as immunohistochemical studies.

4 Conclusion and Outlook

In the present study the molecular structure and tissue expression of two major myelin genes of trout CNS was described. Thereby a striking structural similarity between trout IP and the major mammalian PNS myelin gene Po was revealed. In the case of 36K, on the other side, no significant homology with any of the known mammalian myelin genes could be detected. By virtue of its deduced amino acid sequence, predicted secondary structure and genomic arrangement 36K is suggested as a novel member of the reductase/epimerase/oxidoreductase multigene family. The potential substrate and physiological role of this putative myelin enzyme is still enigmatic.

Since fish oligodendrocytes similar as mammalian Schwann cells readily remyelinate axons after injury, while the regenerative potential of mammalian oligodendrocytes is only poorly developped, the question arises as to the molecular basis for this phenomenon. As a first step into this direction the molecular probes described above should allow to follow expression of 36K and IP genes in the crush-injured trout optic nerve to elucidate the reaction of oligodendrocytes to injury in a regenerating CNS fiber tract at the level of nucleic acids. In summary the data presented here provide a framework for studying in detail the molecular mechanisms controlling myelin gene expression during development and regeneration in the fish CNS. First of all these investigations will be focused on the identification and characterization of nuclear proteins that interact with putative cis-regulatory elements in the 36K and IP gene promoter.

5 Acknowledgements

The authors would like to thank Mrs. B. Flenker, Mrs. U.Mädler and Mrs. M. Düwel for excellent technical assistance. The investigations described here were supported the Deutsche Forschungsgemeinschaft (grants Je 115/4-3, SFB 171/C13).

6 References

1. Breathnach, R., Chambon, P. (1981) Organization and expression of eucaryotic split genes coding for proteins. Annu. Rev. Biochem. **50**, 349-383.
2. Chou P.Y., Fasman G.D. (1978) Prediction of the secondary structure of proteins from their amino acid sequence. Adv. Enzymol. **47**, 45-148.
3. D'Urso, D., Brophy, P.J., Staugaitis, S.M., Gillespie, C.S., Frey, A.B., Stempak, J.G., and Colman D.R. (1990) Protein zero of peripheral nerve myelin: biosynthesis, membrane insertion and evidence for homotypic interaction. Neuron **2**, 449-460.
4. Frohman, M.A., Dush, M.K:, Martin, G.R. (1980) Rapid production of full-length cDNAs from rare transcripts: amplification using a single gene-specific oligonucleotide primer. Proc. Natl. Acad. Sci. USA **85**, 8998-9002.

5. Jeserich G. (1983) Protein analysis of myelin isolated from the CNS of fish: developmental and species comparisons. Neurochem. Res. **8**, 957-969.
6. Jeserich G. and Waehneldt T.V. (1986a) Bony fish myelin: evidence for common major structural glycoproteins in central and peripheral myelin of trout. J. Neurochem. **46**, 525-533.
7. Jeserich G. and Waehneldt T.V. (1986b) Characterization of antibodies against major fish CNS myelin proteins: immunoblot analysis and immunohistochemical localization of 36K and IP2 proteins in trout nerve tissue. J. Neurosci. Res. **15**, 147-158.
8. Jeserich G. and Waehneldt T.V. (1987) Antigenic sites common to major fish myelin glycoproteins (IP) and to major tetrapod PNS myelin glycoprotein (P_0) reside in the amino acid chains. Neurochem. Res. **12**, 825-829.
9. Jeserich G., Müller A., and Jacque C. (1990) Developmental expression of myelin proteins by oligodendrocytes in the CNS of trout. Dev. Brain Res. **51**, 27-34.
10. Kim, J.G., Hudson. L.D. (1992) Novel member of the zinc finger superfamily: a C_2-HC finger that recognizes a glia-specific gene. Mol. Cell. Biol. 12, 5632-5639.
11. Kyte J. and Doolittle R.F. (1982) A simple method for displaying the hydropathic character of a protein. J. Mol. Biol. **157**, 105-132.
12. Labesse, G., Vidal-Cros, A., Chomilier, J., Gaudry, M., Mornon, J.-P. (1994) Structural comparison lead to the definition of a new superfamily of NADP(H)-accepting oxidoreductases: the single-domain reductases/epimerases/dehydrogenases (the 'RED' family). Biochem. J. **304**, 95-99.
13. Lees, M.B., Brostoff, S.W. (1984) Proteins of myelin, in *Myelin*, (Morell,P., ed.) pp 197-224, Plenum Press, New York.
14. Lemke G. and Axel R. (1985) Isolation and sequence of a cDNA encoding the major structural protein of peripheral myelin. Cell **40**, 501-508.
15. Lemke, G., Lamar, E., Patterson, J. (1988) Isolation and analysis of the gene encoding myelin protein zero. Neuron **1**, 73-83.
16. Rossmann, M.G., Liljas; A., Bränden, C.-J., Banaszak, L.J. (1975) Evolutionary and structural relationships among dehydrogenases. In P.D. Boyer (ed.), The Enzymes, Vol 11, Part A, Academic Press, New York, pp 61-102.
17. Saavedra R.A., Fors L., Aebershold R.H., Arden B., Horvath S., Sanders J., and Hood L. (1989) The myelin proteins of the shark brain are similar to the myelin proteins of the mammalian peripheral nervous system. J. Mol. Evol. **29**, 149-156.
18. Salzer, J.L:, Holmes, W.P., Colman, D.R. (1987) The amino acid sequence of the myelin-associated glycoproteins: homology to the immunoglobulin gene superfamily. J. Cell Biol. **104**, 957-965.
19. Schliess F. and Stoffel W. (1991) Evolution of the myelin integral membrane proteins of the CNS. Biol. Chem. Hoppe-Seyler **372**, 865-874.
20. Schneider-Schaulies J., von Brunn A., and Schachner M. (1990) Recombinant peripheral myelin protein P_0 confers both adhesion and neurite outgrowth-promoting properties. J. Neurosci. Res. **27**, 286-297.

21. Stratmann, A., Jeserich, G. (1995) Molecular cloning and tissue expression of a cDNA encoding IP1 - a Po-like glycoprotein of trout CNS myelin. J. Neurochem. **64**, 2427-2436.
22. von Heijne G. (1986) A new method for predicting signal sequence cleavage sites. Nucleic Acids Res. **14**, 4683-4690.
23 Waehneldt T.V., Matthieu J.-M., and Jeserich G. (1986) Appearance of myelin proteins during vertebrate evolution. Neurochem. Int. **9**, 463-474.
24. Wells C.A., Saavedra R.A., Inouye H., and Kirschner D.A. (1993) Myelin P_0-glycoprotein: predicted structure and interactions of extracellular domain. J. Neurochem. **61**, 1987-1995.
25. Williams A.F. (1987) A year in the life of the immunoglobulin super-family. Immunol. Today **8**, 298-303.
26. Wong, M.H., Filbin, M.T. (1994) The cytoplasmic domain of the myelin Po protein influences the adhesive interactions of its extracellular domain. J. Cell Biol. **126**, 1089-1097.
27. Zhang K. and Filbin M.T. (1994) Formation of a disulfide bond in the immunoglobulin domain of the myelin P_0 protein is essential for its adhesion. J.Neurochem. **63**, 367-370.

Transcriptional Controls in the Oligodendrocyte Lineage

Lynn D. Hudson[1], Jin G. Kim[1], Claudia Wiese[1], Da-Lin Yao[2], Xia Liu[2], Henry deF. Webster[2], Denes v. Agoston[3], and Regina Armstrong[4]

[1] Laboratory of Developmental Neurogenetics, NINDS, NIH, Bethesda, MD 20892, USA
[2] Laboratory of Experimental Neuropathology, NINDS, NIH, Bethesda, MD 20892, USA
[3] Laboratory of Developmental Neurobiology, NICHD, NIH, Bethesda, MD 20892, USA
[4] Department of Anatomy and Cell Biology, USUHS, Bethesda, MD 20814-4799, USA

1 An Introduction to Transcription Factors in Oligodendrocytes

Transcriptional controls are operative at all stages in the life cycle of an oligodendrocyte: in the commitment of progenitor cells to the oligodendrocyte lineage, in the coordinated regulation displayed by mature oligodendrocytes of the array of genes that encode myelin proteins and enzymes for myelin lipid synthesis, and in the activation of the myelin program in cells subjected to a demyelinating insult. An appreciation of the molecular events that underly the pinnacle of oligodendrocyte achievement, the compact myelin sheath, requires an understanding of how signalling molecules produced by neuronal and non-neuronal cells transmit information to the transcription factors that ultimately determine the fate and/or regulate myelin production of oligodendrocytes. The specific transcription factors that are involved in each of these events are just beginning to be identified. One of the best characterized factors in precursors of myelinating cells is a POU-homeodomain protein named SCIP for suppressed-cyclicAMP-inducible protein (Monuki et al., 1989; also referred to as Tst-1 or Oct-6). The expression of SCIP in proliferating oligodendrocyte progenitors and subsequent down-regulation of SCIP expression upon differentiation originally suggested that this transcription factor may be tied to the proliferative status of developing oligodendrocytes or the early decision-making steps in this lineage (Collarini et al., 1992). This suggestion was supported by a number of studies that have documented the role that SCIP plays in myelinating cells of the peripheral nervous

Molecular Signaling and Regulation in Glial Cells
A Key to Remyelination and Functional Repair

system. The most relevant ones are transgenic experiments with mice expressing a dominant-negative antagonist of SCIP in Schwann cells, in which the developmental program of Schwann cell differentiation is disrupted, resulting in severe perturbations of the myelin program in the peripheral nervous system (Weinstein et al., 1995).

Another category of transcription factors that may influence the pattern of gene expression in oligodendrocytes are the ubiquitously expressed factors, such as Sp1 (Henson, 1994; Janz and Stoffel, 1993), NFI (Amemiya et al., 1992; Inoue et al., 1990; Zhang and Miskimins, 1993), CREB (Sato-Bigbee et al., 1994) and the thyroid hormone receptor (THR) (Farsetti et al., 1991; Bogazzi et al., 1994). Despite their widespread distribution, these transcription factors cannot be dismissed as unimportant regulators in the oligodendrocyte lineage. Factors like NFI and Sp1 may either compete for binding of glial-specific transcription factors, as hinted at by the ability of Sp1 to bind to the purine-rich region found in myelin promoters (Henson, 1994), or act in conjunction with one of the transcriptional activators, as suggested by the observation that an NFI site is closely coupled with a second activator site in several glial-specific genes (Amemiya et al., 1992). Thus, the concentration of these ubiquitously distributed factors in the oligodendrocyte nucleus may be critical in setting the levels of transcriptional activity of myelin genes. Moreover, some of the ubiquitously distributed transcription factors are directly activated by hormones/cytokines that are known to affect the survival, proliferation or myelinating capacity of oligodendrocytes. An example is the thyroid hormone receptor, a member of the nuclear hormone receptor family that can bind to the myelin basic protein (MBP) and proteolipid protein (PLP) genes and activate their transcription when the ligand (thyroid hormone) is present (Farsetti et al., 1991; Bogazzi et al., 1994). Thyroid hormone has a dual action on the oligodendrocyte lineage: it elevates both the number of mature oligodendrocytes and the average amount of myelin produced (Almazan et al., 1985; Barres et al., 1994). Both of these events are reflected in the enhanced expression of myelin-specific genes in thyroid hormone treated oligodendrocytes (Amur et al., 1984; Shanker et al., 1984, 1987; Tosic et al., 1992), by a mechanism in which the hormone binds to the thyroid hormone receptor which is complexed to a myelin gene promoter (Farsetti et al., 1991; Bogazzi et al., 1994).

The transcription factors mentioned above are the ones that were either already identified as transcription factors in other systems (e.g. the ubiquitous factors) or were originally cloned by their homology to known transcription factors (e.g. SCIP). These approaches to finding transcription factors that influence oligodendrocytes are therefore limited in their scope, as only a small subset of the genomes's array of transcription factors have been characterized to date. To identify novel transcription factors that affect oligodendrocytes, we have employed the gene encoding the most abundant myelin protein in the central nervous system, myelin proteolipid protein (PLP), as the bait to fish out transcription factors that bind myelin-specific genes and thereby dictate lineage decisions and the extent of myelination in oligodendrocytes (Kim and Hudson, 1992). We report here on the characterization of one of the transcription factors that has been isolated by this strategy, named MyTI for Myelin Transcription factor I.

2 A Novel Class of Transcription Factors Expressed in the Developing Nervous System

2.1 MyTI and family members

MyTI represents a new class of zinc finger transcription factors in which the metal cation is bound by three cysteine and one histidine residue in an unusual Cys-Cys-His-Cys arrangement (Kim and Hudson, 1992). A total of six zinc fingers is present in MyTI, split into two widely-separated clusters of these DNA-binding domains. This arrangement is of interest as it suggests that a single molecule of MyTI could bind simultaneously to two discrete sites within the PLP gene and thereby contribute to the bending of DNA at the locus. As the process of bending has been implicated in the transcriptional activation of a variety of genes, it is possible that MyTI is likewise involved in the activation of the locus for transcription.

There is little homology between MyTI and previously identified members of the zinc finger superfamily. A search for transcription factors that are related to MyTI, which was carried out by screening brain cDNA libraries for homology under conditions of reduced stringency, has yielded a clone that displays considerable conservation of the zinc finger DNA binding domains and the same arrangement of these domains in two clusters (J.G.K., unpublished results). The gene encoding this MyTI family member is a unique, single-copy one that maps to a chromosome distinct from MyTI (C.W., unpublished results). Both MyTI and the family member are expressed in the developing nervous system, although the MyTI family member is not found in the oligodendrocyte, astrocyte or microglial lineages as determined by Northern blot analysis (J.G.K. and R.A., unpublished results). It is likely that the MyTI family member is expressed in subsets of neurons, particularly motor neurons, as this population is stained by antibodies directed against the highly conserved zinc finger domain that do not discriminate between MyTI and the MyTI family member (R.A., D.A. and X. L., unpublished results).

2.2 Target genes of MyTI

MyTI was cloned by virtue of its recognition of a site in the PLP promoter region that was required for the full activity of the promoter in transfected glial cells (Berndt et al., 1991). This site, referred to as a "cis element" because it is located within a gene and is subject to binding by a "trans-acting" or transcription factor, is also present in the promoter region of a number of myelin genes (Figure 1).

Thus, PLP, MBP, myelin-associated glycoprotein (MAG), protein zero (P0) and 2',3'-cyclic nucleotide 3'-phosphodiesterase (CNP) are among the candidate target genes for the action of MyTI. By binding to a common cis element that is found in many myelin gene promoters, MyTI may act as a trans-acting factor to coordinate the transcription of these genes.

To identify all possible target sites of DNA binding by MyTI, we also carried out a selective amplification and binding assay with a library of degenerate oligonucleotides. MyTI was found to prefer to bind to G-rich oligonucleotides, and a relatively broad and long consensus site of 10 bases emerged from a comparison of over fifty high affinity binding sites (J.G.K., unpublished results). Multiple recognition sites for MyTI are present in the PLP gene: in addition to the binding site at position -271 (with respect to the initiation of transcription), there exists sites further upstream at positions -481, -901 and -913. These upstream sites may interact with additional MyTI molecules, or may serve as binding sites for one of the two clusters of zinc fingers within a MyTI molecule already bound to the PLP promoter. Analogous to the situtation at the PLP locus, each of the other known myelin gene promoters contain multiple sites that fit the MyTI consensus, an arrangement that reinforces the possibility that MyTI binding to widely separated sites in a myelin gene promoter may promote the looping out, or bending of the intervening DNA. Further experiments to document which of the potential target genes are actually bound by MyTI *in vivo* are underway.

Table 1. MyTI Binding Site in the PLP Promoter is Found in Other Myelin Genes

PLP	A	G	T	T	G	G	A	A	G	T	T	T	-250
MAG	A	G	G	T	G	G	A	G	G	T	G	A	-226
MBP	T	G	T	G	G	G	A	A	G	T	C	C	-640
PO	T	C	C	T	G	G	A	A	G	T	T	T	-550
CNP	A	G	G	T	G	G	C	A	G	C	T	G	-600
CONSENSUS:				T	G	G	A	A	G	T			

The DNA sequences shown are from Berndt et al., 1992 (PLP), Nakano et al., 1991 (MAG), Miura et al., 1989 (MBP), Lemke et al., 1988 (PO), and Monoh et al., 1993 (CNP).

3 MyTI expression during development

MyTI is found at the earliest stages of the oligodendrocyte lineage that can be defined in cultures, and continues to be expressed as the cells differentiate along the oligodendrocyte pathway and synthesize messenger RNA for PLP, one of the presumptive targets of MyTI action (Figure 1; Armstrong et al., 1995). MyTI protein is located in discrete domains within the nucleus of oligodendrocyte precursors. A shift in MyTI localization from the nucleus to the cytoplasm occurs after PLP gene products begin to accumulate in oligodendrocytes, an event which is quickly followed by the disappearance of MyTI (Armstrong et al., 1995).

MyTI expression *in vivo* appears to mimic the pattern observed *in vitro*. In the rat forebrain at postnatal day three, MyTI immunoreactivity is evident in the nuclei of cells of the subventricular zone, the presumptive origin of oligodendrocyte precursors (Armstrong et al., 1995). The MyTI protein is also present in cells that have migrated out of the subventricular zone and appear to be in the initial stages of myelin protein production, as evidenced by the confinement of PLP to the cell body of these cells. No MyTI protein is detectable in the fully myelinated tracts. The presence of MyTI throughout the early development of oligodendrocyte progenitors, and the fact that the expression extends to the period when young oligodendrocytes begin to synthesize myelin proteins, suggests that this transcription factor may play a role in both the determination of cell fate and in the initial coordination of myelin gene expression.

4 MyTI expression in a demyelinating model

Demyelination, which can be induced by a variety of conditions in animal models, is typically associated with a transient decline in the expression of myelin genes,

---

Figure 1 MyTI expression in developing oligodendrocytes
MyTI immunoreactivity is shown for cells at distinct developmental stages in the oligodendrocyte lineage. Very early oligodendrocyte precursors, pre-O-2A cells, which appear as clusters of small round cells (A), exhibit nuclear MyTI immunoreactivity (B) as do bipolar O-2A progenitors (C,D) and multipolar oligodendrocyte precursors (E,F). Cells that have differentiated into oligodendrocytes elaborate a network of thin branched processes (G,I) while continuing to exhibit MyTI immunoreactivity (H). More mature oligodendrocytes lose MyTI immunoreactivity (I,J). Phase contrast micrographs (A,C,E,G,I) are shown for the same fields as the corresponding immunofluoresence micrographs (B,D,F,H,J, respectively). Cultures shown were prepared from neonatal rat brain. A,B: primary mixed glial cell cultures after 11 days *in vitro*. C,D: Secondary cultures grown for 3 days in defined medium containing mitogens (platelet-derived growth factor and basic fibroblast growth factor) to amplify the O-2A progenitor population and prevent differentiation. E-J: Secondary cultures grown in defined medium, without mitogens, to allow differentiation for 1 (E,F), 3(G,H) or 4 (I,J) days. Scale bar, 50 µm.

followed by an upsurge in the transcription of myelin genes during the period of remyelination (Yao et al., 1995a, 1995b). An important question is whether this elevated transcription of myelin genes originates primarily from the resident progenitor cell population, which undergoes proliferation and subsequent maturation following injury, or from mature oligodendrocytes that may respond to a demyelinating insult by either directly increasing myelin gene transcription or by "dedifferentiating" to a precursor cell whose progression through the lineage recapitulates the events in normal development. We have found that in a cryogenic demyelinating model in the rat, MyTI-positive cells identified by in situ hybridization are numerous in the period preceding remyelination; normal adult animals, in contrast, display few MyTI-positive oligodendrocyte lineage cells (D.-L. Y., H.d.W. and L.D.H., unpublished results). Similar results were observed by immunocytochemistry in a model of experimental autoimmune encephalitis (EAE) (X.L., D.-L. Y, H.d.W. and L.D.H., unpublished results). These results suggest that the pool of oligodendrocyte precursors does increase following injury, confirming observations with other markers of immature oligodendrocytes (Komoly et al, 1992). Whether these MyTI positive cells reflect proliferation of the precursor population or "dedifferentiated" oligodendrocytes is not known. However, if the resident population of mature oligodendrocytes are the cells that form most of the newly synthesized myelin sheaths, the re-emergence of MyTI expression indicates that these cells probably undergo a certain amount of dedifferentiation, to at least the stage of a "young" oligodendrocyte that is simultaneously expressing both PLP and MyTI, prior to their synthesis of new myelin sheaths.

The greatly enhanced expression of MyTI in the demyelinating models suggests that this transcription factor may be required for the upregulation of myelin genes following injury, in addition to the role that this transcription factor may have in determining the fate of an oligodendroglial precursor cell in normal development and the possible role that MyTI may play in regulating myelination by initially coordinating the expression of a group of myelin-specific genes. Transgenic experiments designed to answer whether this one transcription factor is instrumental at multiple stages in the production and maintenance of myelin sheaths are underway.

5 Prospectus

This foray into the transcription factors expressed in the oligodendrocyte lineage highlights one of the themes of gene expression in higher organisms. At each locus, a complex array of transcription factors acts in combination to set the overall level of transcriptional activity. While a single transcription factor may be instrumental in determining the transcriptional activity of one or more genes, no one transcription factor acts in isolation. Another facet of transcriptional control that was highlighted by MyTI is the possibility that a single transcription factor can have an impact on several stages/processes within a cell lineage, influencing not only fate decisions, but also the daily regulation of gene activity and the response to special events such as injury to the nervous system. The challenge of the next decade is not simply to identify the constellation of transcription factors that are necessary for specifying an oligodendrocyte and regulating the extent of myelination, but to assess the individual contributions that these transcription factors make in establishing the transcriptional activity of myelin genes.

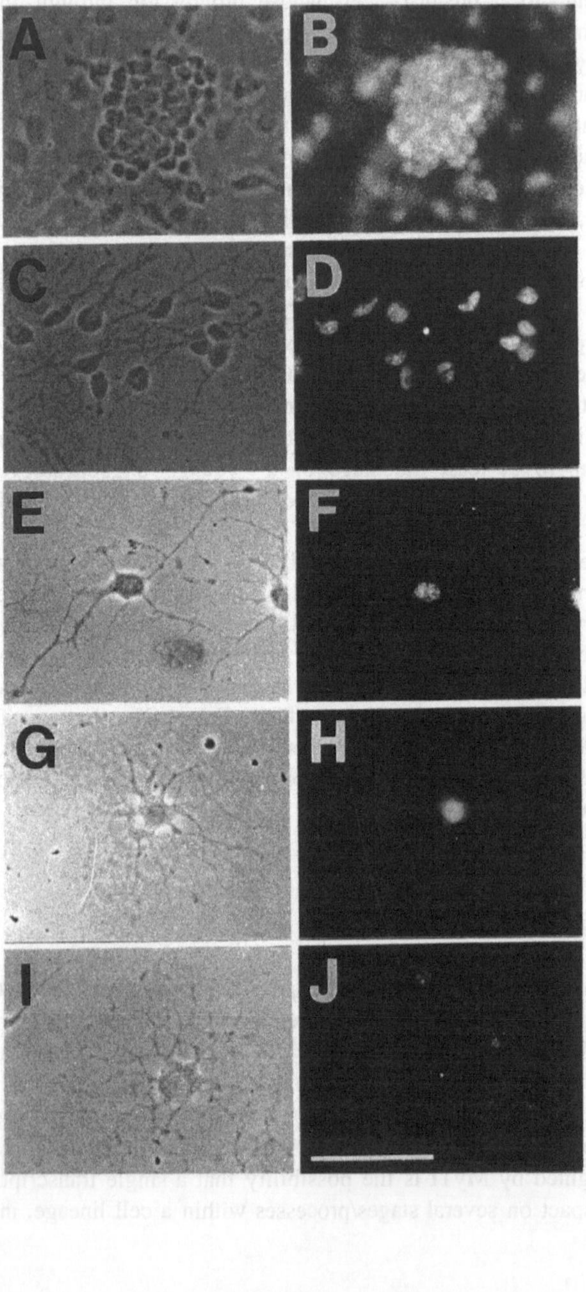

References

Amemiya, K., Traub, R., Durham, L., Major, E. 1992. Adjacent nuclear factor-1 and activator protein binding sites in the enhancer of the neurotropic JC virus. J. Biol. Chem. 267:14204-14211.

Amur, S.G., Shanker, G., Pieringer, R.A. 1984. Regulation of myelin basic protein (arginine) methyltransferase by thyroid hormone in myelinogenic cultures of cell dissociated from embryonic mouse brain. J. Neurochem 43:494-498.

Armstrong, R., Kim, J., Hudson, L.D. 1995. Expression of Myelin Transcription Factor I (MyTI), a "Zinc-Finger" DNA-Binding Protein, in Developing Oligodendrocytes. Glia. 14:303-321.

Barres, B.A., Lazar, M.A., Raff, M.C. 1994. A novel role for thyroid hormone, glucocorticoids and retinoic acid in timing oligodendrocyte development. Development 120:1097-1108.

Berndt, J., Kim, J.G., Hudson, L.D. 1992. Identification of *cis*-regulatory elements in the myelin proteolipid protein (PLP) gene. J. Biol. Chem. 267:14 730-14 737.

Bogazzi, F., Hudson, L.D., Nikodem, V.M. 1994. A novel heterodimerization partner for thyroid hormone receptor. J. Biol. Chem. 269:11683-11686.

Collarini, E.J., Huhn, R., Marshall, C.J., Monuki, E.S., Lemke, G., Richardson, W.D. 1992. Down-regulation of the POU transcription factor SCIP is an early event in oligodendrocyte differentiation. Development 116:193-200.

Farsetti, A., Mitsuhashi, T., Desvergne, B., Robbins, J., Nikodem, V.M. 1991. Molecular basis of thyroid hormone regulation of myelin basic protein gene expression in rodent brain. J. Biol. Chem. 266:23226-23232.

Henson, J.W. 1994. Regulation of the glial-specific JC virus early promoter by the transcription factor Sp1. J. Biol. Chem. 269:1046-1050.

Inoue, T., Tamura T.-A., Furuichi, T., Mikoshiba, K. 1990. Isolation of complementary DNAs encoding a cerebellum-enriched Nuclear Factor I family that activates transcription from the mouse myelin basic protein promoter. J. Biol. Chem. 265:19065-19070.

Janz, R., and W. Stoffel. 1993. Characterization of a brain-specific Sp1-like activity interacting with an unusual binding site within the myelin proteolipid promoter. *Biol. Chem. Hoppe-Seyler* 374:507-517.

Kim, J. G., and L. D. Hudson. 1992. Novel member of the zinc finger superfamily: A C2-HC finger that recognizes a glia-specific gene. Mol. Cell. Biol. 12:5632-5639.

Komoly, S., Hudson, L.D., Webster, H.deF., Bondy, C.A. 1992. Insulin-like growth factor I gene expression is induced in astrocytes during experimental demyelination. Proc. Natl. Acad. Sci. U.S.A. 89:1894-1898.

Lemke, G., Lamar, E., Patterson, J. 1988. Isolation and analysis of the gene encoding peripheral myelin protein zero. Neuron 1:73-83.

Miura, M., T. Tamura, A. Aoyama, and K. Mikoshiba. 1989. The promoter elements of the mouse myelin basic protein gene function efficiently in NG108-15 neuronal/glial cells. *Gene* 75:31-38.

Monoh, K., T.Kurihara, Y. Takahashi, T. Ichikawa, T. Kumanishi, S. Hayashi, S. Minoshima, and N. Shimizu. 1993. Structure, expression and chromosomal localization of the gene encoding human 2',3'-cyclic nucleotide 3'-phosphodiesterase. Gene 129:297-301.

Monuki, E.S., Weinmaster, G., Kuhn, R., Lemke, G. 1989. SCIP: a glial POU domain gene regulated by cyclic AMP. Neuron 3:783-793.

Nakano, R., Fujita, N., Sato, S., Inuzuka T., Sakimura, K., Ishiguro, H., Mishina, M., Miyatake T. 1991. Structure of the mouse Myelin-Associated Glycoprotein gene. BBRC 178:282-290.

Sato-Bigbee, C., Chan, E.L.P., Yu, R.K. 1994. Oligodendroglial cyclic AMP response element-binding protein: a member of the CREB family of transcription factors. J. Neurosci. Res. 38:621-628.

Shanker, G., Campagnoni, A.T., Pieringer, R.A. 1987. Investigations on myelinogenesis in vitro: developmental expression of myelin basic protein mRNA and its regulation by thyroid hormone in primary cerebral cell cultures from embryonic mice. J. Neurosci. Res. 17:220-224.

Shanker, G., Rao, G.S., Pieringer, R.A. 1984. Inestigations on myelination in vitro: regulation of 5'-nucleotidase activity by thyroid hormone in cultures of dissociated cells from embryonic mouse brain. J. Neurosci. Res. 11:263-270.

Tosic, M., Torch, S., Comte, V., Dolivo, M., Honegger, P., Matthiew, J.-M. 1992. Triiodothyronine has diverse and multiple stimulating effects on expression of the major myelin protein genes. J. Neurochem. 59:1770-1777.

Weinstein, D.E., Burrola, P.G., Lemke, G. 1995. Premature Schwann cell differentiation and hypermyelination in ice expressing a targeted antagonist of the POU transcription factor SCIP. Mol. Cell. Neurosci. 6:212-229

Yao, D.L., Liu, X., Hudson, L., Webster, H., deF. 1995. Insulin-like growth factor I treatment reduces demyelination and up-regulates gene expression of myelin-related proteins in experimental autoimmune encephalomyelitis. Proc. Natl. Acad. Sci. 92:6190-6194.

Yao, D.L., West, N.R., Bondy, C.A., Brenner, M., Hudson, L.D., Zhou, J., Collins, G.H., Webster, H., deF. 1995. Cryogenic spinal cord injury induces astrocytic gene expression of insulin-like growth factor I and insulin-like growth factor binding protein 2 during myelin regeneration. Jour. Neurosci. Res. 40:647-659.

Zhang, X., Miskimins, R. 1993. Binding at an NFI site is modulated by cyclic AMP-dependent activation of myelin basic protein gene expression. J. Neurochem. 60:2010-2017.

Regulation of MBP mRNA translocation in oligodendrocytes

Shashi Amur-Umarjee and A.T. Campagnoni

Mental Retardation Research Center, Neuropsychiatric Institute, UCLA School of Medicine, Los Angeles, California 90095, USA

Keywords. RNA transport, glia, neurons, myelin, cell-cell interaction, PDGF

1 Introduction:

Protein targeting in cells is generally achieved by transporting the newly synthesized protein to the appropriate subcellular compartments. In some instances, this specific targeting is achieved by the transport of the corresponding mRNA to specific subcellular compartments. In the central nervous system, mRNAs corresponding to MAP 2 , tau, BC1, Calcium-calmodulin dependent protein kinase, oxytocin and vasopressin are translocated specifically to axons or dendrites in neurons (1) while in oligodendrocytes, MBP and carbonic anhydrase mRNAs are translocated into the processes (2, 3) and in astrocytes, GFAP mRNA is transported (4). The translocation of MBP mRNA has been shown to be developmentally regulated both *in vivo* and *in vitro* (5, 6) and factors regulating the translocation process are not yet completely understood. While studying the translocation of MBP mRNA in oligodendrocytes using a mixed culture system, we observed that translocation of MBP mRNA occurred in only about 25% of the oligodendrocytes in which it was expressed (6). However, we observed that translocation of this mRNA occurred in almost all cells in enriched oligodendrocyte cultures (7). Since about 70% of the cell population in the mixed cultures is made up of astrocytes that form a bed layer on which the oligodendrocytes grow, it appeared possible that astrocytes might influence molecular events in oligodendrocytes either through secreted factors or through cell-cell contact. On replating the purified oligodendrocytes on purified astrocytes, we did observe a profound inhibition of MBP mRNA translocation in the oligodendrocytes. In this report, we show that the inhibition caused by astrocytes was due to cell-cell contact and not due to soluble factors secreted by astrocytes. We also report here that soluble factors secreted by neurons can relieve the astrocytic inhibition and that PDGF can mimic the effect of neuronal factors suggesting that PDGF is important in the regulation of MBP mRNA translocation.

Molecular Signaling and Regulation in Glial Cells
A Key to Remyelination and Functional Repair

2. MATERIALS AND METHODS:

2.1 Preparation of mouse brain primary cultures

Primary cultures were prepared from neonatal BALB/cByJ mouse cerebral hemispheres by mechanical dissociation. The details of the procedure have been described earlier (8). The cells were seeded on chamber slides for performing *in situ* hybridization and on flasks for astrocyte and oligodendrocyte isolation.

2.2 Preparation of enriched oligodendrocytes and astrocytes

Flasks containing primary cultures were shaken at 14 DIV at 200 rpm for 1 hour to remove microglia and then shaken overnight at 280 rpm to separate oligodendrocytes and astrocytes. The details of this procedure have been described earlier also (7).

2.3 Preparation of neuronal cultures

Cerebral hemispheres were dissected out from 15 day-embryonic BALB/cByJ mice and neuron-enriched cultures were prepared and grown in defined medium exactly as we have described previously (9). Neuron-conditioned medium was collected at 7 DIV.

2.4 *In situ* hybridization histochemistry

Cells were fixed in freshly prepared paraformaldehyde solution after washing with phosphate buffered saline and processed for *in situ* hybridization. Digoxigenin-labeled cDNA probes were prepared using MBP cDNA described previouly (6). The non-radioactive *in situ* hybridization method has been described before (6). In brief, cells were permeabilized by treating with ethanol and HCl and then the digoxigenin-labeled probe was added and the probe was allowed to hybridize to the specific mRNA by incubating overnight at 42^0 C. After washing off the excess probe, alkaline phosphatase-conjugated digoxigenin antibody was added and the mRNA bound to the probe and the antibody was detected by addition of organic phosphate and a dye (nitroblue, tetrazolium salt).

2.5 Astrocyte reconstitution experiments

Enriched oligodendrocytes were added back to astrocytes so as to give the same proportion as seen in primary cultures (25% oligodendrocyes and 75% astrocytes) and the details have been given in our earlier publication (7).

Astrocyte-conditioned medium was collected from astrocytes that were remaining in the flasks after the shake-off procedure and astromatrix was prepared from lysing astrocytes with water according the procedure of Rome et al.(9).

2.6 Addition of conditioned media

The oligodendrocytes were plated on astrocytes in medium containing 10% FCS and the medium was changed to defined medium (DMEM:F12 [1:1] containing 5 μg/ml insulin, 50 μg/ml transferrin and 30 nM selenium after 6-7 hours. When conditioned media were added, the medium was removed and a mixture of defined medium: conditioned medium (1:1) was added.

3. Results

The mouse brain primary culture system used in this study has been extensively characterized previously (6, 8). Briefly, these cultures consist mainly of astrocytes and oligodendrocytes with microglia being a minor component. Neurons survive only for a week to ten days under the culture conditions employed. Large, flat astrocytes form a bed layer and small, round oligodendrocytes grow on top of this layer and often form elaborate networks of processes and sheets. Considerable numbers of mature oligodendrocytes are found in these cultures after 14 days *in vitro* (DIV) as evidenced by immunocytochemistry and *in situ* hybridization for MBP, proteolipid protein (PLP) and their mRNAs.

MBP mRNA gets translocated into the processes in oligodendrocyte processes while CNP, MAG and PLP mRNAs remain associated with oligodendrocyte cell somas. Interestingly, in mixed glial cultures, the translocation of MBP mRNA occurred only in 25% of the oligodendrocytes expressing the MBP gene (6). In contrast, when we examined the distribution of MBP mRNA in enriched oligodendrocytes, we observed that MBP mRNA was transported into the processes of almost all oligodendrocytes (7). This suggested that the movement of MBP mRNA in oligodendrocytes may be inhibited by astrocytes.

3.1 Astrocytes inhibit MBP mRNA translocation *in vitro*

Substantial levels of MBP mRNA were found in the processes of almost all oligodendrocytes found in the gaps between astrocytes. This observation was consistent with our notion that astrocytes might inhibit MBP mRNA movement. In order to directly test this possibility, we prepared enriched oligodendrocytes and replated them on astrocytes and examined the distribution of MBP mRNA after 3-4 DIV by *in situ* hybridization. In enriched oligodendrocytes, galactocerebroside (Figure 1a) as well as MBP mRNA (Figure 1b) were distributed in the cell somas, processes and in sheet-like structures. In the reconsituted system, the replated oligodendrocytes resembled oligodendrocytes in mixed glial cultures and grew well on the astrocytes, elaborated processes and expressed galactocerebroside (GC) (Figure 1c). However, astrocytes inhibited the translocation of MBP mRNA almost completely (Figure 1d). Thus, the inhibition could not have been caused because either that oligodendrocytes did not have processes or were not healthy or had dedifferentiated to a GC -ve stage. The distribution of MBP was examined by immunocytochemistry using a polyclonal antibody to MBP. The results showed that the distribution of MBP polypeptide in oligodendrocytes was also influenced by plating the oligodendrocytes on astrocytes. It was interesting that the translocation of MBP mRNA was inhibited by at least 90% while that of MBP polypeptide was inhibited by about 75%. These observations suggest that the translocation of MBP mRNA is the major, if not the sole mode of MBP transport.

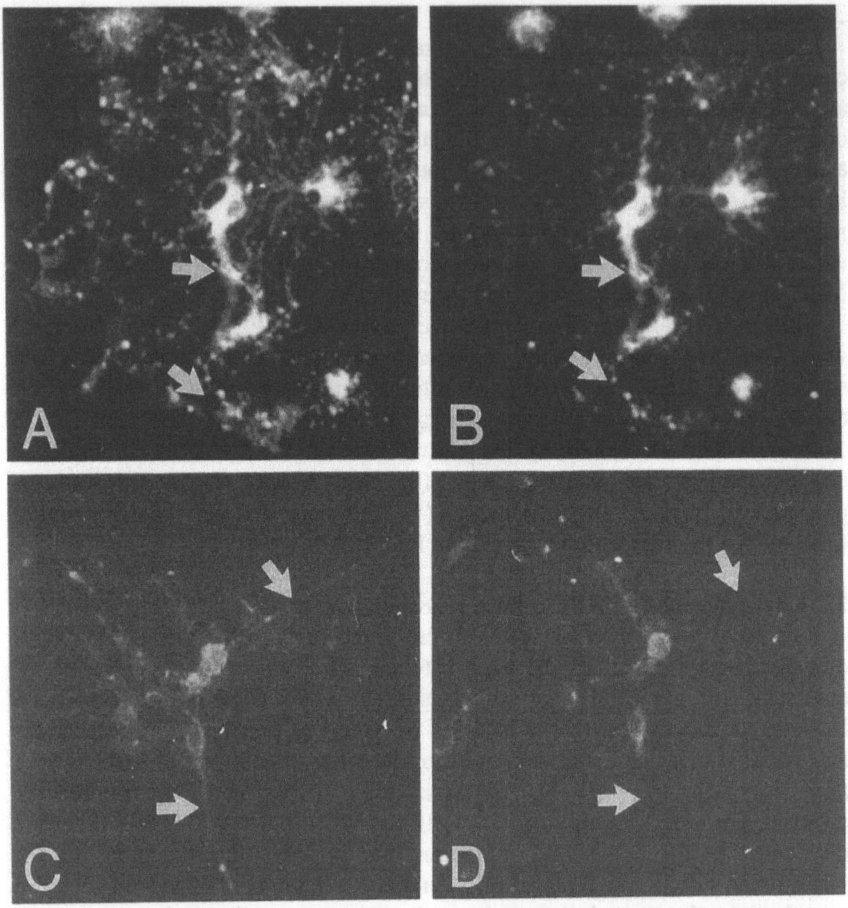

Figure 1. Distribution of MBP mRNA and GC in enriched oligodendrocytes grown in the absence (A, B) or presence (C, D) of astrocytes studied by combined in situ hybridization and immunofluorescence methods. GC (A) as well as MBP mRNA (B) were found to be distributed in the cell soma as well as processes of purified oligodendrocytes (shown with arrows). Purified oligodendrocytes plated on astrocytes elaborated processes as observed with GC labeling (C, marked with arrows) , but failed to translocate MBP mRNA in the processes (D, position of processes is marked with arrows).

3.2 How do astrocytes inhibit MBP mRNA translocation?

In order to investigate the mechanism by which astrocytes inhibit MBP mRNA translocation in oligodendrocytes, we examined the effect of astrocyte-conditioned medium (ACM) on the transport process. The soluble factors secreted by astrocytes did not have any adverse effect on the translocation of MBP mRNA in oligodendrocytes. Thus, astrocytes exert their inhibitory effect through a mechanism different from secretion of factors. The astromatrix, by itself or in combination with ACM did not influence MBP mRNA translocation in oligodendrocytes. In contrast, paraformaldehyde-fixed astrocytes which presumably are intact, but metabolically inert, inhibited the movement of MBP mRNA into oligodendrocyte processes. These findings suggest that the astrocytic inhibition of MBP mRNA translocation is mediated through cell-cell contact and that live astrocytes are not necessary for causing the inhibition.

Table 1. Effect of various cell types on MBP mRNA translocation

Additions	Inhibition of MBP mRNA translocation
Enriched oligodendrocytes (Oligo)	No
Oligo + Astrocytes (Astro)	Yes
Oligo + Astro-conditioned medium (ACM)	No
Oligo + Astromatrix	No
Oligo + Astromatrix + ACM	No
Oligo + Fixed Astrocytes	Yes
Oligo + 3T3 fibroblasts	No
Oligo + SKN-SH neuroblastoma cells	No
Oligo + Immortalized astrocytes (BAS 8.2)	Yes

The inhibitory effect appears to be specific to astrocytes since mouse 3T3 fibroblasts as well as neuroblastoma cells did not inhibit the translocation of MBP mRNA while astrocytes and immortalized astrocytes inhibited the mRNA movement.

3.3. Do neurons play a role in regulating MBP mRNA translocation ?

Interestingly, an almost complete inhibition of MBP mRNA movement into oligodendrocyte processes is observed in the oligodendrocytes replated on astrocytes while about 75% inhibition of the transport process was observed in mixed glial cultures. No inhibition of MBP mRNA has been observed *in vivo*. If neurons did relieve the inhibition due to astrocytes, the results can be explained. Since a few neurons survive in mixed glial cultures, the neurons may be able to relieve the inhibition partially and since no neurons remain in the reconstituted sysem, there is an almost complete inhibition of MBP mRNA translocation. To test this hypothesis, we collected neuron-conditioned medium (NCM) from neuron-enriched cultures grown for 7 DIV from 15-day embryonic BALB/cByJ mice and added to the reconstituted mixture of astrocytes and oligodendrocytes. The added NCM relieved the astrocytic inhibition of MBP mRNA movement proving that factors released by neurons are capable of regulating MBP mRNA translocation *in vitro*.

Conditioned media were also collected from SKN-SH neuroblastoma cells, PC-12 cells and from dorsal root ganglion neurons. The conditioned media from the SKN-SH cells and PC-12 cells did not alleviate the inhibition due to astrocytes whereas the conditioned medium collected from dorsal root ganglion cells was able to relieve the inhibition due to astrocytes. These results suggest that neurons isolated from cerebral hemispheres and from dorsal root ganglia secrete the factor (s) responsible for reversing the astrocytic inhibition whereas the two cell lines used probably do not secrete the factor (s) needed for the alleviation of astrocytic inhibition.

Since fixed astrocytes also inhibited the translocation of MBP mRNA, we examined the effect of added neuronal factors on the astrocytic inhibition. Neuronal factors failed to relieve the astrocytic inhibition of the mRNA translocation. These results suggested that the effect of the neuronal factors on the alleviation of the astrocytic inhibition is mediated through astrocytes. It appeared possible to us that neuronal factors may alter the composition of the astrocytic surface including concentrations of the adhesion molecules. Thus, we treated the astrocytes with NCM and fixed them with paraformaldehyde and plated oligodendrocytes on these treated astrocytes. These NCM-treated astrocytes also inhibited MBP mRNA movement.

3.4. Which of the factors secreted by neurons is responsible for the alleviation of the astrocytic inhibition of MBP mRNA translocation?

In order to identify the factor (s) responsible for relieving the astrocytic inhibition, various growth factors were added to the reconstituted system of oligodendrocytes and astrocytes and the effects tested after 3 DIV. Among the growth factors tested only PDGF was capable of mimicking the effect of NCM (11). The various isoforms of PDGF, AA, AB and BB were then tested. We found that both PDGF- AB and -BB isoforms were equally capable of mimicking the effect of NCM while PDGF-AA isoform had no effect on the inhibition of translocation of MBP mRNA in oligodendrocytes. Interestingly, astrocytes in culture also secrete the A chains of PDGF, but neither PDGF-AA nor ACM could relieve the inhibition due to astrocytes. Thus, only the neuron-derived forms of PDGF are capable of relieving the astrocytic inhibition.

4. Discussion

Our studies have demonstrated that astrocytes inhibit the translocation of MBP mRNA into oligodendrocyte processes *in vitro*. Consistent with these findings, astrocytes have been shown to inhibit myelination of neuronal axons by oligodendrocytes and Schwann cells *in vitro* (12, 13). Seemingly, in contrast to these observations, astrocytes are believed to have beneficial effects on oligodendrocyte development by providing the right matrix for the migration of oligodendrocyte progenitor cells (14) and by supporting the differentiation of oligodendrocyte progenitors to 04 +ve immature oligodendrocytes (15). However, it has also been shown that astrocytes facilitate the differentiation of oligodendrocyte progenitors only to the 04 +ve immature stage, but not to O1 +ve more mature oligodendrocyte stage (15). The picture that emerges from these findings is that of astrocytes regulating oligodendrocyte differentiation, by promoting the migration and proliferation of oligodendrocyte progenitor cells and subsequent differentiation up to the formation of immature oligodendrocytes and in later stages of inhibiting the formation of mature oligodendrocytes capable of synthesizing myelin in the absence of other regulatory influences..

Astrocyte-secreted factors did not inhibit MBP mRNA translocation suggesting that the inhibition of the mRNA movement was not mediated by factors secreted by astrocytes. The "astromatrix" also did not inhibit MBP mRNA translocation, but the paraformaldehyde-fixed astrocytes did inhibit the mRNA movement, which suggested to us that intact, not necessarily live, astrocytes could inhibit MBP mRNA translocation. These results further suggest that adhesion molecules on astrocytes might be involved in this inhibition process.

The mechanism by which astrocytes inhibit MBP mRNA translocation in oligodendrocytes is unclear. We have shown that cytoskeletal elements are involved in the translocation of MBP mRNA and that microtubules, and not microfilaments, mediate the movement of MBP mRNA into oligodendrocyte processes (16). Thus, it is possible that the adhesion molecules on astrocytes could interact with adhesion molecules on oligodendrocytes that may interact with a membrane-cytoskeleton-linker in oligodendrocytes and could cause an inhibition of MBP mRNA translocation. Consistent with this suggestion, astrocytes have been shown to express several recognition molecules such as N-CAM, G-CAM, CD44, laminin, N-cadherin and integrin receptors and N-CAM has been shown to bind to spectrin, a membrane-cytoskeleton-linker (for references, see 7).

Under normal circumstances, MBP mRNA is translocated into oligodendrocyte processes *in vivo* (2, 5), even though presumably, substantial interaction of oligodendrocytes with astrocytes is continuously occurring. A possible explanation is that neurons may relieve the inhibition due to astrocytes *in vivo* and since neurons are essentially absent in the primary cultures around the time the mRNA translocation occcurs and entirely absent in the reconstituted system, the astrocytic block on the mRNA movement is not removed. Consistent with this theory, several reports that demonstrate that neurons regulate gene expression in oligodendrocytes and in astrocytes (17-20) are available. Our observations that factors secreted by neurons could alleviate the astrocytic inhibition further support the theory. We have also found that PDGF-AB and/ -BB are capable of mimicking the effects of NCM. Reports are available in the literature that neurons synthesize both A and B chains of PDGF whereas astrocytes synthesize only A chains of PDGF (21, 22)suggesting that these factors could be the factors in NCM that are responsible for the alleviation of the astrocytic inhibition of MBP mRNA translocation. The mechanism of alleviation of astrocytic inhibition of MBP mRNA translocation of NCM and PDGF remains to be understood.

It is also possible that astrocytes inhibit maturation of oligodendrocytes and thereby inhibit MBP mRNA translocation. In support of this theory, we have observed that fewer oligodendrocytes in mixed or reconstituted astrocyte-oligodendrocyte cultures have elaborate processes and sheets compared to those grown in the absence of astrocytes. However, the translocation of MBP mRNA in these cells is inhibited. Since the presence of GC, MBP mRNA and protein in oligodendrocytes is generally believed to represent a "mature" oligodendrocyte, the absence of MBP mRNA translocation in these "mature" oligodendrocytes in the presence of astrocytes, may define a unique stage in the differentiation of oligodendrocytes. This is true in the case of *jimpy* oligodendrocytes also (23, 24) where the translocation of MBP mRNA occurs to a significantly lesser extent and the oligodendrocytes are considered to be less mature.

We can not rule out the possibility that the astrocytes in culture behave like "reactive" astrocytes and the inhibition we have observed represents a pathological state where astrogliosis has occurred and neurons are damaged. Based on our findings, we would suggest that a complex interaction between astrocytes, oligodendrocytes and neurons influences oligodendrocyte differentiation and myelination

References

1. Wilhelm, J. E. and Vale, R. D. (1993) RNA on the move: The mRNA localization pathway. J. Cell Biol. 123, 269-274.
2. Trapp, B. D., Moench, T.,Pulley, M., Barbosa, E., Tennekoon, G. and Griffin, J. (1987) Spatial segregation of mRNA encoding myelin-specific proteins. Proc. Natl. Acad. Sci. USA 84, 7773-7777.
3. Ghandour, M. S. and Skoff, R. P. (1991) Double- lebeling in situ hybridization analysis of mRNAs for carbonic anhydrase II and myelin basic protein: Expression in developing cultured glial cells. Glia 4, 1-10.
4. Rataboul, P., Faucon Biguet, N., Vernier, P., De Vitry, F., Boularand, S., Privat, A. and Mallet, J. (1988) Identification of a human glial fibrillary acidic protein cDNA: a tool for the molecular analysis of reactive gliosis in the mammalian central nervous system. J. Neurosci. Res. 20, 165-175.
5. Verity, A. N. and Campagnoni, A. T. (1988) Myelination and its underlying mechanisms. Regional expression of myelin protein genes in the developing mouse brain: In situ hybridization studies. J. Neurosci.Res. 21, 238-248.
6. Amur-Umarjee, S. G., Hall, L. and Campagnoni, A. T. (1990) Spatial distribution of mRNAs for myelin proteins in primary cultures of mouse brain. Dev. Neurosci. 12, 263-272.
7. Amur-Umarjee, S., Phan, T. and Campagnoni, A. T. (1993) Myelin basic protein mRNA translocation in oligodendrocytes is inhibited by astrocytes in vitro. J. Neurosci.Res. 36, 99-110.
8. Amur-Umarjee, S. G., Dasu, R. and Campagnoni, A. T. (1990) Temporal expression of myelin-specific components in neonatal mouse brain cultures: Evidence that 2, 3-cyclic nucleotide 3-phosphodiesterase appears prior to galactocerebroside. Dev. Neurosci. 12, 251-262.
9. Schlinger, B. A., Amur-Umarjee, S., Campagnoni, A. T. and Arnold, A. P. A. T. (1995) 5β-reductase and other androgen-metabolizing enzymes in primary cultures of developing zebra finch telencephalon. J.

Neuroendocrinol. 7, 187-192.
10. Rome, L. H., Bullock, P. N., Chiappelli, F., Cardwell, M., Adinolfi, A. M. and Swanson, D. (1986) Synthesis of a myelin-like membrane by oligodendrocytes in culture. J. Neurosci. Res. 15, 49-65.
11. Amur-Umarjee, S. and Campagnoni, A. T. (1995) The inhibition of myelin basic protein (MBP) mRNA translocation in oligodendrocytes by astrocytes is reversed by neuron conditioned medium and by platelet-derived growth factor (PDGF) *in vitro*. J. Neurochem. 64, S46D.
12. Rosen, C.L., Bunge, R. P., Ard, M. D. and Wood, P. M. (1989) Type 1 astrocytes inhibit myelination by adult rat oligodendrocytes in vitro. J. Neurosci. 9, 3371-3379.
13. Guenard, V., Gwynn, L. A. and Wood, P. M. Astrocytes inhibit schwann cell proliferation and myelination of dorsal root ganglion neurons in vitro. J. Neurosci. 14, 2980-2992.
14. Aloisi, F., Agresti, C., D'Urso, D. and Levi, G. (1988) Differentiation of bipotential glial precursors into oligodendrocytes is promoted by interaction with type-1 astrocytes in cerebellar cultures. Proc. Natl. Acad. Sci. USA 85, 6167-6171.
15. Keilhauer, G., Meier, D. H., Kuhlmann-Krieg, S., Nieke, J. and Schachner, M. (1985) Astrocytes support incomplete differentiation of an oligodendrocyte precursor cell. EMBO J. 4, 2499-2504.
16. Amur-Umarjee, S. and Campagnoni, A. T. (1993) Myelin basic protein mRNA translocation in oligodendrocytes involves microtubules and is inhibited by astrocytes *in vitro*. In S. Salvati (ed.), A multidisciplinary approach to myelin diseases.II, Plenum Press, New York.
17. Macklin, W. B., Weill, C. L. and Deininger, P. L. (1986) Expression of myelin proteolipid and basic protein mRNAs in cultured cells. J. Neurosci. Res. 16, 203-217.
18. Hardy, R. and Reynolds, R. (1993) Neuron-oligodendroglial interactions during central nervous system development. 36, 121-126.
19. Dutly, F. and Schwab, M. E. (1991) Neurons and astrocytes influence the development of purified 0-2A progenitor cells. Glia 4, 559-571.
20. Steward, O., Torre, E. R., Tomasulo, R. and Lothman, E. (1991) Neuronal activity up-regulates astroglial gene expression. Proc. Natl. Acad. Sci. USA 88, 6819-6823.
21. Yeh, H-J, Ruit, K. G., Wang, Y-X, Parks, W. C., Snider, W. D. and Deuel, T. F. (1991) PDGF A-chain gene is expressed by mammalian neurons during development and in maturity. Cell 64, 209-216.
22. Sasahara, M., Fries, J. W. U., Raines, E. W., Gown, A. M., Westrum, L. E., Frosch, M. P., Bonthron, D. T., Ross, R. and Collins, T. PDGF B-chain in neurons of the central nervous system, posterior pituitary, and in a transgenic model. Cell 64, 217-227.
23. Verity, A. N., Levine, M. S. and Campagnoni, A. T. (1990) Gene expression in the *jimpy* mutant: Evidence for fewer oligodendrocytes expressing myelin protein genes and impaired translocation of myelin basic

protein mRNA. Dev. Neurosci. 12, 359-372.

24. Shiota, C., Ikenaka, K. and Mikoshiba, K. (1991) Developmental expression of myelin protein genes in dysmyelinating mutant mice: Analysis by nuclear run-off transcription assay, in situ hybridization, and immunohistochemistry. J. Neurochem. 56, 818-826.

Molecular cloning of new basic helix-loop-helix proteins from the mammalian central nervous system

Moritz Rossner, Angelika Bartholomä, Markus Schwab, Klaus-Armin Nave

Zentrum für Molekulare Biologie (ZMBH), Universität Heidelberg,
Im Neuenheimer Feld 282, D-69120 Heidelberg

Introduction

The development of neuronal and glial cells in the mammalian central nervous system requires the coordinate expression of cell type-specific subsets of genes. When development has been completed, fully differentiated cells maintain their characteristic properties by continously integrating cell-intrinsic and -extrinsic signals which converge, ultimately, at the level of gene transcription. This integration is executed by the combined activity of cell type-specific and ubiquitous transcription factors, but the underlying combinatorial code is little understood. In such regulatory systems, the repression of some genes is as important as specific gene activation. Autoregulative circuits, which include both transcriptional activators and repressors, are thought to stabilize the phenotype of each cell. To better understand the molecular mechanisms which underlie normal development and differentiation of the nervous system, it is essential to identify the key players in gene transcription. This also applies to the understanding of regenerative processes, which are thought to recapitulate, at least in part, the steps of normal development.

A growing number of transcription factors, which belong to the *helix-loop-helix* (*HLH*) superfamily of DNA-binding proteins, have been shown to determine cell fate decisions and differentiation in *Drosophila* embryonic development and, outside the central nervous system, in mammals. This group of transcription factors is structurally defined by the *HLH* domain, a 60-70 amino acid long region of the protein which mediates protein dimerization and is frequently located adjacent to a *b*asic domain responsible for DNA binding (thus referred to as bHLH proteins). In general, hetero- or homodimers of bHLH proteins are capable of DNA-binding and transcriptional activation (for review see Benezra and Weintraub, 1991; Jan and Jan, 1993). One of the best analyzed examples of a bHLH protein is *myoD*, prototype of the myogenic subfamily which controls determination and differentiation of myoblasts from mesodermal stem cells (Weintraub et al., 1991; Lassar and Münsterberg, 1994). A similar role is played by E2A gene products which control B-cell development and by SCL (stem cell leukemia), a protein required for erythroid differentiation in the hematopoetic system (Zhuang et al., 1994; Shivdasani et al., 1995). In Drosophila, proteins of the *achaete-scute* and the *enhancer-of-split* complex regulate the earliest

Molecular Signaling and Regulation in Glial Cells
A Key to Remyelination and Functional Repair

development of the nervous system and their mammalian homologues appear to be involved in early neurogenesis (Campos-Ortega, 1994; Guillemot et al., 1993).
The superfamily of bHLH proteins has been further subdevided into different classes, based on structural and functional considerations. *Class A* proteins are functional transcriptional activators after heterodimerization. *Class B* proteins form homo- or heterodimers and share an additional a-helical domain, the leucine zipper (LZ), adjacent to the bHLH region. Hence, they are also known as bHLH-LZ proteins (Beckmann and Kadesch, 1990). Only a few proteins from *Class B* are involved in cell type-specific gene expression. Members of this subfamiliy are generally involved in cell cycle control or general housekeeping functions; some of them also interact with other LZ-bearing transcription factors (Prendergast and Ziff, 1992). Finally, *Class C* proteins are transcriptional repressors, which control developmental processes by interacting with the same promoters activated by *Class A* bHLH proteins (Ohsako, S. et al., 1994). A different kind of repressor molecules are dominant-negative *(dn) HLH* proteins which lack a basic region. Proteins such as Id or emc (in Drosophila) form heterodimers with *Class A* proteins that are then unable to bind to DNA (Benezra et al, 1990).
Because of the involvement of *Class A* and *Class C* bHLH proteins in cellular differentiation processes outside the central nervous sytem, we have directed a search for new members of these subfamilies expressed in the mammalian brain.

Experimental Strategy

Whereas the sequence identity of different bHLH proteins in one subfamily is high, even between distantly related species, the overall degree of bHLH sequence identity between unrelated subfamilies is less than 50% (Benezra et al., 1990; Murre and Baltimore, 1992). To identify cDNAs encoding new bHLH proteins of the mammalian nervous system, we have optimized a "low-stringency" reverse transcriptase-polymerase chain reaction (RT-PCR) technique. Each primary amplification product was accurately size-selected, reamplified, size-selected again, and subcloned for DNA sequence analysis. The approach is based on the following assumptions:

1. Fully degenerate PCR primers (up to 8000-fold degenerate) can be designed which correspond in sequence to clusters of the most conserved amino acids in the basic region and helix-2 of known bHLH proteins. Subfamilies were chosen based on their proven involvement in cell fate determation and differentiation (*achaete-scute* /MASH; *enhancer-of-split/ hairy* /HES; SCL,NSCL; MyoD, myogenin, myf-5).
2. We assumed that total cDNA prepared from adult rat brain is a starting point for this RT-PCR approach, because bHLH proteins which are involved in terminal neuronal and glial differentiation should be expressed throughout life.
3. When PCR amplifications are performed under conditions that allow mismatches between oligonucleotide primers and their cDNA targets, new bHLH subfamilies should be also identifiable. We specifically utilized the partial enzymatic activity of Taq Polymerase at a low temperature (37°- 40°C) and

allowed long annealing times (up to 10 min per cycle) in order to obtain temperature-stable primer extension products.

Neuronal helix-loop-helix protein -1 (NEX-1)

The first successfully identified cDNA, encoding a new bHLH protein (later termed NEX-1), hybridized on Northern blots to two mRNAs of approximately 2.6 kb and 2.2 kb in size. These transcripts were highly abundant in adult rodent brain, but undetectable in liver and kidney. The complete cDNA sequence encoded a polypeptide of 337 amino acids in length and has a relative molecular mass of 37,000. Structural features of NEX-1 include, in addition to the bHLH domain, a glutamic acid-rich domain as a possible transcriptional activator (Ptashne, 1988), a proline-rich carboxyl terminus, and numerous possible sites for protein phosphorylation (Bartholomä and Nave, 1994). A data base search revealed no amino acid sequence homology to known proteins except for the 57 residue bHLH region itself (Figure 1a). Recently two proteins highly homologoues to NEX-1 were identified by Weintraub and colleagues (Lee et al., 1995) and Akazawa et al. (1995), demonstrating that NEX-1 defines a novel subfamily of neural bHLH proteins.

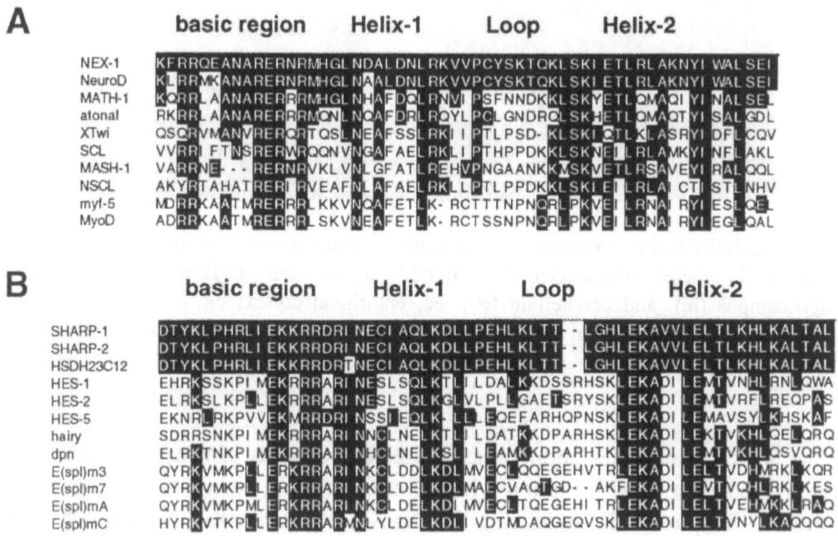

Figure 1: Sequence alignment of the conserved basic helix-loop-helix region (bHLH domain) **A,** NEX-1 defines a new subfamily of neuronally expressed bHLH proteins which also includes NeuroD/Beta2 and MATH-1. **B,** SHARP-1 and -2 are related to Drosophila *enhancer-of-split* (e-spl) and *hairy* proteins.

Transcription of the NEX-1 gene is absolutely brain-specific. On Northern blots, NEX-1 mRNA was not detectable before embryonic day 14. The highest steady-state level of NEX-1 mRNA was measured in the first postnatal week, after which transcripts decreased about four-fold in abundance but remained at a stable level into adulthood. By in situ hybridization at embryonic day 17, NEX-1 mRNA was most abundant in the dorsal telencephalon, including the developing neocortex, hippocampus anlage, and olfactory bulb (Bartholomä and Nave, 1994). NEX-1 transcripts were only detected in outwardly migrating cells and cells positioned in the cortical plate, presumablly postmitotic neurons. Remarkably, NEX-1 was strongly expressed in subsets of mature neurons of the neocortex, the entorhinal cortex, the amygdala, cerebellum and hippocampus (Figure 2).

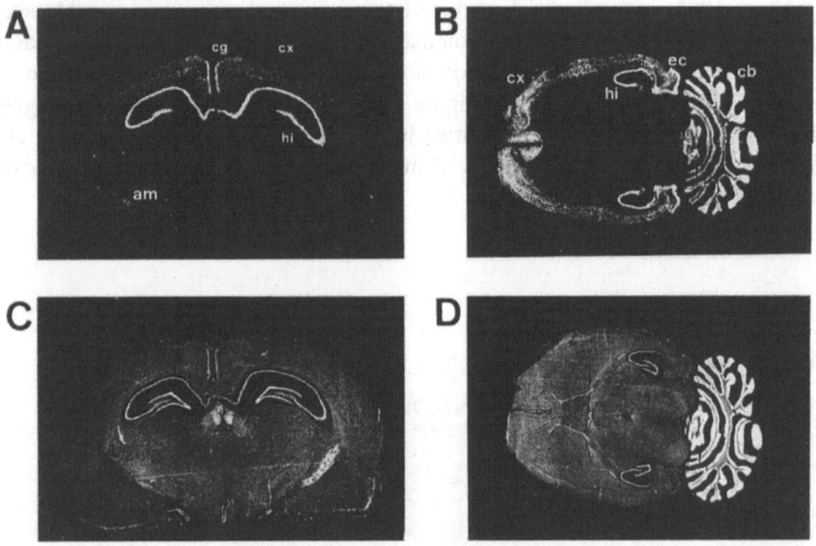

Figure 2: NEX-1 gene expression in the adult rat brain revealed by an situ hybridization with 35-S labeled oligonucleotide. Highest signals are obtained in cortex (cx), hippocampus (hi), and cerebellum (cb). ec, entorhinal cortex; cg, cingulate gyrus. **A,** frontal section; **B,** horizontal section; **C,** and **D,** thionin blue counter-staining.

The NEX-1 gene spans approximately 3 kb and consists of two exons which are interrupted by a 1200 bp intron with the entire protein coding region located on exon 2. To pursue the possibility that NEX-1 stimulates the transcription of its own gene (autoactivation is a critical function of myogenic bHLH proteins), we fused the human CMV promoter to the NEX-1 cDNA for cotransfection experiments. A NEX-1 target gene was generated by linking its own promoter to the fire fly luciferase gene. Overexpression of NEX-1 in heterologous cells (e.g.

PC12 cells) consistantly stimulated the basal activity of the NEX-1 promoter, an effect even more pronounced when E12, a heterodimerization partner, was coexpressed. The NEX-1 promoter contains several E-box sequences (5'-CANNTG-3') as potential target sites, one of which was identified as a binding site for NEX-1/E12 bHLH protein dimers (Bartholomä and Nave, unpublished results). The target genes for NEX-1 are unknown, but the gene for growth-associated protein GAP-43 is a good candidate, based on an overlapping expression pattern in the CNS and its transactivation by NEX-1/E12 in heterologous cells (Bartholomä et al., 1995). The function of NEX-1 in neuronal differentiation and in the adult brain is still unknown. Ongoing experiments focuss on the phenotype of transgenic mice in which the NEX-1 gene has been disrupted through homologous recombination in embryonic stem cells (Schwab and Nave, unpublished results).

Enhancer-of *S*plit/*Ha*iry *R*elated *P*rotein-1 and -2 (Sharp-1 and -2)

In a second screen for bHLH proteins with primers that correspond to *Class C* proteins, we have amplified a new bHLH transcript from the adult rat, termed SHARP-1 (for Enhancer-of *S*plit/*Ha*iry *R*elated *P*rotein-1). A low-stringency screen of a cDNA library made from CG4 glial cells with the SHARP-1 PCR probe led to the identification of a closely related transcript, termed SHARP-2. Moreover, a data base search with the SHARP-1 and 2 sequences revealed a previously unrecognized human expressed sequence tag (EST) with considerable homology within the bHLH region.

Cloning and sequencing of full-length cDNAs for SHARP-1 revealed an open reading frame of 441 amino acids (predicted molecular weight 49 kD) with the bHLH region located at the N-terminus (Rossner and Nave, unpublished results). An alanine-rich stretch at the SHARP-1 C-terminus might confer a transcriptional repressor function (Licht et al., 1990) and is in line with presumed negative regulatory role of enhancer of split and hairy proteins in Drosophila development.

However, a comparison of the deduced SHARP-1 amino acid sequence with the already known *Class C* proteins revealed that SHARP-1 may be the prototype of a new subgroup of bHLH proteins (see Figure 1b). Both, SHARP-1 and -2 contain the so called "orange"-domain (Dawson et al., 1995) but lack the C-terminal "WRPW"-domain that is common to all other *Class C* bHLH proteins (Paroush et al., 1994). Both regions of homology, located outside the bHLH domain, have been implicated in the activity of *Class C* proteins as transcriptional repressors (Dawson et al., 1995).

By Northern blot analysis, the SHARP-1 mRNA is approximately 4.5 kb in size, with highest expression in the adult rat brain and moderate expression in lung, muscle and heart. No expression was detected in liver, kidney, gut, uterus and testis. Different from all other bHLH proteins in the mammalian nervous system, expression of SHARP-1 in the rat CNS increases continually throughout prenatal and postnatal development until it peaks in the adult brain. In situ hybridizations of sections of adult rat brains reveal expression of the SHARP-1 gene in defined brain regions. Transcription of the gene is strongest in the cerebellum, the dentate gyrus, and the CA1 field of the hippocampus (Rossner and Nave, unpublished results). In cultured PC12 cells, transcription of the SHARP-1 gene is rapidly induced following application of NGF. Thus, although the function of SHARP-1 is presently unknown, PC12 cells may provide a convenient system to study its role in neuronal differentiation.

Discussion

In an attempt to identify regulatory genes of neuronal and glial differentiation, we have cloned several new basic helix-loop-helix (bHLH) proteins from the adult rodent brain. These proteins share significant structural homology to a group of transcription factors that have properties of "master" regulators outside the nervous system (Weintraub et al., 1991): upon ectopic overexpression, these bHLH proteins can alter the phenotype of some (but not all) host cells. Best studied in this respect are the cell-specific myogenic determination factors such as myoD; a similar potential have bHLH proteins in Drosophila which determine cell fate in embryonic development. Although some bHLH proteins have previously been associated with early events of neurogenesis (Lo et al., 1991; Guillemot et al., 1993; Akazawa et al., 1992; Duncan et al., 1992; Bernard et al., 1992; Feder et al., 1993; Begley et al., 1992), we conclude from our studies and those of others (Bartholomä and Nave, 1994; Lee et al., 1995; Akazawa et al.,1995; Rossner and Nave, unpublished) that bHLH proteins are involved in the entire spectrum of neuronal differentiation and, most likely, in specific functions of the mature nervous system.

NEX-1 defines a new subfamily of neuronally expressed mammalian bHLH proteins (Bartholomä and Nave, 1994). A protein termed MATH-2 (Shimizu et al., 1995) is identical to NEX-1. Additional members of this family include neuroD (Lee et al., 1995) and MATH-1 (Akazawa et al., 1995), all of which show a distant relationship to the Drosophila *atonal* gene product. By Northern blot analyses and in situ hybridizations, the NEX-1 gene is specifically expressed in the central nervous system, starting with the generation of neurons at the end of the second week of gestation. This late induction suggests that NEX-1 serves a function downstream from the events that determine neuronal cell fate in the central nervous system. The developmental time course and unambiguous identification of its mRNA in single hippocampal pyramidal cells of the CA4 field reveals NEX-1 as a neuronal protein. Maximal expression in the rat brain occurs in the first postnatal week, closely following the time-course of synaptogenesis in the cerebral cortex. The overall pattern of NEX-1 transcripts in the adult rat brain suggest that a major function of this protein is shared by a diverse group of fully committed CNS neurons which are in the stage of terminal differentiation or fully mature.

Possible target genes of NEX-1 are genes required for neuronal differentiation steps known to peak in the first postnatal week, e.g. dendritic arborization and synaptogenesis, taking place soon after neurons have positioned themselves in the cortical plate. NEX-1 gene is then sustained in several adult brain regions, such as the hippocampus, the amygdala, entorhinal cortex, cerebellum, and neocortex. These areas have been repeatedly associated with long-term functional changes of synaptic connections and synaptic strength as well as learning and memory formation. NEX-1 is a transcription factor which heterodimerizes with E12 proteins and which is tied into positive autoregulation. Positive autoregulatory loops may serve as an "epigenetic memory" of past cell fate decisions and is likewise a feature of the myogenic determination factors (Thayer et al., 1989).

Whereas expression of the NEX-1 gene is specific to the central nervous system, both SHARP-1 and SHARP-2 show developmentally regulated expression patterns that are not restricted to the nervous system but show regional specificity within the CNS. The highest expression level of SHARP-1 transcripts are detected in the cerebellum, in the dentate gyrus and in the CA1 region of the hippocampus formation. SHARP -2 transcripts were found mainly in the neocortex, in the entorhinal cortex, and in the CA1 region of the hippocampus. Both proteins are closely related and members of the *Class C* of bHLH proteins. These serve as transcriptional repressors in Drosophila by genetically interacting with *Class A* bHLH proteins (Jan and Jan, 1993). By suppressing specific genes in different regions of the vertebrate central nervous system, SHARP-1 and -2 may contribute to define specific properties of a subset of neuronal cell types.

In summary, a reverse genetic approach has been successfully used to clone several new genes for neuronal transcription factors that belong to the superfamily of bHLH proteins. These newly identified proteins are developmentally and tissue-specifically under tight control in the mammalian central nervous system. We speculate that they provide a missing link in coupling general and early transcriptional control mechanisms to neuron-specific gene expression and neuronal functions.

References

Akazawa, C., Ishibashi, M., Shimizu, C., Nakanishi, S., Kageyama, R. (1995) A mammalian helix-loop-helix factor structurally related to the product of Drosophila proneural gene atonal is a positive transcriptional regulator expressed in the developing nervous system. J. Biol. Chem. 270, 8730-8738.

Bartholomä, A. and Nave, K.-A. (1994) NEX-1: a novel brain-specific helix-loop-helix protein with autoregulation and sustained expression in mature cortical neurons. Mech. of Dev. 48, 217-228.

Bartholomä, A., Schwab, M., Gass, P., Nave, K.-A. (1995) The brain-specific Helix-Loop-Helix protein NEX-1: A neuronal activator of GAP43? Soc. Neurosci. Abstracts. 21, 1290.

Beckmann, H.L. and Kadesch, T. (1990) TFE3: a helix-loop-helix protein that activates transcription through the immunoglobulin enhancer muE3 motif. Genes & Dev. 5, 1057-1066.

Begley, C. G., Lipkowitz, S., Gobel, V., Mahon, K. A., Bertness, V., Green, A. R., Gough, N. M. and Kirsch, I. R. (1992) Molecular characterization of NSCL, a gene encoding a helix-loop-helix protein expressed in the developing nervous system. Proc. Natl. Acad. Sci. USA 89, 38-42.

Benezra, R., Davis, R. L., Lockshon, D., Turner, D. and Weintraub, H. (1990) The protein Id: a negative regulator of helix-loop-helix DNA binding proteins. Cell 61, 49-59.

Bernard, O., Drago, J. and Sheng, H. (1992) L-myc and N-myc influence lineage determination in the central nervous system. Neuron 9, 1217-1224.

Campos-Ortega, J.A. (1994) Cellular interactions in the developing nervous system of Drosophila. Cell 77, 969-975.

Dawson, S.R., Turner, D.L., Weintaub, H., Parkhurst, S.M. (1995) Specificity for the Hairy/Enhancer of split basic Helix-Loop-Helix (bHLH) proteins maps outside the bHLH domain and suggests two seperable modes of transcriptional repression. Mol.Cell.Biol. 15, 6923-6931.

Duncan, M., DiCicco-Bloom, E. M., Xiang, X., Benezra, R. and Chada, K. (1992) The gene for the helix-loop-helix protein, Id, is specifically expressed in neural precursors. Dev. Biol. 154, 1-10.

Feder, J. N., Jan, L. Y. and Jan, Y. N. (1993) A rat gene with sequence homology to the Drosophila gene hairy is rapidly induced by growth factors known to influence neuronal differentiation. Mol. Cell. Biol. 13, 105-113.

Guillemot, F., Lo, L.-C., Johnson, J. E., Auerbach, A., Anderson, D. J. and Joyner, A. L. (1993) Mammalian achaete-scute homolog 1 is required for the early development of olfactory and autonomic neurons. Cell 75, 463-476.

Jan,Y.N. and Jan, L.Y. (1993) HLH proteins, fly neurogenesis, and vertebrate myogenesis. Cell 75, 827-830.

Lassar, A. and Munsterberg, A. (1994) Wiring diagrams: regulatory circuits and the control of skeletal myogenesis. Curr-Opin-Cell-Biol. 6(3): 432-42

Lee, J.E., Hollenberg, S.M., Snider, L., Turner, D.L., Lipnick, N. and Weintraub, H., (1995) Conversion of Xenopus ectoderm into neurons by NeuroD, a basic helix-loop-helix protein. Science 268, 836-844.

Licht, J.D., Grossel, M. J., Figge, J., Hansen, U.M. (1990) Drosophila Kruppel protein is a transcriptional repressor. Nature 346, 76-79.

Lo, L.-C., Johnson, J. E., Wuenschell, C. W., Saito, T. and Anderson, D. J. (1991) Mammalian achaete-scute homolog 1 is transiently expressed by spatially restricted subsets of early neuroepithelial and neural crest cells. Genes Dev. 5, 1524-1537.

Murre, C. and Baltimore, D. (1992) In Transcriptional Regulation. S. L. McKnight and K. R. Yamamoto, eds. (New York: Cold Spring Harbor Laboratory) pp. 861-879.

Ohsako, O., Hyer, J., Panganiban, G., Oliver, I. and Caudy, M., (1994) Hairy function as a DNA-binding helix-loop-helix repressor of Drosophila sensory organ formation. Genes & Dev. 8, 2743-2755.

Paroush, Z., Finley, R.L., Kidd, T., Wainwright, S.M., Ingham, P.W., Brent, R., Ish-Horowicz, D. (1994) Groucho is required for Drosophila neurogenesis, segmentation, and sex determination and interacts directly with hairy-related bHLH proteins. Cell 79, 805-815.

Prendergast, G.C. and Ziff, E.G. (1992) A new bind for Myc. Trends Genet. 8, 91-96.

Ptashne, M. (1988) How eukaryotic transcriptional activators work. Nature 335, 683-689.

Shimizu, C., Akazawa, C., Nakanishi, S. and Kageyama, R. (1995) MATH-2, a mammalian helix-loop-helix factor structurally related to the product of *Drosophila* proneural gene *atonal*, is specifically expressed in the nervous system. Eur.J.Biochem. 229, 239-248.

Shivdasani, R.A.; Mayer, E.L.; Orkin, S.H. (1995) Absence of blood formation in mice lacking the T-cell leukaemia oncoprotein tal-1/SCL. Nature 373, 432-434.

Thayer, M. J., Tapscott, S. J., Davis, R. L., Wright, W. E., Lassar, A. B. and Weintraub, H. (1989) Positive autoregulation of the myogenic determination gene MyoD1. Cell 58, 241-248.

Weintraub, H., Davis, R., Tapscott, S., Thayer, M., Krause, M., Benezra, R., Blackwell, T. K., Turner, D., Rupp, R. and Hollenberg, S. (1991) The myoD gene family: nodal point during specification of the muscle cell lineage. Science 251, 761-766.

Zhuang, Y., Soriano, P. and Weintraub , H. (1994) The helix-loop-helix gene E2A is required for B cell formation. Cell 79, 875-884.

Part 3 Neural Repair Mechanisms in the CNS

Cellular Responses to Injury

Microglial and astroglial activation patterns in neural graft rejection and neuronal and axonal degeneration.

Finsen B., Pedersen E.B., Lehrmann E., Jensen M.B., Aznar S. and Zimmer J.

Department of Anatomy and Cell Biology, Institute of Medical Biology, University of Odense,Winsløwparken 19, DK-5000 Odense C, Denmark

Abstract. There are many parallels in the reactions of brain astroglial and microglial cells to immune mediated intracerebral graft rejection and to non-immune ischemic and excitotoxic neuronal and axonal degeneration. There are, however, also significant differences, as some types of brain injury only induce a partial expression of the full repertoire of microglial activation markers. This suggests a tight, stimulus-dependent regulation of the microglial activation. Based on our work on the hippocampus, this review provides an overview of the microglial and astroglial activation and reaction patterns and potential functions of these cells in relation to neural graft rejection and neural degeneration and regeneration.

Key words. Brain, hippocampus, regeneration, ischemia, excitotoxin, macrophages, lymphocytes, perivascular cells, CD4 antigen, MHC antigens, TGF-ß1.

1 Introduction

The success of clinical restorative use of brain grafts depends on refinement of the surgical procedures, the procurement and quality of donor tissue, and the establishment of immunological stable conditions, with indefinite survival of maybe even xenografts. In basic experimental studies we have focused on the ability of mouse hippocampal xenografts to survive and integrate in the hippocampal region of newborn (70) and adult rats (10), and pointed to the role of donor and host microglial cells and host macrophages in the presentation of donor antigens to host T-cells (9,11). For immunosuppression we have studied both Cyclosporin A (10), and the additive effect of Azathioprine and steroids on xenograft survival and cell infiltration (46).

One concern for neural grafts in humans is of course also the mechanisms underlying the progressive neurodegenerative disorders, like Parkinson's and Huntington's diseases. Non-specific inflammatory mechanisms involving glial cells may contribute to the degenerative processes (52), and autopsi material from patients with Parkinson's and Alzheimer's diseases (35,69) displays the same types

Molecular Signaling and Regulation in Glial Cells
A Key to Remyelination and Functional Repair

of glial pathological changes as observed after experimental ischemic and excitotoxic or traumatic injuries in rats (8,12,23,24,25,31).

2 Neural graft rejection

2.1 Effector cells and effector mechanisms

The cellular infiltrate of neural grafts undergoing rejection mainly consists of macrophages and T-cells, while B-cells only are present in very low numbers. Both CD8+ and CD4+ cells are numerous (4,11,42). For these cells lymphocyte-depletion studies have shown, that CD4+ cells are essential for initiating the rejection of both xeno- and allogeneic CNS grafts (41,66). CD8+ T-cells on the other hand play an important role in the later stages of rejection, but are unable by themselves to initiate rejection (40). The role of T-cells as effector cells is further substantiated by the findings that administration of antibodies against the T-cell growth factor IL-2 significantly can enhance both xeno- and allograft survival (21,65). The population of CD4+ cells does, however, not only comprise T-cells, but also high numbers of macrophages and reactive microglial cells (Figs. 1 and 2). Among these at least the macrophages can exert effector functions by delayed type of hypersensitivity mechanisms, which has been implicated in the rejection of extracerebral non-neuronal xenografts (54). Other effector mechanisms such as killing by Natural Killer cells and donor-specific antibodies may also take place in the solid grafts.

2.2 Antigen presention and immunostimulation

All available data on intracerebral graft rejection support the concept that foreign antigens are presented to CD4+ host T-cells by MHC antigen class II+ host or donor antigen presenting cells (APCs) (11,21,41,65,66). This leads to cytokine secretion, attraction and proliferation of T-cells and macrophages, with induction of MHC antigens on host and donor brain cells. This facilitates further activation of immune effector cells, with graft rejection ensuing. Both astrocytes and microglial cells can express MHC antigen class II *in vitro*, but in functional lymphocyte-activation assays microglial cells are significantly more efficient than astrocytes, although less efficient than brain macrophages and thymic epithelial cells (13,57). The generally observed restriction of MHC antigen class II expression to microglial cells and macrophages *in vivo* (11,48) also points to these cells as the key candidates for antigen presentation.

2.2.1 Microglial precursor cells

The APCs triggering rejection could in principle be of both donor and host origin. In this context an interesting cell is the microglial precursor cell. This cell type is

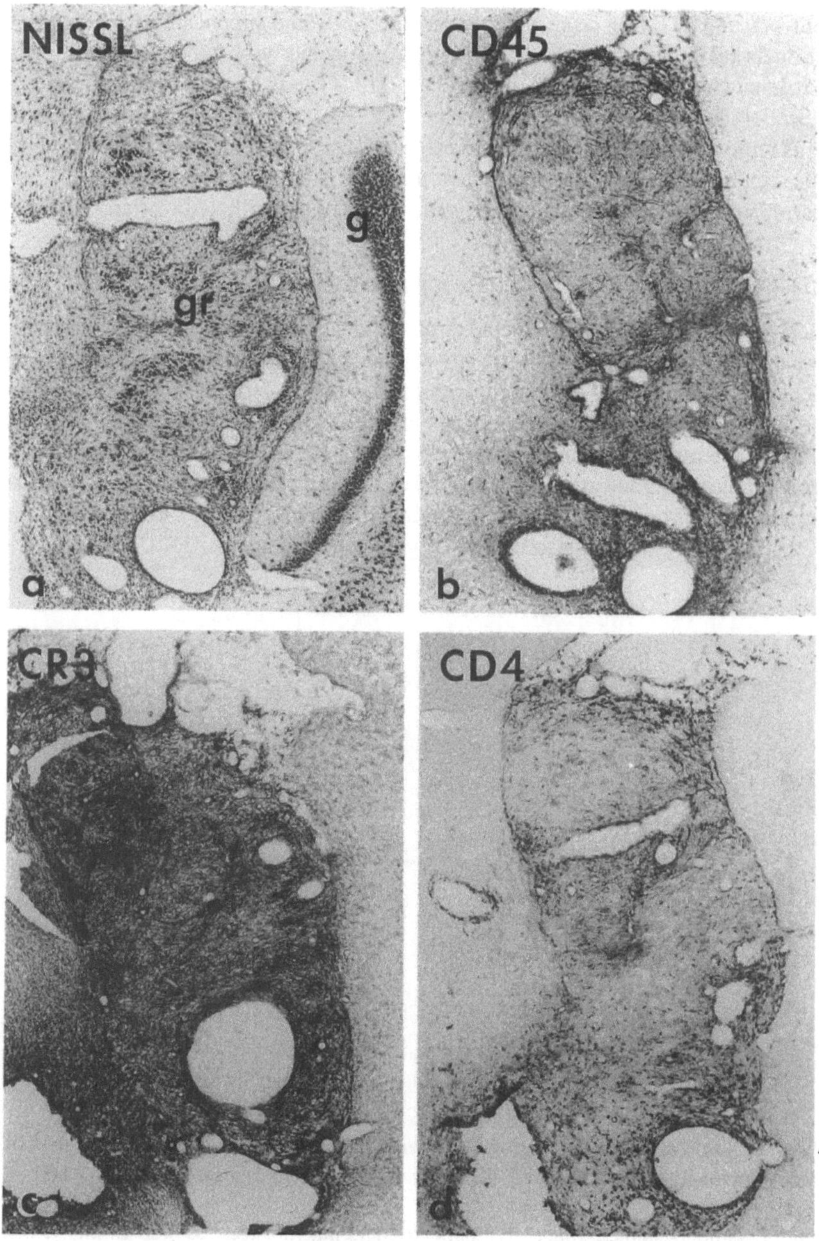

Fig. 1. Cryostate sections (20 μm) of 3 weeks old mouse hippocampal xenograft (gr) located in the subarachnoid space adjacent to the host hippocampus. (g, dentate granule cell layer). The Nissl staining (a) shows a partial preservation of the xenograft cell layers, while CD45 immunohistochemistry (b) displays large numbers of intensily

almost absent in embryonic day 15 old rat and mouse hippocampal donor tissue, but present in fairly large numbers in the CNS and hippocampus of newborn animals ((3) and unpublished observations). Together with other factors this may explain why very immature xenogeneic brain tissue is less prone to undergo rejection than more mature donor tissue (70). The microglial precursor cells might thus represent the MHC antigen class I+ donor cells, which Hickey (19) detected in the central lymphoid organs in his allograft studies. These cells might also participate in antigen presentation *in situ* to T-cells patrolling the brain.

2.2.2 Host microglial cells, macrophages and dendritic cells

Besides antigen presentation by the direct route, indirect antigen presentation could be operative by means of host microglial cells, infiltrating macrophages and dendritic cells, as well as macrophages and dendritic cells in the regional lymph nodes and spleen. Recent findings of cells with ultrastructural (30) and immunophenotypical characteristics (58) akin to dendritic cells within intracerebral grafts undergoing rejection have indicated, that circulating host dendritic cells are implicated in brain graft rejection. Similarly, data obtained by Lund and collegues (32,47,48) suggest, that also host microglial cells can act as APCs. Although the

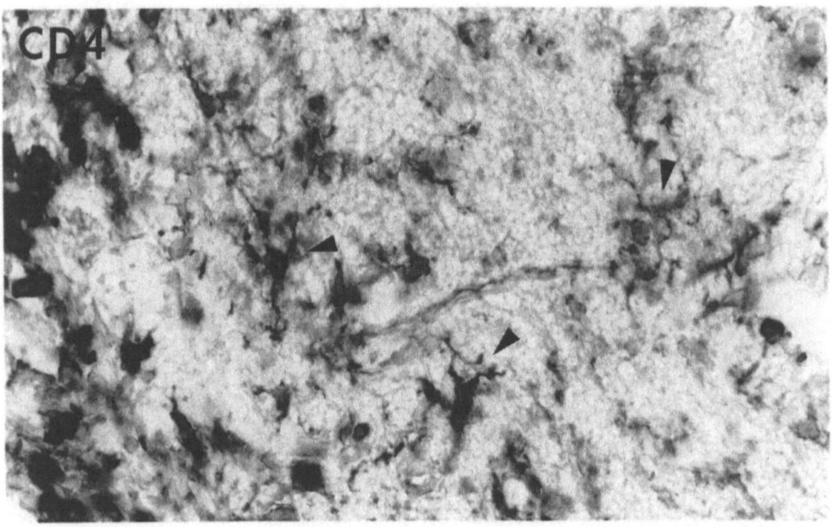

Fig. 2. High magnification of CD4+ xenograft-infiltrating reactive microglial-like cells (arrowheads), and rounded, distinctly stained T-cells and macrophages. x 600.

Fig. 1. cont.: - stained, round lymphocytes and macrophages, as well as many, more lightly stained, irregularly shaped microglial-like cells. The reaction pattern of microglial cells and macrophages is also visualized by CR3 (c), and CD4 (d) immunohistochemistry. The CD4 antigen induction includes a large number of microglial-like cells (see Fig. 2). x 45.

embryonic mouse retina is devoid of microglial precursor cells when grafted into the optic tectum of newborn rats, it becomes populated in an organotypic manner by host microglial cells within a few weeks (47). A large number of these retinal xenografts undergo spontaneous rejection even after several months of survival (32), just as they can be induced to undergo acute rejection (32,48). In both instances, the ensuing graft rejection is preceded by microglial expression of MHC antigen class I and II within the grafts, again suggesting a role of these cells in antigen presentation.

As for newborn retinal xenografts (47), we demonstrate here in this paper, that embryonic Kyoto-Wistar rat hippocampal tissue grafted into MHC antigen class I histoincompatible brains of adult rats of the PVG strain becomes populated with host microglial cells (Fig. 3). The same process is likely to occur in mouse hippocampal xenografts after grafting to the brains of adult rats. After xenografting the microenvironment of the T-cell infiltrated grafts is, however, so, that immigrating microglial cells and macrophages differentiate into the macrophage phenotype rather than into ramified microglial cells (cp. Figs. 1 and 2 with Fig. 3).

2.2.3 Perivascular cells

Other bone marrow derived cells of interest in CNS graft rejection are the perivascular cells. These cells constitutively express ED2 and MHC antigen class I and to a lesser extent MHC antigen class II (16), and are capable of antigen presentation to T-cells (20). When we recently studied the occurrence of host perivascular cells within grafts undergoing rejection, we were unable, however, to find $ED2^+$ cells in the xenografts. This indicates to us, that the role of perivascular cells in effecting graft rejection is miminal, but it does not exclude that donor or host perivascular cells may be involved in triggering the rejection reaction.

3 Anterograde axonal degeneration and neuronal death

3.1 Experimental models of hippocampal lesioning

Studies of selective neuronal and axonal degeneration is greatly facilitated by the highly ordered and laminar, cellular and connective organization of the hippocampus. Transient global cerebral ischemia produced by the four-vessel occlusion method induces a rapid degeneration of dentate hilar neurons followed by a delayed degeneration of hippocampal CA1 pyramidal cells (25,50). In contrast, intracerebroventricular injections of the glutamate agonist kainic acid (KA) preferentially affects the hilar neurons and CA3 pyramidal cells (39). More distinct axonal degeneration can be produced by lesioning of the perforant path (PP) projection, which induces a distantly located and distinctly laminar axonal and terminal degeneration in the outer parts of the dentate molecular layer (24,33). Axonal degeneration can also be induced by colchicine-injection into the fascia dentata leading to anterograde degeneration of the hippocampal mossy fibers ((12) and references therein), or KA-induced degeneration of the CA3 pyramidal cells

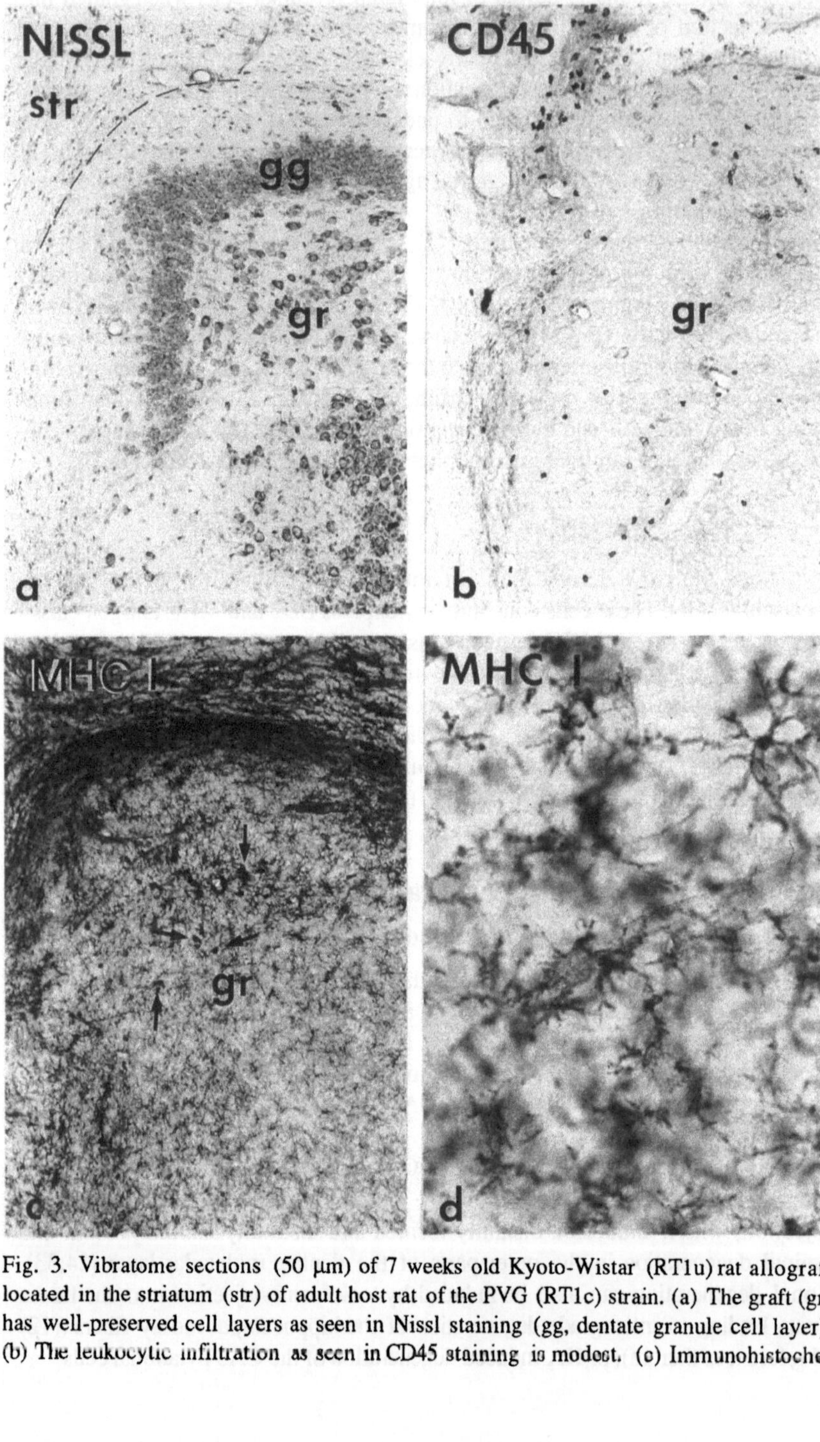

Fig. 3. Vibratome sections (50 µm) of 7 weeks old Kyoto-Wistar (RT1u) rat allograft located in the striatum (str) of adult host rat of the PVG (RT1c) strain. (a) The graft (gr) has well-preserved cell layers as seen in Nissl staining (gg, dentate granule cell layer). (b) The leukocytic infiltration as seen in CD45 staining is modest. (c) Immunohistoche-

leading to anterograde degeneration of Schaffer collaterals and commissural fibers to stratum radiatum and oriens of CA1 (39). Using this panel of neuronal and axonal lesion models, we have performed a detailed and comparative analysis of the micro- and astroglial activation and reaction patterns to different and graded types of degeneration *in vivo* (8,12,24,26,31). An extension of these observations to the level of cytokine mRNA (31) and protein expression will provide valuable, general clues to the mechanisms of glial activation, and hence provide new insight in the pathophysiological significance of astro- and microglial cell reactions.

3.2 Micro- and astroglial activation and reaction patterns

3.2.1 Microglial cells

The studies of the microglial cell reactions (8,12,23,24,25), have demonstrated, that these cells display a stereotyped and graded activation pattern. The initial signs of activation consist of partial retraction and coarsening of the cellular processes, increased immunohistochemical and histochemical staining for the constitutively expressed C3 receptor and the plasmalemmal enzyme nucleoside diphosphatase (NDPase)(25), induction of MHC antigen class I (8) - and as demonstrated recently after ischemia - induction of mRNA encoding the anti-inflammatory, multifunctional cytokine TGF-ß1 (31). In the absence of additional stimuli microglial cells may revert from this initial stage of activation to their resting state within a couple of days. Upon further activation a major proportion of the microglial cells begins to express CD4 and CD45 antigen (23), while maintaining the high levels of the abovementioned activation markers, including TGF-ß1 mRNA (31). When maximally activated microglial cells may also express MHC antigen class II (8,12). A report of a similar progressive induction of immunomolecules has been presented by Streit and collaborators (38). Along with the graded expression of activation markers, microglial cells are induced to proliferate and migrate at a very early stage, just as they may become phagocytotic in the presence of degenerating cells or abundant axonal debris (5,11,23,61). The actual mode and the extent of morphological transformation is determined by the cells and anatomical structures undergoing degeneration, emphasizing once again the plastic nature of the microglial cell. Weeks to months after the initial injury, when all debris has been removed, the distribution and morphology of microglial cells have adapted to the rearranged histoarchitecture, and surface antigen and gene expression have returned to their resting states (8,26). The normalization of cell numbers implies that cells at some stage must be eliminated, which could be brought about either by necrosis, apoptosis or by a combination hereof.

Fig. 3. cont.: - mistry with host MHC antigen class I specific antibody (OX27, Serotec, UK) shows that the graft is infiltrated by lightly activated host-derived microglial cells. These are shown in higher magnification in (d). The distintcly stained cells marked by arrows in (c), probably correspond to the CD45+ leukocytes in (b). x 90 (a-c) and x 600 (d).

3.2.2 Astroglial cells

Characterization of the astroglial reactions to injury has been complicated by the more limited range of markers specific for normal and reactive astrocytes. The classical marker proteins are still the cytoskeletal intermediary filament proteins glial fibrillary acidic protein (GFAP), abundant both in normal and reactive astrocytes, and vimentin (VIM), which is induced in reactive astrocytes (24,25), but also present in smaller amounts in macrophages and reactive microglial cells (69). Signs of astroglial activation classically includes hypertrophy, increased GFAP immunoreactivity, induction of VIM immunoreactivity, and proliferation (7,25,55). Other characteristics of the activated astrocytes include an altered expression of enzymes involved in the neurotransmitter metabolism, as exemplified by glutamine synthetase (2), reduced expression of glutamate transporter proteins (J. Storm-Mathisen, personal communication), increased production of extracellular matrix proteins (45), and expression of various cytokines and cytokine receptors (discussed later). Astrocytes are usually described to react later to pertubations and injury than microglial cells (7,24). However, by being linked together by gap-junctions in the brain, and having their cellular processes next to synaptic specializations, the astrocytes must inevitably be influenced by ionic, transmitter and metabolic changes in the CNS just as early as the microglial cells. The fact that their mode of reaction is different probably reflects their prime function in maintaining the homeostasis of the microenvironment, securing normal neuronal function (18). Emphasis should therefore be placed on the observation, that early blood-brain barrier changes and the early signs of activation of both astroglial and microglial cells all were detected at 24 hours (or even earlier) after ischemia (8,14,25,37,55). We believe, that the advent of more sensitive functional astrocytic markers eventually will demonstrate simultaneous activation of astrocytes and microglial cells.

3.3 Anterograde axonal degeneration

Transection of axonal pathways with anterograde axonal degeneration, induces a microglial response within the terminal fields within 24 hours. This includes expression of microglial activation markers up to the level of MHC antigen class I, which is seen in all our models. Thereafter the reactions differ. In the KA-induced Schaffer collateral denervated CA1 there is no further activation of microglial cells (8), while the microglial cells in the PP-denervated dentate molecular layer are induced to the stage of expressing CD45 antigen (23), but not MHC antigen class II (23). In contrast to this, the full repertoire of microglial immunomolecules including MHC antigen class II is induced when the mossy fibers in the hippocampus degenerate following colchicine destruction of the dentate granule cells (12). The different activation patterns of microglial cells cannot be accounted for by differen-ces in the density of the degenerating synapses or the degree of myelination of the lesioned pathways. Both the Schaffer collaterals and the commissural fibers to CA1 and the perforant path fibers to the dentate molecular layer are thin myelinated fibers with small boutons en passage, and in both cases around 80% or more of the synapses in the terminal areas are degenerating. The mossy fibers are unmyeli-nated, but they carry extremely large terminal boutons ((12) and references therein).

In response to anterograde axonal degeneration following PP-lesioning astrocytes express increased GFAP mRNA levels already at 24 hours (60), and display signs of hypertrophy, increased GFAP immunoreactivity and induction of VIM immunoreactivity at day 2 (23,53). They begin to proliferate around postlesional day 4, which is about 2 days later than microglial cells (7). It is in axonal degeneration paradigms like these that astrocytes have been shown to express EGF-receptors (44), and to change their immunohistochemical content of glutamine synthetase (2).

3.4 Transient global cerebral ischemia

Evidence is accumulating, that the ischemia-induced raise in extracellular glutamate and potassium levels, together with acidosis, increases the neuronal sensitivity to subsequent, normal neuronal firing and subtle changes in the microenvironment to such an extent that this in certain types of neurons can lead to delayed neuronal degeneration (62). As the microenvironment is maintained by glial cells, these cells may play a key role in relation to the delayed degeneration of the CA1 pyramidal neurons which is initiated day 2-4 after ischemia (25). It is therefore of great interest that the microglial and astroglial cells begin to display the classical signs of early cellular activation from minutes to a few hours after the ischemic insult (8,25,37,14,55). We have in our studies shown, that there is an early, transient activation of both micro- and astroglial cells in *all* ischemia-exposed areas, including also areas with no subsequent neuronal degeneration, such as the dentate molecular layer, the CA3 region and the neocortex. Starting from minutes to a few hours after the ischemic episode, *microglial cells* throughout the brain begin to retract their processes, increase their CR3 and NPDase reactivity (26), and to express MHC antigen class I and TGF-ß1 mRNA (8,31)(Fig. 4). These reactions are transient and disappear in brain regions with no subsequent neuronal degeneration. In regions with ongoing and delayed neuronal degeneration the microglial reactions escalate to the level of CD45 (8) and CD4 (38) antigen expression, further enhancement of TGF-ß1 mRNA expression (Fig. 4)(31), and eventually for some cells expression of MHC antigen class II (8,38,31). The early activation of *astroglial cells* was most strikingly visualized by their uptake of extravasated plasmaconstituents (26,28,55), whereas hypertrophy and intense immunoreactivity for GFAP and VIM characterized the later and stronger activation of astrocytes in CA1 (26,58).

The early activation of both types of glial cells may be ascribed to a widespread leakage of plasmaproteins into the neuropil, changes in pH and extracellular ion concentrations, as well as to a systemic stress response following the electrocauteri-

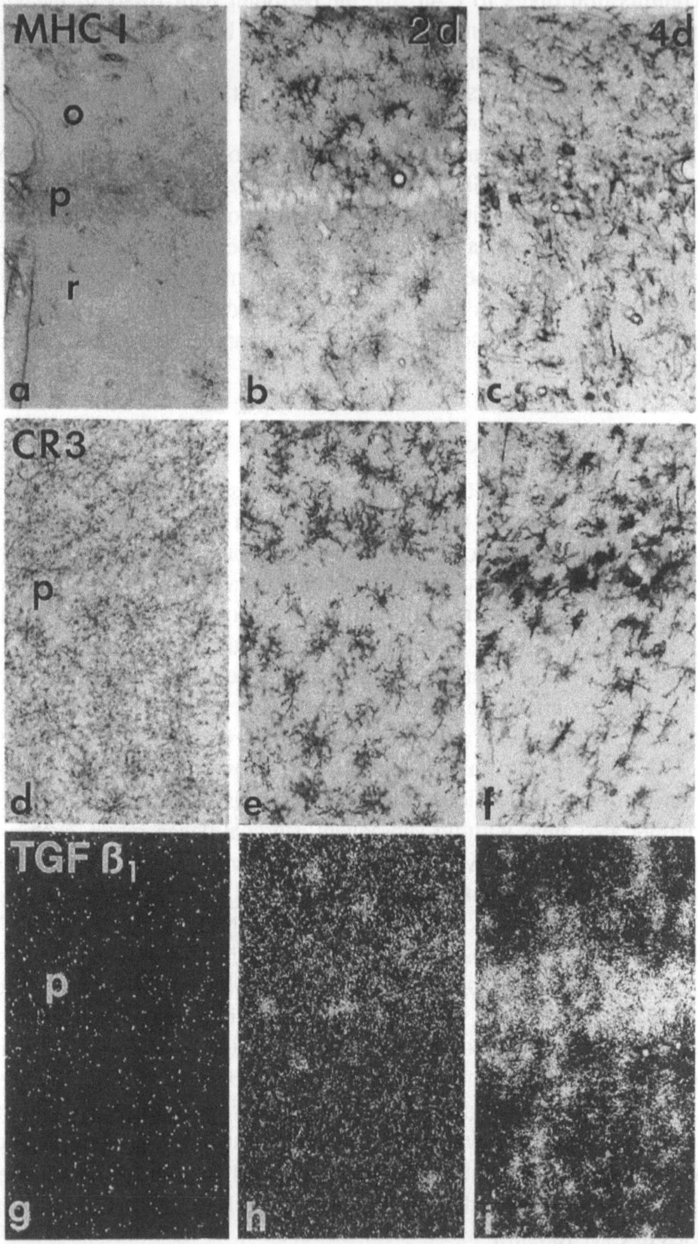

Fig. 4. Early, transient, and late, degeneration-induced microglial activation patterns in transient, global cerebral ischemia. All photomicrographs display the same sector of CA1, showing the pyramidal cell layer (p), the inner third of stratum radiatum (r), and stratum oriens (o). The upper and middle horisontal rows display immunohistochemical

zation of the vertebral arteries ((8) and references therein). One characteristic feature of the delayed degeneration of the CA1 pyramidal neurons is an extensive accumulation of reactive microglial cells and/or macrophages in the degenerating neuronal cell layer (Fig. 4), substantiating the optimal anatomical position of these cells for direct cell-cell interaction with the sensitized neurons at an early stage of the degenerative process. Another characteristic is the morphological transformation of microglial cells to "rod cells" in the adjacent stratum radiatum. This is most manifest around postischemic day 7, when the dendrites in this layer are in the process of profound degeneration (25).

4 Cytokines and cellular interactions

Cytokines are potent, diffusible or cell bound factors, which act in low, picomolar concentrations. They are pleiotrophic factors, which in addition to their own direct effects, can modulate the effect of other factors in a dose- and timedependent manner. This means that even very low levels of a cytokine and very small numbers of cytokine-producing cells may have dramatic effects on cells in the microenvironment. The induction of TGF-ß1 mRNA in microglial cells at the earliest stages of brain pertubation, suggests that TGF-ß1, if translation to protein has occurred, may exert its modulatory effect on other cytokines and extracellular matrix components at all, early as well as late, timepoints after lesioning. In the case of *anterograde axonal degeneration*, a few other cytokines have recently been demonstrated in addition to TGF-ß1 mRNA (36). IL-1 protein (6) and insulin-like growth factor-1 (IGF-1) mRNA (17) have been located to activated microglial cells, and bFGF to activated astrocytes (15). There is general agreement that all four factors play a role in the postlesional axonal sprouting processes (6,15,17,36). In addition to the microglial expression of TGF-ß1 mRNA in the

Fig. 4. cont.: - stainings for MHC antigen class I (a-c) and CR3 (d-f), with dark field photomicrographs of isotopic *in situ* hybridization for TGF-ß1 mRNA in the bottom row (g-i). The first vertical column represents the resting state of the microglial cells, showing MHC antigen class I (a) and TGF- ß1 mRNA (g) negative, but CR3 (b) positive ramified microglial cells in the normal hippocampus. The second column shows the microglial activation state 2 days after ischemia and is representative of both the early generalized microglial activation (see text) and of the initial stages of microglial activation preceeding the more pronounced degeneration-specific reactions, shown in the third column. Note in (e) the partial retraction and coarsening of the microglial processes, in (b) and (g) the microglial expression of MHC antigen class I and TGF-ß1 mRNA, respectively. The third column shows the microglial activation 4 days after ischemia. At this time the pyramidal cells are, as reflected by the accumulation of CR3 (f), MHC antigen class I (c) and TGF-ß1 mRNA (i) expressing microglial cells/macrophages within the cell layer, beginning to undergo degeneration. The immunohistochemical stainings were carried out on 50 µm thick, free floating vibratome sections (8,25) and the *in situ* hybridization on 16 µm cryostate sections (31). x 135.

postischemic hippocampus (31,64), microglial cells have also been shown to express IL-1ß mRNA (69). Other studies have reported the induction of bFGF protein and mRNA in astroglial cells (63). Interestingly, both TGF-ß1 (29,49) and astrocyte-derived factors (27,33), including bFGF, have been shown to reduce hypoxic neuronal degeneration in newborn rats, and to enhance neurite outgrowth *in vitro*. The observation that treatment of stroke-lesioned rats with IL-1 receptor antagonists reduces the area of infarction (51), indicates a potential role of IL-1 in antagonizing the neurotrophic effects of TGF-ß1 and bFGF.

Regarding *in situ* cytokine production during *neural graft rejection*, the situation becomes even more complex, as there additionally will be T-cell derived cytokines present. Among the most important are the Th_1-cell derived immunostimulatory cytokines IL-2 and interferon-γ (IFN-γ), which activates $CD8^+$ cytotoxic T cells and macrophages/microglia, and the Th_2-cell derived cytokines IL-4 and IL-10, which antagonizes the effect of IL-2 and IFN-γ and overall have an anti-inflammatory effect (43). Other cytokines to be considered are again the macrophage/microglial derived cytokines IL-1 as well as IL-6 and tumour necrosis factor-α (TNF-α), which all have been detected during extracerebral graft rejection (reviewed by Chandler (1)) and in EAE-animals (67). Macrophage-depletion and cytokine-intervention studies have unequivocally proven the detrimental effect of infiltrating macrophages and macrophage/microglia-derived products in the course of EAE (22,59). These findings are highly relevant in relation to both neural graft rejection and ischemic- and excitototin-induced neuronal degeneration.

5 Concluding remarks

The *in vivo* activation of microglial and astroglial cells occurs in a graded manner, which can be demonstrated in simple lesion paradigms. The increasing number of *in situ* hybridization studies on glial cytokine expression indicates, that this graded activation can be extended to the level of cytokine mRNA, and perhaps protein, expression. The pleiotrophic factor TGF-ß1 clearly belongs to a class of easily induceable cytokines, which may be present in its active form and modulate the effects of other cytokines, cell adhesion molecules, and extracellular matrix components at all postlesional timepoints. *In vivo* intervention studies of the effects of this and other cytokines and cytokine antagonists, the use of transgenic animals and knock-out experiments will provide further insight into the mechanism by which individual cytokines influence neural degeneration and regeneration, and CNS graft rejection.

6 Acknowledgements

The coauthors on the articles forming the basis for this review are thanked for their valuable contributions. The studies were supported by the Danish State

Biotechnology Programme, the Danish MRC, the Lundbeck Foundation, the Universities of Aarhus and Odense, the Carlsberg Foundation, P. Carl Petersens Foundation, Desireé and Niels Ydes Foundation, Direktør Leo Nielsens Foundation, The Foundation to the Advancement of Medical Sciences, the Novo Foundation and a number of other Danish Foundations.

7 References

1. Chandler, C. and Passaro, E. (1993) Transplant rejection. Arch. Surg. 128, 279-283.
2. Condorelli, D.F., Dell'Albini, P., Kaczmarek, L., Messina, L., Spampinato, G., Avola, R. Messina, A. and Giuffrida Stella, A.M. (1990) Glial fibrillary acidic protein messenger RNA and glutamine synthetase activity after nervous system injury. J. Neurosci. Res. 26, 251-257.
3. Dalmau, I., Finsen, B.R., González, B., Castellano, B., Tønder, N. and Zimmer, J. (1992) Astroglial and microglial cells in the developing rat hippocampus. A histochemical and immunocytochemical study. Eur. J. Neurosci. Suppl. 5, 28.
4. Date, I., Kawamura, K. and Nakashima, H. (1988) Histological signs of immune reactions against allogeneic solid fetal neural grafts in the mouse cerebellum depend on the MHC locus. Exp. Brain. Res. 73, 15-22.
5. del Rio-Hortega, P. and Penfield, W. (1919) Cerebral cicatrix. The reaction of neuroglia and microglia to brain wounds. Bull. Johns Hopkins Hospital 41, 278-303.
6. Fagan, A.M. and Gage, F.H. (1990) Cholinergic sprouting in the hippocampus: A proposed role for IL-1. Exp. Neurol. 110, 105-20.
7. Fagan, A.M. and Gage, F.H. (1994) Mechanisms of sprouting in the adult central nervous system: Cellular responses in areas of terminal degeneration and reinnervation in the rat hippocampus. Neurosci. 4, 705-25.
8. Finsen, B.R., Jørgensen, M.B., Diemer, N.H. and Zimmer, J. (1993) Microglial MHC antigen expression after ischemic and kainic acid lesions of the adult rat hippocampus. Glia 7, 41-49.
9. Finsen, B., Oteruelo, F. and Zimmer, J. (1998) Immunocytochemical characterization of the cellular immune response to intracerebral xenografts of brain tissue. Prog. Brain. Res. 78, 261-70.
10. Finsen, B., Poulsen, P.H. and Zimmer, J. (1988) Xenografting of fetal mouse hippocampal tissue to the brain of adult rats: Effects of Cyclosporin A treatment. Exp. Brain Res. 70, 117-133.
11. Finsen, B.R., Sørensen, T., Castellano, B., Pedersen, E.B. and Zimmer, J. (1991) Leukocyte infiltration and glial reactions in xenografts of mouse brain tissue undergoing rejection in the adult rat brain. A light and electron microscopical immunocytochemical study. J. Neuroimmunol. 32, 159-83.
12. Finsen, B.R., Tønder, N., Xavier, G.F., Sørensen, J.C. and Zimmer, J. (1993) Induction of microglial immunomolecules by anterogradely degenerating mossy fibres in the rat hippocampal formation. J. Chem. Neuroanat. 6, 267-275.
13. Ford, A.L., Gooodsall, A.L., Hickey, W.F. and Sedgwick, J.D. (1995) Normal adult ramified microglia separated from other central nervous system

macrophages by flow cytometric sorting. Phenotypic differences defined and direct ex vivo antigen presentation to myelin basic protein-reactive CD4+ T cells compared. J. Immunol. 154, 4309-21.

14. Gehrmann, J., Bonnekoh, P., Miyazawa, T., Hossmann, K.-A. and Kreutzberg, G. (1992) Immunocytochemical study of an early microglial activation in ischemia. J. Cereb. Blood Flow Metab. 12, 257-69.

15. Gómez-Pinilla, F., Lee, J.W.-K., Cotman, C.W. (1992) Basic FGF in adult rat brain: Cellular distribution and response to entorhinal lesion and fimbria-fornix transection. J. Neurosci. 12, 345-355.

16. Graeber, M.B., Streit, W. J. and Kreutzberg, G.W. (1989) Identity of ED2-positive perivascular cells in rat brain. J. Neurosci. Res. 22, 103-106.

17. Guthrie, K.M., Nguyen, T. and Gall, C.M. (1995) Insulin-like growth factor-1 mRNA is increased in deafferented hippocampus: spatiotemporal correspondence of a trophic event with axon sprouting. J. Comp. Neurol. 352, 147-160.

18. Hertz, L. (1991) Neuronal-astrocytic interactions in brain development, brain function and brain disease. In: Timiras PS et al, eds. Plasticity and Regeneration of the Nervous System. New York: Plenum Press, 143-59.

19. Hickey, W.F. (1992) Tolerance of CNS allografts: a hypothesis about a persistent enigma. Restor. Neurol. Neurosci. 4(3), 38.

20. Hickey, W.F. and Kimura, H. (1988) Perivascular microglial cells of the CNS are bone marrow-derived and present antigen in vivo. Science 139, 290-3.

21. Honey, C.R., Clarke, D.J., Dallman, M.J. and Charlton, H.M. (1990) Human neural graft function in rats treated with anti-interleukin II receptor antibody. NeuroReport 1, 247-249.

22. Huitinga, I., Fuuls, S.R., Jung, S., Van-Rooijen, N., Hartung, H.P. and Dijkstra, C.D. (1995) Macrophages in T cell line-mediated, demyelinating, and chronic relapsing experimental autoimmune encephalomyelitis in Lewis rats. Clin. Exp. Immunol. 100, 344-51.

23. Jensen, M.B., Finsen, B.R., González, B. and Zimmer, J. (1989) Cellular occurrence of MHC- and leukocytic antigens in response to anterograde axonal degeneration in the rat fascia dentata. Eur. J. Neurosci. Suppl 2, 71.27.

24. Jensen, M.B., Gonzalez, B., Castellano, B. and Zimmer, J. (1994) Microglial and astroglial reactions to anterograde axonal degenertion: a histochemical and immunocytochemical study of the adult rat fascia dentata after entorhinal perforant path lesions. Exp. Brain Res. 98, 245-60.

25. Jørgensen, M.B. and Diemer, N.H. (1982) Selective neuron loss after cerebral ischemia in the rat: Possible role of transmitter glutamate. Acta Neurol. Scand. 66: 536-46.

26. Jørgensen, M.B., Finsen, B.R., Jensen, M.B., Castellano, B., Diemer, N.H. and Zimmer, J. (1993) Microglial and astroglial reactions to ischemic and kainic acid-induced lesions of the adult rat hippocampus. Exp. Neurol. 120, 70-88.

27. Kinoshita, A., Yamada, K., Kohmura, E. and Hayakawa, T. (1990) Effect of astrocyte-derived factors on ischemic brain edema induced by rat MCA-occlusion. APMIS 98, 851-7.

28. Klatzo, I. (1987) Pathophysiological aspects of brain edema. Acta Neuropathol. (Berl) 72, 236-9.

29. Klempt, N.D., Sirimanne, E., Gunn, A.J., Kleempt, M., Singh, K., Williams, C. and Gluckman, P.D. (1992) Hypoxia-ischemia induces transforming growth factor $ß_1$ mRNA in the infant rat brain. Mol. Brain Res. 13, 93-101.

30. Lawrencc, J.M., Morris, R.J., Wilson, D.J. and Raisman, G. (1990) Mechanisms of allograft rejection in the rat brain. Neurosci. 37, 431-62.

31. Lehrmann, E., Kiefer, R., Finsen, B., Diemer, N.H., Zimmer, J. and Hartung, H.-P. (1995) Cytokines in cerebral ischemia: expression of transforming growth factor beta-1 (TGF-1ß) mRNA in the post-ischemic adult rat hippocampus. Exp. Neurol. 131, 114-123.

32. Lund, R.D., Rao, K., Kunz, H.W. and Gill. III T.J. (1988) Instability of neural xenografts placed in neonatal rat brains. Transplant. 46, 216-223.

33. Matthews, D.A., Cotman, C. and Lynch, C. (1976) An electron microscopic study of lesion-induced synaptogenesis in the dentate gyrus of the adult rat. I. Magnitude and time course of degeneration. Brain Res. 115: 1-21.

34. Mattson, M.P. and Rychlik, B. (1990) Glia protect hippocampal neurons against excitatory amino acid-induced degeneration: involvement of fibroblast growth factor. Int. J. Devl. Neurosci. 8, 399-415.

35. McGeer, P.L., McGeer, E.G., Kawamata, T., Yamada, T. and Akiyama, H. (1991) Reactions of the immmune system in chronic degenerative neurological diseases. Can. J. Neurol. Sci. 18, 376-9.

36. Morgan, T.E., Nichols, T.R., Pasinetti, G.M. and Finch, C.E. (1993) TGF-ß1 mRNA increases in macrophage/microglial cells of the hippocampus in response to deafferentation and kainic acid neurodegeneration. Exp. Neurol. 120, 291-301.

37. Morioka, T., Kalehua, A.N. and Streit, W.J. (1991) The microglial reaction in the rat dorsal hippocampus following transient forebrain ischemia. J. Cereb. Blood Flow Metab. 11, 966-73.

38. Morioka, T., Kalehua, A.N. and Streit, WJ. (1992) Progressive expression of immunomolecules on microglial cells in rat dorsal hippocampus following transient forebrain ischemia. Acta Neuropathol. (Berl) 83, 149-57.

39. Nadler, J.V., Perry, B.W., Gentry, C. and Cotman, C.W. (1980) Degeneration of hippocampal CA3 pyramidal cells induced by intraventricular kainic acid. J Comp. Neurol. 192: 333-59.

40. Nicholas, M.K. and Arnason, B.G.W. (1990) A role for $CD8^+$ T lymphocytes late in the rejection of intraventricular fetal neocortical fragment allografts in the mouse. In: Long Term Consequences of Barrier Dysfunction for the Brain (Proceedings of Eric K Fernström Symposium Pathophysiology of the Blood Brain Barrier).

41. Nicholas, M.K., Chenelle, A.G., Brown, M.M, Stefansson, K. and Arnason, B.G.W. (1990) Prevention of neural allograft rejection in the mouse following in vivo depletion of L3T4+ but not Lyt-2+ T lymphocytes. Prog. Brain Res. 87, 161-7.

42. Nicholas, M.K., Sagher, O., Hartley, J.P., Stefansson, K. and Arnason, B.W.G. (1988) A phenotypic analysis of T lymphocytes isolated from the brain of mice with allogeneic neural transplants. Prog. Brain Res. 78, 249-59.

43. Nickerson, P., Steurer, W., Steiger, J., Zheng, X., Steele, A.W. and Strom T. B (1994) Cytokines and the Thl/Th2 paradigm in transplantation. Curr. Biol. 6, 757-764.

44. Nieto-Sampedro, M., Gómez-Pinilla, F., Knauer, D.J. and Broderick, J.T. (1988) Epidermal growth factor receptor immunoreactivity in rat brain astrocytes. Response to injury. Neurosci. Lett. 91, 276-282.
45. Niquet, J., Jorquera, I., Ben-Ari, Y. and Represa, A. (1994) Proliferative astrocytes may express fibronection-like protein in the hippocampus of epileptic rats. Neurosci. Lett. 180, 13-16.
46. Pedersen, E.B., Poulsen, F.R., Zimmer, J. and Finsen, B. (1995) Prevention of mouse-rat brain xenograft rejection by a combination therapy of Cycloporin A, Prednisolone and Azathioprine. Exp. Brain Res. In press.
47. Perry, V.H. and Lund, R.D. (1989) Microglia in retinae transplanted to the central nervous system. Neurosci. 31, 453-62.
48. Pollack, I.F., Lund, R.D. and Rao, K. (1990) MHC antigen expression and cellular response in spontaneous and induced rejection of intracerebral neural xenografts in neonatal rats. Restor. Neurol. Neurosci. 1: 347-60.
49. Prehn, J.H.M., Backhauss, C. and Krieglstein, J. (1993) Transforming growth factor β_1 prevents glutamate neruotoxicity in rat neocortical cultures and protects mouse neocortex from ischmeic injury in vivo. J. Cereb. Blood Flow Metab. 13, 521-525.
50. Pulsinelli, W.A., Brierly, J.B. and Plum, F. (1982) Temporal profile of neuronal damage in a model of transient forebrain ischemia. Ann .Neurol. 11: 491-8.
51. Relton, J.K. and Rothwell, N.J. (1992) Interleukin-1 receptor antagonist inhibits ischaemic and excitotoxic neuronal damage in the rat. Brain Res. Bull. 29, 243-246.
52. Rogers, J., Kirby, L.C., Hempelman, S.R., Berry, D.L., McGeer, P.L., Kaszniak, A.W. et al. (1993) Clinical trial of indomethacin in Alzheimer's disease. Neurol. 43, 1609-11.
53. Rose, G., Lynch, G. and Cotman, C.W. (1976) Hypertrophy and redistribution of astrocytes in the deafferented dentate gyrus. Brain Res. Bull. 1, 87-92.
54. Satake, M., Korsgren, O., Ridderstad, A., Karlsson-Parra, A., Wallgren, A.C. and Möller, E. (1994) Immunolgoical characteristics of islet cell xenotransplantation in humans and rodents. Immunol. Rev. 141, 191-211.
55. Schmidt-Kastner, R., Szymas, J. and Hossmann, K.-A. (1990) Immunohistochemical study of glial reaction and serum-protein extravasation in relation to neuronal damage in rat hippocampus after ischemia. Neurosci. 38, 527-40.
56. Schwartz, J.P., Nishiyama, N., Wilson, D. and Taniwaki, T. (1994) Receptor-mediated regulation of neuropeptide gene expression in astrocytes. Glia 1, 185-90.
57. Sedgwick, J.D., Mössner, R., Schwender, S. and ter Meulen, V. (1991) Major histocompatibiligy complex-expressing nonhematopoietic astroglial cells prime only CD8+ T lymphocytes: astroglial cells as perpetuators but not initiators of CD4+ T cell responses in the central nervous system J. Exp. Med. 173, 1235-46.
58. Sloan, D.J., Brenan, M., Harrison, C.J., Wood, M.J.A., Puklavec, M. and Charlton, H.M. (1992) Regulation of the immune response within the CNS II: Migration of MRC OX-62 positive cells into rejecting neural allografts. Restor. Neurol. Neurosci. 4:149.
59. Sommer, N., Löschmann P.-A., Northoff, G.H., Weller, M., Steinbrecher, A., Steinbach, J.H., Lichtenfels, R., Meyermann, R., Riethmüller, A., Fontana, A.,

Dichgans, J. and Martin, R. (1995) The antidepressant rolipram suppresses cytokine production and prevents autormmune encephalomyelitis. Nature Med. 1, 244-248.

60. Steward, O., Torre, E.R., Phillipi, L.L. and Trimer, P.A. (1990) The process of rcinnervation in the dentate gyrus of adult rats: time course of increases in mRNA for glial fibrillary acidic protein. J. Neurosci. 10, 2373-2384.

61. Streit, W.J., Graeber, M.B. and Kreutzberg, G.W. (1988) Functional plasticity of microglia: a review. Glia 1, 301-7.

62. Szatkowski, M. and Attwell, D. (1994) Triggering and execution of neuronal death in brain ischaemia: two phases of glutamate release by different mechanisms. TINS, 17, 359-365.

63. Takami, K., Iwane, M., Kiyota, Y., Miyamoto, M., Tsukuda, R. and Shiosaka, S. (1992) Increase of basic fibroblast growth factor immunoreactivity and its mRNA level in rat brain following transient forebrain ischemia. Exp. Brain Res. 90, 1-10.

64. Wießner, C., Gehrmann, J., Lindholm, D., Töpper, R., Kreutzberg, G.W. and Hossmann, K.A. (1993) Expression of transforming growth factor-ß1 and interleukin-1ß mRNA in rat brain following transient forebrain ischemia. Acta Neuropathol. (Berl.) 86, 439-446.

65. Wood, M.J.A., Sloan, D.J., Dallman, M.J. and Charton, H.M. (1993) Specific tolerance to neural allografts induced with an antidboy to the interleukin 2 receptor. J. Exp. Med. 177, 97-603.

66. Wood, M.J.A., Sloan, D.J., Wood, K.J. and Charlton, H.M. (1992) The role of CD4+ and CD8+ T lymphocytes in the rejection of neural xenografts. Restor. Neurol. Neurosci. 4, 149.

67. Woodroofe, M.N. and Cuzner, M.L. (1993) Cytokine mRNA expression in inflammatory multiple sclerosis lesions: Detection by non-radioactive in situ hybridization. Cytokine 5, 583-588.

68. Yamada, T., Kawamata, T., Walker, D.G. and McGeer, P.L. (1992) Vimentin immunoreactivvity in normal and pathological human brain tissue. Acta Neuropathol. (Berl.) 84, 157-162.

69. Yabuuchi, K., Minami, M., Katsumata, S, Yamazaki, A. and Satoh, M. (1994) An in situ hybridization study on interleukin-1ß mRNA induced by transient forebrain ischemai in the rat brain. Mol. Brain Res. 26, 135-142.

70. Zimmer, J., Finsen, B., Sørensen, T. and Poulsen, P.H. (1988) Xenografts of mouse hippocampal tissue. Exchange of laminar and neuropeptide specific nerve connections with the host rat brain. Brain Res. Bull. 20, 369-79.

The Effect of Myelin Disruption on Spinal Cord Regeneration

Hans S. Keirstead[1] and John D. Steeves[2]

[1] MRC Cambridge Centre for Brain Repair and Department of Clinical Veterinary Medicine, University of Cambridge, Cambridge, U.K., CB2 2PY
[2] Department of Zoology Anatomy and Surgery, University of British Columbia, Vancouver, B.C., Canada, V6T 1Z4

Keywords: spinal cord injury, regeneration, CNS, myelin sheath, GalC

1. Introduction

The adult central nervous system (CNS) of higher vertebrates has a very limited capacity for regeneration following injury. CNS regeneration in higher vertebrates is primarily restricted to monoaminergic fibres (7), unmyelinated cholinergic axons (7) primary olfactory axons (4), and neurosecretory fibres (14). With these exceptions, damage to the adult CNS most often results in aborted attempts at regeneration followed by degeneration (7, 45). However, some CNS axons will anatomically regenerate and make functional synaptic connections if peripheral nerve segments containing Schwann cells are grafted to a site of CNS injury (1, 13, 45). Similarly, axotomized CNS neurons will extend axons *in vivo* through implants of fetal (8, 9, 34), or genetically modified (20, 65) CNS tissue. These findings indicate that adult CNS neurons retain intrinsic growth programs which enable long-distance axonal regeneration in the presence of a favorable extraneuronal environment. Nevertheless, the intrinsic neuronal determinants of regeneration may differ in distinct populations of CNS neurons (19, 35, 63).

The poor regenerative capacity of the mature CNS of higher vertebrates may be accounted for in great part by the paucity of growth promoting factors and the presence of growth inhibitory factors within the extraneuronal environment, a situation reciprocal to that of the developing CNS (eg. 25). Strategies to evoke substantial regeneration of the adult CNS are precluded by information addressing the relative contribution of these various factors to the regenerative potential of injured adult neurons. Many of the events underlying adult axonal regeneration are a recapitulation of axonal development (29, 64), thus an investigation of the regenerative capacity of the injured embryonic spinal cord should identify the principal developmental events which prohibit functional regeneration of adult CNS axons.

Documentation of spinal cord development and\or the regenerative capacity of the injured embryonic CNS is difficult to perform in viviparous mammals, due to maternal complications. The avian egg, however, provides for easy and uncomplicated access to the embryonic chick at any stage of development. Chickens undergo a relatively accelerated rate of neurodevelopment, such that their CNS (in particular those brainstem and spinal systems concerned with

Molecular Signaling and Regulation in Glial Cells
A Key to Remyelination and Functional Repair

locomotion) can be considered fully developed and adult-like in function at the time of hatching (23, 24, 27, 28, 43, 53, 54). It is well established that the structure and function of the vertebrate brainstem and spinal cord has remained virtually unchanged throughout evolutionary history (48), and studies of the development and organization of avian locomotor systems indicate that they are similar to all other vertebrates (2, 5, 43, 55, 56, 57, 58, 60, 61, 67, 68). Following injury to the adult spinal cord, birds suffer the same motor deficits as a mammal (18, 55, 68, 69).

2. Regeneration of the Early Embryonic Spinal Cord

Studies of developing chick embryos have indicated the ability to functionally recover from spinal cord injury is lost at a critical point during the 21 day developmental period (27, 28, 43). Transections of the thoracic spinal cord prior to embryonic day 13 (E13) result in complete neuroanatomical regeneration and physiological recovery (27, 28, 62, 66). If the thoracic spinal cord is transected on E13 or E14, the repair of descending supraspinal pathways progressively diminishes, resulting in little or no functional recovery (27, 28, 54). Spinal cord transections on E15 result in no neuroanatomical regeneration or functional recovery and the chick, upon hatching, is as paralyzed as a bird or mammal transected as an adult (27, 28, 31, 54, 55, 68).

Thus, there appears to be a transition during development from a permissive to a restrictive period for spinal cord repair. Investigations of neural changes that coincide with this developmental transition may provide insights into the lack of repair following injury to the adult spinal cord.

2.1 Developmental Onset of Myelination

It is noteworthy that the developmental onset of myelination in the embryonic chick spinal cord has been reported to begin around E13 (6, 26, 38). Using immunohistochemical markers to document the developmental appearance of myelin basic protein (MBP), myelin-associated glycoprotein (MAG), 2',3'-cyclic nucleotide 3'-phosphodiesterase (CNP) and galactocerebroside (GalC), we have independently confirmed these findings, noting that the onset of myelination is coincident with the transition from a permissive to a restrictive period for spinal cord repair (Figure 1; 31, 32). MBP, MAG, CNP and GalC are myelin-specific markers and, along with proteolipid protein (PLP), constitute the major myelin markers evident during CNS myelination (26, 40).

The onset of myelination in the developing chick spinal cord begins following the completion of axonal projections from the brainstem to the lumbar levels of the spinal cord. Stepping movements can be evoked by brainstem stimulation as early as E7 (42). Brainstem-spinal projections are complete to the cervical spinal cord by E8 and are complete to the lumbar spinal cord prior to E12 (43). Myelination of the spinal cord begins in the ventrolateral funiculi shortly afterwards on E13 and proceeds in a rostral-caudal direction with development (31). Thus, the developmental onset of chick spinal cord myelination occurs after brainstem-spinal neurons have completed their projections and have made functional synaptic connections with their targets. These results are in agreement with observations of myelin development in the rat and mouse spinal cord, although myelination in these species begins perinatally, and is completed within the first three weeks of neonatal life (21, 50).

Myelin-associated proteins have been identified in the rat spinal cord that inhibit the neuroanatomical regrowth of axotomized coticospinal fibres (10, 49). At least some of this inhibitory activity has been found to reside in 35kD and 250kD proteins isolated from rat CNS white matter (3). Neutralization of these myelin-associated proteins with specific antibodies results in partial neuroanatomical regeneration of axotomized adult corticospinal fibres, although functional synapse formation with spinal cord neurons has not yet been documented (49). In addition, MAG has recently been shown to inhibit neurite outgrowth from certain cell types (37, 41).

To assess the inhibitory role of myelin in regeneration of brainstem-spinal projections, we utilized an immunological method of delaying the developmental onset of myelination until well into the restrictive period for spinal cord repair. The assessment of spinal cord repair following a subsequent restrictive period transection (eg. E15) then serves as a direct test of whether myelin inhibits neuroanatomical regeneration and functional recovery.

3. Developmental Myelin-Suppression

Suppression of the onset of myelination in the developing chick embryo was initiated by the direct intraspinal injection of heterologous serum complement proteins plus a myelin/oligodendrocyte-specific, complement-binding antibody. GalC is the major sphingolipid produced by oligodendrocytes and is expressed upon differentiation of oligodendrocytes at E10-E11, 2-3 days prior to myelin formation (6, 32). GalC is highly conserved across species (46) and the polyclonal antibody that we use has been shown to cross-react with chicken and exhibit specificity for oligodendrocytes within the CNS (32, 46). Antisera against GalC have been shown to: 1) suppress myelin development *in vitro* (15, 16, 17, 22, 30), 2) Suppress myelin development *in vivo* (31, 32, 47), and 3) reversibly alter CNS myelin structure *in vivo* (guinea pig optic nerve: 44, 52; cat optic nerve: 11, 12; rat spinal cord: 39).

We have found that a single injection of GalC antibody plus serum complement proteins into the thoracic spinal cord of an E9-E12 chick embryo (i.e. around the time of oligodendrocyte differentiation) delayed the onset of spinal cord MBP, CNP and MAG immunoreactivity until E17 (Fig. 1; 31, 32). Toluidine blue stained transverse sections of myelin-suppressed E15 spinal cords revealed a complete absence of myelin (32). Importantly, myelin development following myelin-suppression was shown to be fully compensatory. Equally effective inhibition of myelin development can be obtained by utilizing the complement-fixing, oligodendrocyte-specific 04 antibody (59) in place of the GalC antibody (32).

Several findings suggest that oligodendrocyte cell bodies survive our myelin-suppression protocol. The proposed mechanism of anti-GalC-induced demyelination *in vitro* involves microtubule disassembly of oligodendrocyte processes mediated by an influx of extracellular calcium (17). Using a specific MBP cDNA probe, we have observed that MBP gene expression is not down regulated in myelin-suppressed animals (32). Furthermore, electron microscopic studies of myelin disrupted hatchling chick and adult rat spinal cords have identified oligodendrocyte cell bodies within the lesion (see below). These findings suggest that oligodendrocytes survive the myelin-suppression protocol and are a likely origin of subsequent myelin.

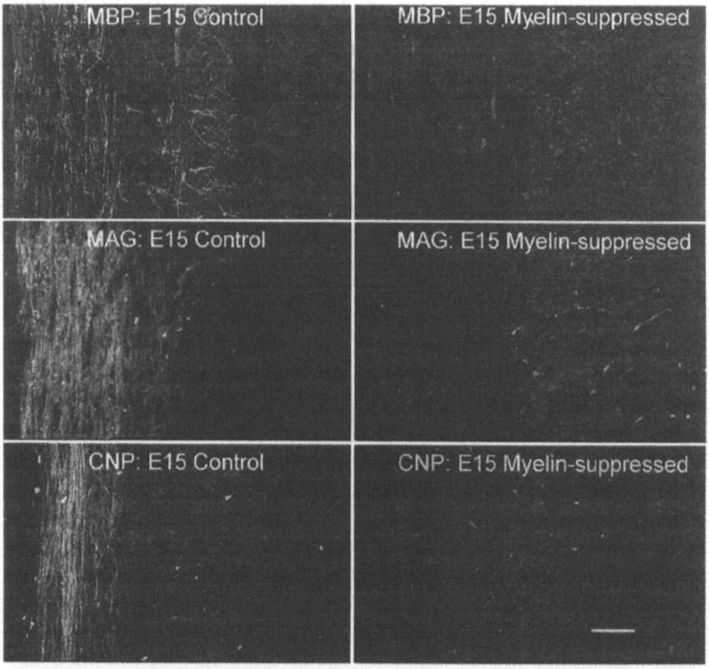

Figure 1. Immunostaining of embryonic chick spinal cords. In unoperated control spinal cords, MBP, MAG and CNP are present at E15 in development (left side). These myelin markers are absent at E15 in animals subjected to developmental myelin-suppression. (Bar=100um).

3.1 Regeneration of the Late Embryonic Spinal Cord

A thoracic spinal cord transection as late as E15 (i.e. during the normally restrictive period for repair) in myelin-suppressed embryos resulted in 90-100% neuroanatomical regeneration and functional recovery (Figs. 2;31). This is in contrast to normally-myelinated control embryos transected on E15, which showed no neuroanatomical evidence of regeneration and were completely incapable of voluntary locomotion after hatching. The number and distribution of retrogradely-labelled brainstem-spinal neurons in myelin-suppressed, E15-transected animals was indistinguishable from untransected control embryos, indicating virtually complete neuroanatomical regeneration (Fig. 2). This neuroanatomical regeneration was also equivalent to that observed in chicks transected during the permissive period for repair (when myelin has yet to appear; 27, 28, 54). The locomotor functions of myelin-suppressed, E15-transected chicks were also equivalent to untransected control animals, or animals transected during the permissive period for repair (Fig. 2; 31).

These findings demonstrate that developmental myelin-suppression extends the permissive period for functional repair and, to our knowledge, constitute the first observation of functional CNS recovery following therapeutic intervention. The results confirm and extend the proposition that the presence of CNS myelin contributes to the inhibition of neuronal repair after an adult CNS injury (10, 31, 45, 49). These data are also supported by the recent evidence for neuronal repair prior to the developmental onset of myelination in neonatal opposums (64), and the functional repair/regeneration in animals that do not have myelinated CNS fibre tracts (eg. lamprey; 36).

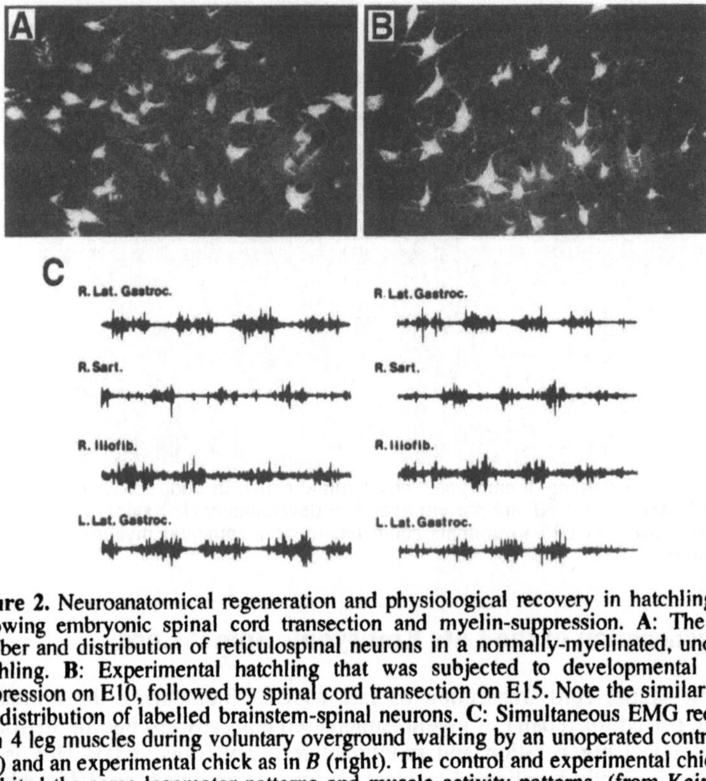

Figure 2. Neuroanatomical regeneration and physiological recovery in hatchling chicks following embryonic spinal cord transection and myelin-suppression. **A**: The normal number and distribution of reticulospinal neurons in a normally-myelinated, unoperated hatchling. **B**: Experimental hatchling that was subjected to developmental myelin-suppression on E10, followed by spinal cord transection on E15. Note the similar number and distribution of labelled brainstem-spinal neurons. **C**: Simultaneous EMG recordings from 4 leg muscles during voluntary overground walking by an unoperated control chick (left) and an experimental chick as in *B* (right). The control and experimental chicks both exhibited the same locomotor patterns and muscle activity patterns. (from Keirstead et al., 1992)

4. Myelin Disruption in the Mature Spinal Cord

The neuroanatomical regeneration and functional recovery observed in myelin-suppressed embryonic chicks prompted the development of a protocol for removing myelin from the mature chick spinal cord. We observed that a similar immunological treatment does not result in demyelination, but instead produces a severe alteration of myelin structure within the hatchling chick spinal cord (33).

The transient disuption of spinal cord myelin in the hatchling chick was achieved by infusing serum complement proteins plus GalC or 04 antibodies directly into the low thoracic spinal cord via an osmotic mini-pump. The subsequent state of spinal cord myelin was assessed with MBP immunohistochemistry and electon microscopy. Within 24 hours of treatment, several segments of the spinal cord on either side of the infusion site showed prominent myelin disruption, characterized by an 'unravelling' of myelin lamella and a separation of the myelin sheath from the axon (Fig. 3). MBP immunoreactivity was undetectable within this region of myelin disruption. Myelin disruption can be maintained for as long as the complement and antibody are infused and the animals have no untoward side effects due to the treatment (14 days being the longest infusion period tested). Upon termination of the treatment, MBP immunoreactivity reappears in the disrupted region within one week, indicating myelin recovery.

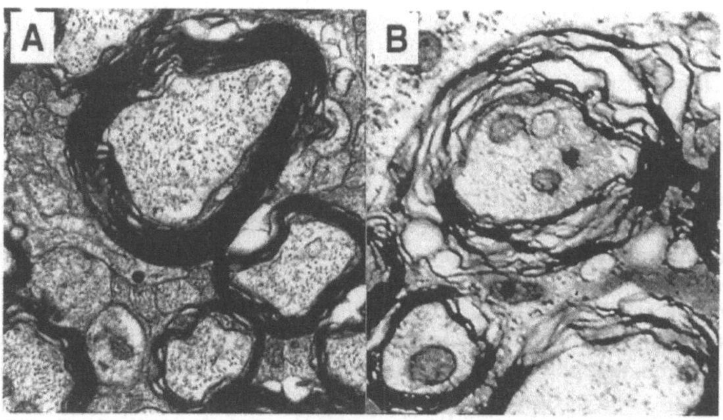

Figure 3. Electron micrographs of the hatchling chick spinal cord. **A**: Normally-myelinated, unoperated spinal cord showing compacted myelin. **B**: Myelin disrupted spinal cord showing uncompacted myelin; the major myelin markers are not detectable within regions of myelin disruption. (from Keirstead et al., 1995)

Oligodendrocyte cell bodies were identified by their morphology within the region of myelin disruption, indicating that they survived the immunological protocol. The rapid onset of myelin recovery following myelin disruption is also consistent with the survival of oligodendrocytes. Preliminary data using a transgenic mouse line where oligodendrocytes can be unequivocally identified (due to a LacZ reporter gene coupled to the MBP promotor region) also suggested that the oligodendrocyte cell bodies survived the immunological protocol (H.S. Keirstead and A.C. Peterson, unpublished data). Astrocytes were identified by their morphology within the region of myelin disruption, and did not appear to be hypertrophic.

As mentioned above, the alterations of myelin structure revealed by ultrastructural analysis were accompanied by a loss of immunoreactivity for MBP. We have also observed a loss of MAG and PLP immunoreactivity in immunologically treated spinal cords. This indicates that a lack of immunoreactivity for myelin markers does not necessarily indicate a loss of myelin. It is conceivable that the loss of MBP immunoreactivity following myelin disruption was a result of excess GalC or 04 antibodies 'coating' the myelin membranes, effectively masking other epitopes. It is also plausible that the gross alterations of myelin structure were accompanied by alterations in protein structure. Nevertheless, the presence or absence of specific myelin proteins within disrupted myelin should be confirmed with appropriate western blot analysis.

4.1 Regeneration of the Mature Spinal Cord

Myelin disruption was initiated in the hatchling chick spinal cord in an attempt to rescue previously axotomized brainstem-spinal projections. The immunological protocol was initiated shortly after complete thoracic transection of the spinal cord, and neuroanatomical regeneration of brainstem-spinal projections was assessed 2-4 weeks after spinal cord injury by counting the brainstem neurons retrogradely-labelled by an injection of a retrograde tracer into the lumbar spinal cord (for details see 31, 32, 33). Myelin disruption facilitated approximately 6-19% neuroanatomical regeneration of severed brainstem-spinal neurons (Fig. 4; 33). This assessment of axonal regeneration was based on comparisons with the number of labelled brainstem-spinal neurons observed in untransected or sham-operated animals. Double-labelling studies indicated that true axonal regeneration contributed to the neuroanatomical recovery (33).

Neuroanatomical regeneration of hatchling chicks was accompanied by functional synaptogenesis (33). Although hatchling chicks that had undergone spinal transection and myelin disruption did not show any voluntary signs of locomotor recovery, focal electrical stimulation of a brainstem locomotor region (in the decerebrate animal) evoked rhythmic motor activity in the right and left legs (Fig. 4). These findings indicate that neuroanatomical regeneration of brainstem-spinal projections following spinal cord transection in myelin-disrupted hatchling chicks is accompanied by functional synaptogenesis with 'target' spinal cord interneurons or motoneurons.

These studies indicate that demyelination is not required to render the CNS extraneuronal environment permissive for axonal regeneration. It is pausible that immunological myelin disruption alters proteins other than MBP. Myelin-associated proteins that have been shown to inhibit axonal growth and regeneration (37, 41, 51) may also be altered by immunological myelin disruption. We cannot, however, discount the possibility that immunological myelin disruption facilitates axonal regeneration indirectly by altering the expression or release of other cellular factors within the spinal cord.

The neuroanatomical regeneration of brainstem-spinal projections in transected, myelin disrupted hatchling chicks did not result in any overt behavioral signs of locomotor recovery. It is recognized that adult birds supported on a treadmill are capable of locomotion, mediated by the intrinsic activity of neural networks confined to the lumbar spinal cord (55, 61). Although spontaneous alternating leg movements were often observed in transected, myelin disrupted hatchling chicks, this activity could have been initiated by either intrinsic spinal circuits or descending supraspinal projections. Our brainstem stimulation experiments demonstrated that at least some of the regenerating brainstem-spinal axons were capable of forming functional synapses with target spinal neurons, which could then be activated in a behaviorally appropriate manner.

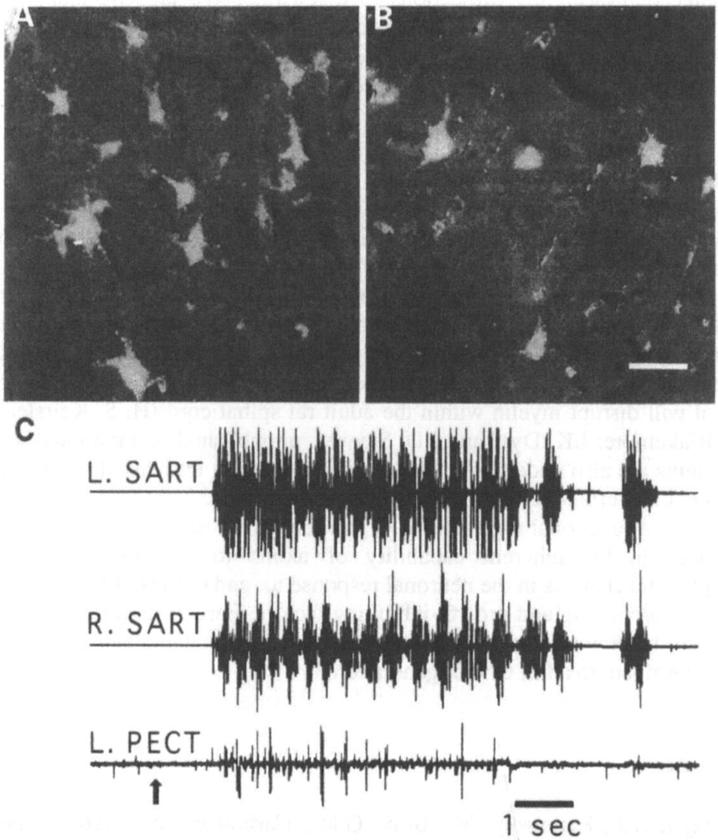

Figure 4. Neuroanatomical regeneration and physiological activity approximately 3 weeks after spinal cord transection and myelin disruption in a hatchling chick. **A**: The normal number and distribution of reticulospinal neurons in a normally-myelinated, unoperated hatchling. **B**: Approximately 6-19% of the normal number of retrograde labelled reticulospinal neurons is evident in experimental animals, indicating partial neuroanatomical regeneration. **C**: Focal electrical stimulation of a brainstem locomotor region evoked alternating stepping activity in the animal referred to in *B*. Electromyographic recordings are from the left and right sartorius and right pectoralis muscles. Brainstem stimulation at 50μA evoked motor activity in both the wings and legs. In this trial, the wings were held to prevent movement artifacts from appearing in the leg EMG records. (from Keirstead et al., 1995)

However, these studies do not allow us to determine whether the brainstem-evoked nerve and muscle activity is only due to direct synaptic connections onto lumbar locomotor interneurons and motoneurons, or also via regenerating propriospinal projections from more rostral regions of the spinal cord. The lack of voluntary locomotor behavior in transected, myelin disrupted hatchling chicks may be due to the failure of regenerating axons to form 'appropriate' synaptic connections. Alternatively, functional locomotor behavior may require a greater percentage of descending projections to regenerate. We propose that the most likely reason for the lack of voluntary locomotor behavior in transected, myelin disrupted hatchling chicks may have been the significant muscular atrophy that occurred over the 3-4 week recovery period.

5. Future Outlook

These studies indicate that CNS myelin is inhibitory to the neuroanatomical regeneration and physiological recovery of transected spinal cord axons in the late-embryonic and hatchling chick. Due to the presence of myelin-associated inhibitors in mammalian CNS white matter (37, 41, 51) it is highly probable that myelin disruption of the mammalian spinal cord would also facilitate a regenerative response. Preliminary studies indicate that a similar immunological protocol will disrupt myelin within the adult rat spinal cord (H. S. Keirstead and W. F. Blakemore; J.K. Dyer and J.D. Steeves, unpublished observations). Further experiments are also underway to determine whether an intrathecal administration route is equally effective in producing myelin disruption.

The lack of functional regeneration in the adult CNS may be due to age-related differences in the inherent capability of axons to regenerate, as well as developmental changes in the neuronal response to, and extraneuronal distribution of, growth-promoting and growth-inhibiting factors. Further research is underway in several laboratories in an effort to identify and manipulate the biological systems pertinent to adult CNS regeneration.

References

1. Aguayo, A.J., Raminsky, M., Bray, G.M., Carbonetto, S., McKerracher, L., Villegas-Perez, M.P., Vidal-Sarz, M. and Carter, D.A. (1991) Degenerative and regenerative responses of injured neurons in the central nervous system of adult mammals [Review]. Philosophical Transactions of the Royal Society of London- Series B: Biological Sciences. 331: 337-343.
2. Armstrong, D.M. (1986) Supraspinal contributions to the initiation and control of locomotion in the cat. Prog. Neurobiol. 26: 273-361.
3. Bandtlow, C., Zachleder, T. and Schwab, M.E. (1990) Oligodendrocytes arrest neurite growth by contact inhibition. J. Neurosci. 10: 3837-3848.
4. Barber, P.C. (1981) Axonal growth by newly-formed vomeronasal neurosensory cells in the normal adult mouse. Brain Res. 216: 229-237.

5. Bekoff, A. (1976) Ontogeny of leg motor output in the chick embryo: a neural analysis. Brain Res. 106: 271-291.
6. Bensted, J.P.M., Dobbing, J., Morgan, R.S., Reid, R.T.W. and Payling Wright, G. (1957) Neuroglial development and myelination in the spinal cord of the chick embryo. J. Embryol. Exp.Morphol. 5: 428-437.
7. Bjorklund, A., Katzman, R., Stenevi, U. and West, K.A. (1971) Development and growth of axonal sprouts from noradrenaline and 5-hydroxytryptamine neurons in the rat spinal cord. Brain Res. 31: 21-33.
8. Bjorklund, A. and Stenevi, U. (1979) Regeneration of monoaminergic and cholinergic neurons in the mammalian central nervous system. [Review] Physiol. Rev. 59: 62-100.
9. Bjorklund, A. (1991) Neural transplantation- an experimental tool with clinical possibilities. [Review] TINS. 14: 319-322.
10. Caroni, P. and Schwab. M.E. (1988) Antibody against myelin-associated inhibitor of neurite growth neutralizes nonpermissive substrate properties of CNS white matter. Neuron 1: 85-96.
11. Carroll, W.M., Jennings, A.R. and Mastaglia, F.L. (1984) Experimental demyelinating optic neuropathy induced by intraneural injection of galactocerebroside antiserum. J. Neurol. Sci. 65: 125-135.
12. Carroll, W.M., Jennings, A.R. and Mastaglia, F.L. (1985) Immunocytochemical study of the glial cell response in antibody-mediated optic nerve demyelination. Neurosci. Lett. Suppl. 19: S49.
13. David, S. and Aguayo, A.J. (1981) Axonal elongation into peripheral nervous system "bridges" after central nervous system injury in adult rats. Science. 214: 931-933.
14. Dellman, H.P. (1973) Degeneration and regeneration of neurosecretory systems [Review]. Int. Rev. Cytol. 36: 215-315.
15. Dorfman, S.H., Fry, J.M. Silberberg, D.H. (1979) Antiserum induced myelin inhibition *in vitro* without complement Brain Res. 177: 105-114.
16. Dubois-Dalcq, M., Niedieck, B. and Buyse, M. (1970) Action of anti-cerebroside sera on myelinated tissue cultures. Pathol. Eur. 5: 331-347.
17. Dyer, C.A. & Benjamins, J.A. (1990) Glycolipids and transmembrane signaling: antibodies to galactocerebroside cause an influx of calcium in oligodendrocytes J. Cell Biol. 111: 625-633.
18. Eidelberg, E., Story, J.L., Walden, J.G. and Meyer, B.L. (1981) Anatomical correlates of return of locomotor function after partial spinal cord lesions in cats. Brain Res. 42: 81-88.
19. Fawcett, J.W. (1992) Intrinsic neuronal determinants of regeneration. Trends Neurosci. 15:5-8.
20. Fisher, L.J. and Gage, F.H. (1993) Grafting in the mammalian central nervous system. [Review] Physiol. Rev. 73: 583-616.
21. Foran, D.R. and Peterson, A.C. (1992) Myelin acquisition in the central nervous system of the mouse revealed by an MBP-Lac Z trangene. J. Neurosci. 12: 4890-4897.
22. Fry, J.M., Weissbarth, S., Lehrer, G.M. and Bornstein, M.B. (1974) Cerebroside antibody inhibits sulfatide synthesis and myelination and demyelinates in cord tissue cultures. Science. 183: 540-542.

23. Glover, J.C. and Petursdottir, G J. (1991) Regional specificity of developing reticulospinal, vestibulospinal and vestibulo-ocular projections in the chicken embryo. Neurobiol. 22:353-376.
24. Glover, J.C. (1993) The development of brain stem projections to the spinal cord in the chicken embryo. Brain Res. Bull. 30: 265-272.
25. Hamburger, V. and Levi-Montalcini, R. (1949) Proliferation, differentiation and degeneration in the spinal cord ganglia of the chick embryo under normal and experimental conditions. J. Exp. Zool. 111: 457.
26. Hartman, B.K., Agrawal, H.C., Kalmbaach, S. and Shearer, W.T. (1979) Development and maturation of central nervous system myelin: comparison of immunohistochemical localization of proteolipid protein and basic protein in myelin and oligodendrocytes. J. Comp. Neurol. 188: 273-290.
27. Hasan, S.J., Nelson, B.H., Valenzuela, J.I., Keirstead, H.S., Schull, S.E., Ethell, D.W. and Steeves, J.D. (1991) Restor. Neurol. Neurosci. 2:137-154.
28. Hasan, S.J., Keirstead, H.S., Muir, G.D. and Steeves, J.D. (1993) Axonal regeneration contributes to repair of injured brainstem-spinal neurons in embryonic chick. J. Neurosci. 13: 492-507.
29. Holder, N. and Clarke, J.D.W. (1988) Is there a correlation between continuous neurogenesis and directed axon regeneration in the vertebrate nervous system? TINS. 11: 94-99.
30. Hruby, S., Alvord, E.C., Jr and Seil, F.J. (1977) Synthetic galactocerebrosides evoke myelination-inhibiting antibodies, Science. 195: 173-175.
31. Keirstead, H.S., Hasan, S.J., Muir, G.D. and Steeves, J.D. (1992) Suppression of the onset of myelination extends the permissive period for the functional repair of embryonic spinal cord. Proc. Natl. Acad. Sci. 89: 11664-11668.
32. Keirstead, H.S., Pataky, D.M., Wisniewska, A.B., Schachner, M. and Steeves, J.D. (1995) *In vivo* immunological suppression of spinal cord myelin development. (submitted)
33. Keirstead, H.S., Dyer, J.K., Sholomenko, G.N., McGraw, J., Delaney, K.R. and Steeves, J.D. (1995) Axonal regeneration and physiological activity following transection and immunological disruption of myelin within the hatchling chick spinal cord. J. Neurosci. (in press)
34. Kromer, L.F., Bjorklund, A. and Stenevi, U. (1981) Regeneration of the septohippocampal pathways in adult rats is promoted by utilizing embryonic hippocampal implants as bridges. Brain Res. 210: 173-200.
35. Li, D., Field, P.M. and Raisman, G. (1995) Failure of axon regeneration in postnatal slice entorhino-hippocampal slice coculture is due to maturation of the axon, not that of the pathway or target. Eur. J. Neurosci. 7: 1164-1171.
36. McClellan, A.D. (1990) Locomotor recovery in spinal-transected lamprey: role of functional regeneration of descending axons from brainstem locomotor command neurons. Neuroscience 37: 781-798
37. McKerracher, L., David, S., Jackson, D.L., Kottis, V., Dunn, R.J. and Braun, P.E. (1994) Identification of myelin-associated glycoprotein as a major myelin-derived inhibitor of neurite growth. Neuron (in press).
38. Macklin, W.B. and Weill, C.L. (1985) Appearance of myelin proteins during development in the chick central nervous system. Dev. Neurosci. 7: 170-178.

39. Mastaglia, F.L., Carrol, W.M. and Jennings, A.R. (1989) Spinal cord lesions induced by antigalactocerebroside serum Clin. Exp. Neurol. 26: 33-44.
40. Morell, P., Quarles, R.H. and Norton, W.T. Formation, structure and biochemistry of myelin. In Siegel, G.J. et al (eds.), Basic Neurochemistry: Molecular, cellular, and medical aspects, 4th ed., New York: Raven Press, 1989, pp. 109-136.
41. Mukhopadhyay, G., Doherty, P., Walsh, F.S., Crocker, P.R. and Filbin, M. (1994) A novel role for myelin-associated glycoprotein as an inhibitor of axonal regeneration. Neuron. 13: 757-767.
42. O'Donovan, M., Sernagor, E., Sholomenko, G., Ho, S., Antal, M. and Yee, W. (1992) Development of spinal motor networks in the chick embryo. J. Exp. Zool. 261: 261-73.
43. Okado, N. and Oppenheim, R.W. (1985) The onset and development of descending pathways to the spinal cord in chick embryo. J. Comp. Neurol. 232:143-161.
44. Ozawa, K., Saida, T., Saida, K., Nishitani, H. & Kameyama, M. (1989) *In vivo* CNS demyelination mediated by anti-galactocerebroside antibody *Acta Neuropathol.* 77: 621-628.
45. Ramon y Cajal, S. (1959) Degeneration and regeneration of the nervous system. (New York: Oxford UP, 1928) Reprint (May R.M., trans., ed.). New York: Hafner.
46. Ranscht, B., Clapshaw, P.A., Price, J., Noble, M. and Seifert, W. (1982) Development of oligodendrocytes and schwann cells studies with a monoclonal antibody against galactocerebroside. Proc. Natl. Acad. Sci. 79: 2709-2713.
47. Rosenbluth, J., Lui, Z., Guo, D. and Schiff, R. (1994) Inhibition of CNS myelin development *in vivo* by implantation of anti-GalC hybridoma cells. J. Neurocytol. 23: 699-707.
48. Sarnat, H. and Netsky, M. (1981) Evolution of the nervous system. New York, NY: Oxford UP.
49. Schnell, L. and Schwab, M.E. (1990) Axonal regeneration in the rat spinal cord produced by an antibody against myelin-associated neurite growth inhibitors. Nature (London) 343: 269-272.
50. Schreyer, D.J. and Jones, E.G. (1982) Growth and target finding by axons of the corticospinal tract in prenatal and postnatal rats. Neuroscience 7: 1837-1853.
51. Schwab, M.E. and Caroni, P. (1988) Oligodendrocytes and CNS myelin are nonpermissive substrates for neurite growth and fibroblast spreading *in vitro*. J. Neurosci. 8: 2381-2393.
52. Sergott, R.C., Brown, M.J., Silberberg, D.H. & Lisak, R.P. (1984) Antigalactocerebroside serum demyelinates optic nerve *in vivo*. J. Neurol. Sci. 64: 297-303.
53. Shiga, T., Kunzi, R. and Oppenheim, R.W. (1991) Axonal projections and synaptogenesis by supraspinal descending neurons in the spinal cord of the chick embryo. J. Comp. Neurol. 305: 83-95.
54. Shimizu, I., Oppenheim, R.W., O'Brian, M. and Schneiderman, A. J. (1990) Anatomical and functional recovery following spinal cord transection in the chick embryo. J. Neurobiol. 21:918-937.

55. Sholomenko, G.N. and Steeves, J.D. (1987) Effects of selective spinal cord lesions on hind limb locomotion in birds. Exp. Neurol. 95: 403-418.
56. Sholomenko, G.N., Funk, G.D. and Steeves, J.D. (1991a) Avian locomotion activated by brainstem infusion of neurotransmitter agonists and antagonists. I. Acetylcholine, excitatory amino acids and substance P. Exp. Brain Res. 85: 659-673.
57. Sholomenko, G.N., Funk, G.D. and Steeves, J.D. (1991b) Avian locomotion activated by brainstem infusion of neurotransmitter agonists and antagonists. II. Gamma-aminobutyric acid. Exp. Brain Res. 85: 674-681.
58. Sholomenko, G.N., Funk, G.D. and Steeves, J.D. (1991c) Locomotor activities in the decerebrate bird without phasic afferent input. Neuroscience. 40: 257-266.
59. Sommer, I. and Schachner, M. (1981) Monoclonal antibodies (01 to 04) to oligodendrocyte cell surfaces: an immunocytological study in the central nervous system. Exp. Biol. 83: 311-327.
60. Steeves, J.D. and Jordan, L.M. (1980) Localization of a descending pathway in the spinal cord which is necessary for controlled treadmill locomotion. Neurosci. Lett. 20: 283-288.
61. Steeves, J.D., Sholomenko, G.N. and Webster, D.M.S. (1987) Stimulation of the pontomedullary reticular formation initiates locomotion in decerebrate birds. Brain Res. 401:205-212
62. Steeves, J.D., Hasan, S.J., Keirstead, H.S., Muir, G.D., Ethell, D.W., Pataky, D.M., McBride, C.B., Rott, M.E. and Wisniewska, A.B. (1993) The embryonic chicken as a model for central nervous system injury and repair. Neuroprotocols. 3: 35-43.
63. Tetzlaff, W., Alexander, S., Miller, F. and Bisby, M. (1991) Response of facial and rubrospinal neurons to axotomy: changes in mRNA expression for cytoskeletal proteins and GAP-43. J. Neurosci. 11: 2528-2544.
64. Treherne, J.M., Woodward, S.K.A., Varga, Z.M. Ritchie, J.M. and Nicholls, J.G. (1992) Restoration of conduction and growth of axons through injured spinal cord of neonatal oppossum in culture. Proc. Natl. Acad. Sci. U.S.A. 89: 432-434.
65. Tuszynski, M.H., Peterson, D.A., Ray, J., Baird, A., Nakahara, Y. and Gage, F.H. (1994) Fibroblasts genetically modified to produce nerve growth factor induce robust neuritic ingrowth after grafting to the spinal cord. Exp. Neurol. 126: 1-14.
66. Valenzuela, J.I., Hasan, S.J. and Steeves, J.D. (1990) Stimulation of the brainstem reticular formation evokes locomotor activity in embryonic chicken, in ovo. Dev. Brain Res. 56: 13-18.
67. Webster, D.M.S. and Steeves, J.D. (1988) Origins of brainstem-spinal projections in the duck and goose. J. Comp. Neurol. 273: 573-583.
68. Webster, D.M. and Steeves, J.D. (1991) Funicular organization of avian brainstem-spinal projections. J. Comp. Neurol. 312: 467-476.
69. Zemlan, F.P., Kow, L.M. and Pfaff, D.W. (1983) Effect of interuption of bulbospinal pathways on lordosis, posture and locomotion. Exp. Neurol. 81: 177-194.

Does Myelin Inhibit Regeneration and Remyelination?

Stephen J. Moorman

Department of Anatomy and Cell Biology, University of North Texas Health Science Center at Fort Worth, Fort Worth, TX 76107

Abstract. Neither regeneration nor remyelination occur readily in the mammalian central nervous systems. Regeneration is thought to be limited by inhibition of neuronal growth cone motility. Likewise, inhibition of oligodendrocyte motility might play a role in limiting remyelination. *In vitro*, inhibition of the neuronal growth cone and oligodendrocytes is indicated by collapse of the fine structure upon application of myelin extract. In both cases, the inhibition is accompanied by an increase in internal free calcium concentration. In addition, neuronal growth cone motility is inhibited by contact with an oligodendrocyte. Contact between two oligodendrocytes also causes an inhibition of oligodendrocyte motility. Both cases of contact-induced inhibition of motility are physiologically similar to the myelin-induced inhibition. These similarities suggest that a variety of inhibitory environmental influences probably exert their effects through common intracellular pathways, even in different cell types.

Key words. calcium, cell-cell contact, cell motility, FURA2, growth cone morphology, multiple sclerosis, myelination, oligodendrocyte

1 Introduction.

My work is motivated by the hypothesis that many of the cellular and molecular mechanisms that have been shown to be important during neuronal development and myelination will also prove to be relevant to neuronal regeneration and remyelination. In the last decade, much has been learned about neurogenesis, axonogenesis, neuronal growth cone navigation, synaptogenesis, and myelination in the developing nervous system. In a number of cases, detailed control mechanisms have been elucidated at the cellular and molecular level. One example is the substantial body of evidence indicating that changes in the level of intracellular free calcium can play an important role in the behavior of neuronal growth cones (see 12). A second example is the identification of myelin molecules responsible for compaction of myelin lamellae (34).

For recovery of function to occur in an injured mammalian central nervous system, many developmental processes will probably be utilized. Regenerating neuronal growth cones must 1) navigate either around or through the site of the

Molecular Signaling and Regulation in Glial Cells
A Key to Remyelination and Functional Repair

injury, 2) recognize and follow an appropriate pathway, and 3) recognize and synapse with their appropriate target cells. For remyelination to occur in the central nervous system after demyelination due to injury or disease such as multiple sclerosis, oligodendrocytes must migrate into demyelinated regions, and recognize and ensheath demyelinated axons. Most of these requirements are a recapitulation of processes that neurons and oligodendrocytes perform with great precision during development.

1.1 Regeneration.

In this paper I concentrate on a neuronal growth cone behavior referred to as growth cone collapse. Growth cones of a variety of neurons have been demonstrated to undergo collapse *in vitro* (see 13). Growth cone collapse is usually quite rapid and is characterized by a cessation of filopodial movements, a general loss of filopodia, and retraction of the fine structure of the growth cone so that the overall surface area of the growth cone decreases. In addition to the growth cone collapse, the distal neurite often retracts. Growth cone collapse can occur spontaneously, but most commonly collapse is the result of contact with specific types of cells (see 13). Cell surface molecules that are potential signals triggering collapse are currently under intensive scrutiny. Although a great deal of progress has been made in understanding the cellular and molecular mechanisms that trigger growth cone collapse in some model systems, an important caveat to all studies in this area is that there is no conclusive evidence that growth cones collapse *in vivo* either during development or during regeneration. It may be some time before such information is available because of technical difficulties in imaging elongating axons *in vivo*.

Growth cone collapse might prevent regeneration in the spinal cord. In support of this idea are the long-standing observations that, although many neurons survive an acute injury and reform growth cones, these growth cones have great difficulty traversing the damaged area (see 33). Studies over the last decade have suggested a possible explanation for these results. Since axons grow readily through grafts of peripheral nerve, the implication is that something specific to the damaged central nervous system is inhibitory to regrowth. By systematically varying the types of cells present, it was concluded that oligodendrocytes were specifically associated with the inhibitory process (2, 4, 5). This inference has been greatly strengthened by two lines of observation. First, an extract of CNS-myelin can elicit the collapse of the growth cones of dorsal root ganglion (DRG) neurons and locus coeruleus neurons *in vitro* (4, 18) but PNS-myelin does not elicit collapse. Second, when CNS-myelin is pretreated with antibodies against two CNS-myelin-specific proteins, collapse of DRG neuronal growth cones is no longer evoked (5). These results suggested the hypothesis that the site of injury in the spinal cord might present an environment that is inhibitory to regenerating growth cones. This hypothesis is also supported by the observation that the same antibodies that prevent the collapse of DRG neuronal growth cones *in vitro* significantly enhance spinal cord regeneration after injury *in vivo* (32).

1.2 Remyelination.

In vitro, cells of the oligodendrocyte lineage are highly motile (24). They extend numerous processes that resemble neurites and end in growth cone-like expansions (7, 17, 38). This resemblance to developing neurons is also seen *in vivo* (36). The neuronal growth cone has been the focus of a large number of studies aimed at understanding the signal transduction mechanisms that underlie the detection and integration of environmental cues that affect the development of axons (see 11). The growing tip of an oligodendrocyte process or oligodendroglial growth cone (oligo-GC), or its homologous structure *in vivo*, is that portion of the oligodendrocyte that transduces the environmental signals that lead to axon recognition and the initiation of myelination. Therefore, the oligo-GC serves as the focus of my studies to determine some of the signal transduction mechanisms that underlie the initiation of myelination and remyelination.

A major impediment to complete recovery from episodes in multiple sclerosis (MS), and related demyelinating diseases, appears to be the inability of the oligodendrocytes to successfully recapitulate the robust myelination that occurs during development. In MS, demyelinating plaques are widespread in the CNS, including the cerebrum, the optic nerve, the brainstem, and the spinal cord. However, the mechanism(s) that initiates demyelination is unknown. The earliest morphological changes seen in oligodendrocytes of demyelinating plaques of MS include degeneration of inner glial loops and inner myelin lamellae (28). The oligodendrocytes appear to survive the initial stages of demyelination but are depleted later (25). Oligodendrocytes are almost absent in the center but numerous around the periphery of chronic demyelinated lesions (26). This suggests that either oligodendrocytes that migrate into the center of a demyelinated region do not survive or that oligodendrocytes are not able to migrate into the center of the demyelinated region.

Remyelination in MS is represented by thinly myelinated fibers at the edge of areas of long-standing disease (27). Remyelinated fibers have disproportionately thin sheaths around large diameter axons, attenuation of myelin, short internodes, and aberrant collections of loops of oligodendroglial cell cytoplasm (23, 25). Oligodendrocytes are numerous in these areas of remyelination (26, 27, 28). This increase in number of oligodendrocytes around the periphery of a lesion might be due to proliferation of mature and immature cells of the oligodendroglial lineage in the area of the demyelination (22).

Recently, antibodies have been shown to promote remyelination in mice infected with Theiler's murine encephalomyelitis virus (TMEV). In many aspects, the TMEV model is indistinguishable from the chronic progressive form of MS (6, 30). Mice chronically infected with TMEV and treated with either of two specific monoclonal antibodies had significantly more CNS remyelination than control animals (15). The identity of the antigen(s) recognized by the antibodies is not known. However, the only cultured cells in which these antibodies recognized an extracellular antigen were from the rat glial cell line 5.5B8 (15). Rat glial cell line

5.5B8 is MBP+, CNP+, A2B5+, and O4+ (3), suggesting that they are cells in the oligodendroglial lineage. These results suggest that the antibodies enhance remyelination by either negating the inhibitory properties of an oligodendrocyte-specific surface protein or by binding to a surface receptor for an inhibitory molecule.

CNS-myelin induces collapse of the oligo-GC *in vitro* (17). Contact between oligodendrocytes has similar inhibitory effects on oligodendrocyte motility *in vitro* (16). In this paper, I will discuss these results within the context of their similarity to the effects of myelin on the neuronal growth cone. It is this similarity that suggests that inhibitory environmental influences might exert their effects through common pathways, even in different cell types. If this were true, similar therapies might enhance regeneration and enhance remyelination.

2 Materials and Methods

2.1 Preparation of Neurons and Oligodendrocytes.

Locus coeruleus neurons and oligodendrocytes were grown as previously described (17, 18). Oligodendrocytes and astrocytes were identified by their characteristic morphology when viewed with phase contrast microscopy (38). The accuracy of visual identification was verified with immunofluorescence. Cells classified as astrocytes had flat cell bodies. In 8 cultures that were fixed, permeabilized, and incubated with an antibody against glial fibrillary acid protein (Boeringer Mannheim) , >95% of the cells classified as astrocytes based on phase contrast microscopy were labeled. Cells classified as oligodendrocytes had small, round, phase bright cell bodies, an eccentric nucleus, and a few short thick, highly branched processes. In 4 cultures incubated with an antibodies against galactocerebroside (24) and O4 (35), >85% of the cells classified as oligodendrocytes based on phase contrast microscopy were labeled. Because these cells are galC+ and O4+, using the terminology of Pfeiffer et al. (21), they are be classified as "immature" oligodendrocytes.

In all of the experiments on neuronal growth cones only growth cones of large multipolar neurons were studied. Neurons in this group had phase bright cell bodies averaging 25.6 μm (+/- 3.63, n = 20) in diameter, and were clearly a distinct population from the next largest class of neurons, which averaged 16.3 μm (+/- 1.7, n = 20) in diameter. Using similar dissection and culture techniques, Masuko, et al. (14) demonstrated that > 98% of neurons with cell bodies larger than 20 μm stained positive for catecholamines. For these reasons, the neurons that were studied are referred to as locus coeruleus neurons.

2.2 Preparation of Myelin.

Extracts of CNS-myelin were prepared from adult rat spinal cords. Extracts of PNS-myelin were prepared from adult rat ischiatic nerve. Both extracts were prepared according to the methods of Norton (20).

2.3 Video-Microscopy.
Video microscopic observations of oligodendrocytes were made as previously described (17, 18). Observations were made of neonatal rat locus coeruleus neurons and oligodendrocytes that had been in primary culture for 3 to 7 days. No more than 10 cells were studied in a single dish, and over 100 different dishes of cells were used for these studies. For all experiments, the cells/dishes were maintained at 37°C during observation.

2.4 Oligodendrocyte Manipulation.
A standard patch clamp microelectrode, with a tip diameter of less than 3 μm, and motorized micromanipulator (Newport/Klinger Inc.) were used to manipulate the oligodendrocytes. To remove an oligodendrocyte from the substratum, the microelectrode tip was placed in contact with the cell body. As the microelectrode was moved the cell body remained adhered to the microelectrode tip and the oligodendrocyte's cell body and processes were pulled off the substratum. Thus, membrane from the cell body of the manipulated oligodendrocyte was attached directly to the microelectrode tip and most of the processes appeared to be hanging from the cell body. Most likely, when this manipulated oligodendrocyte was placed in contact with a neuronal growth cone or another oligodendrocyte, it was membrane of the processes of the manipulated oligodendrocyte that came in contact.

2.5 FURA2 Microscopy.
Cells were loaded with FURA2-AM as previously described (16, 17, 18). FURA2 ratio measurements were made using a photometry system that has also been described previously (16). For each oligodendrocyte or neuronal growth cone studied, an initial series of measurements were made to verify that there was a steady baseline calcium concentration. The final 30 seconds of the baseline data were stored and then the acquisition program was paused. The oligodendrocyte 'ghost' attached to the microelectrode was then brought into contact with the neuronal growth cone or oligodendrocyte being monitored. Data acquisition was then resumed.

The ratios of the amplitude of fluorescence values were converted into free calcium concentrations using the formula of Grynkiewicz et al. (9).

3 Results.
The results are described briefly in the text and summarized in Table 1. Some results, as indicated, have already been published.

3.1 Myelin Causes Inhibition of Motility (17, 18)
Once a stable growth rate had been established, myelin extract was puffed onto the locus coeruleus growth cones (LC-GCs). After a sixty minute period, the myelin was washed off, and the size and position of each growth cone were recorded. By 60 minutes after application of CNS-myelin extract, 12 out of 16 growth cones were collapsed. Neither heat-inactivated CNS-myelin (n=5) nor trypsin-treated CNS-

myelin (n=5) evoked growth cone collapse. In contrast to the effects of CNS myelin, 0 out of 5 LC-GCs collapsed after 60 minutes of contact with an extract of myelin from the adult rat ischiatic nerve (PNS-myelin extract).

When CNS-myelin was applied to oligodendrocytes for 30 minutes, and then washed off, 23 of 25 oligo-GCs had collapsed. Neither heat-inactivated (n = 10) nor trypsin-treated (n = 10) CNS-myelin evoked any change in morphology of the oligo-GC. However, PNS-myelin evoked collapse of the oligo-GC in 8 of 10 cells tested.

3.2 Myelin Induces a Calcium Increase in Both Cell Types (17, 18)
Within six minutes of application of CNS myelin, an elevation in intracellular free calcium was observed in 9 out of 12 LC-GCs studied. On average, the resting intracellular free calcium more than doubled from 94 nM to 237 nM. Treatments that failed to evoke LC-GC collapse also failed to elicit a rise in intracellular free calcium. Thus, neither PNS myelin, heat-inactivated myelin, nor trypsin-treated myelin had any effect on the level of intracellular free calcium.

Application of CNS-myelin extract onto oligodendrocytes consistently caused a large increase in [Ca]i in the oligo-GC. On average the resting intracellular free calcium increased by more than a factor of five, from 71 nM to 431 nM. This increase in [Ca]i in response to myelin did not occur if the myelin was pretreated by incubation at 100°C or by trypsin treatment. However, contact with PNS-myelin evoked a [Ca]i increase in 8 of 10 oligodendrocytes tested. The [Ca]i increase evoked in oligodendrocytes by PNS-myelin, from 83 nM to 437 nM, was not significantly different from that evoked by CNS-myelin.

3.3 Contact With Oligodendrocytes Causes Inhibition of Motility
Spontaneous interactions between LC-GCs and oligodendrocytes resulted in collapse of all LC-GCs (n = 5). To facilitate cell-cell interactions, oligodendrocytes were individually removed from the substrate with a microelectrode. Each 'manipulated' oligodendrocyte was then placed in contact with an LC-GC. After the manipulated oligodendrocyte was placed in contact with the LC-GC, most (19 of 20) LC-GCs had collapsed within 30 minutes. As a control for possible nonspecific contact-mediated effects inducing collapse of the fine structure of the LC-GC, manipulated O2A cells were used in place of oligodendrocytes. In the majority of cases (15 of 20), O2A cells did not induce collapse of LC-GCs.

Initial spontaneous interactions between oligodendrocyte pairs resulted in collapse of the oligo-GC (n = 5). To facilitate cell-cell interactions, oligodendrocytes were individually removed from the substrate with a microelectrode. Each 'manipulated' oligodendrocyte was then placed in contact with a second oligodendrocyte. After the manipulated oligodendrocyte was placed in contact with the oligo-GC, most (31 of 35) oligo-GCs had collapsed within 30 minutes. As a control for possible nonspecific contact-mediated effects inducing collapse of the fine structure of the oligo-GC, manipulated astrocytes were used in place of oligodendrocytes. Contact with astrocytes (n = 10) did not induce any significant change in morphology of the oligo-GC.

3.4 Contact With Oligodendrocytes Induces a Calcium Increase

Contact between an oligodendrocyte and an LC-GC consistently caused a large [Ca]i increase in the LC-GC. On average, the resting [Ca]i increased by more than a factor of 2, from 83 nM to 176 nM (n = 18). The [Ca]i increase occurred rapidly, within 3 seconds or less and returned to precontact levels within 1-2 minutes. As a control for possible nonspecific contact mediated effects inducing a [Ca]i increase, manipulated O2A cells were placed in contact with the LC-GCs. Contact with O2A cells (n = 5) did not induce any significant change in [Ca]i.

Contact between oligodendrocytes consistently caused a large [Ca]i increase in the oligo-GC. On average, the resting [Ca]i increased by more than a factor of 3, from 69 nM to 213 nM (n = 17). The [Ca]i increase occurred rapidly, within 3 seconds or less. Typically, the [Ca]i returned to pre-application levels with a time course that was much slower (1.5 to 2 minutes) than the increase. As a control for possible nonspecific contact-mediated effects inducing a [Ca]i increase in the oligo-GC, manipulated astrocytes were placed in contact with the oligo-GC. Contact with astrocytes (n = 10) did not induce any significant change in [Ca]i.

3.5 Pertussis Toxin Blocks the Contact-induced Inhibition

In the presence of pertussis toxin (0.0015 mg/mL), manipulated contact between an oligodendrocyte and an LC-GC no longer resulted in collapse of the LC-GC (n = 20).

In the presence of pertussis toxin (0.0005 mg/mL), manipulated-contact between two oligodendrocytes no longer induced any change in morphology of the oligo-GC (n = 25).

Table 1. Summary of the results of interactions of locus coeruleus growth cones (LC-GC) and oligodendroglial growth cones (oligo-GC) with the types of myelin or the types of cells indicated. (nd = not done)

	% That Collapsed		[Ca]i Increase (nM)	
	LC-GC	Oligo-GC	LC-GC	Oligo-GC
CNS-Myelin	75	92	143	360
PNS-Myelin	0	80	0	354
Heat Inactivated CNS-Myelin	0	0	0	0
Trypsin-Treated CNS-Myelin	0	0	0	0
Oligodendrocyte (Spontaneous)	100	100	nd	nd
Oligodendrocyte (Manipulated)	95	89	93	144
Astrocyte (Manipulated)	0	25	0	0
Oligodendrocyte + Pertussis Toxin	15	0	nd	nd

4 Discussion.

The previous studies of neuronal growth cone collapse suffer two significant drawbacks in extrapolating their relevance to spinal cord injury. First, most of the experiments used a model system, neurons from the peripheral nervous system, rather than CNS neurons. I felt that it was important to study neurons whose cell bodies lie in the CNS and whose axons would probably be damaged by a spinal cord injury. I chose to study growth cones of locus coeruleus neurons from the neonatal rat brainstem. One reason for this choice is that these cells can be readily identified in cell culture (6). In addition, the noradrenaline produced by locus coeruleus neurons and released in the lumbar region of the spinal cord has well defined effects on spinal cord reflexes (10, 19) and locomotion (8). Therefore, it should be possible to monitor recovery of function due to regeneration of locus coeruleus neurons *in vivo*.

The second drawback the previous studies, including our past studies, suffer is that there is no *a priori* reason to assume that the neuronal growth cone will respond to contact with cells that produce inhibitory molecules in the same manner that the neuronal growth cone responds to contact with each purified inhibitory molecule. The neuronal growth cone integrates all the environmental influences. This is particularly relevant since the cellular mechanisms that underlie contact with one purified inhibitory myelin molecule (1) and the cellular mechanisms that underlie contact with crude myelin extract (18) appear to differ. During regeneration after an injury to the CNS, the neuronal growth cone might contact both myelin debris and oligodendrocytes. Contact with crude myelin extract can be considered a good model for neuronal growth cone contact with myelin debris. Until now, the cellular mechanisms that underlie neuronal growth cone contact with oligodendrocytes have remained uninvestigated.

Molecules in myelin and on the surface of oligodendrocytes induce collapse of LC-GCs and collapse of the oligo-GC. Since PNS-myelin has no effect on LC-GCs but causes collapse of the oligo-GC, different myelin molecules are probably involved in inducing collapse in the two different cell types. However, the two cell types respond to contact with these inhibitory molecules in very similar ways. The results of the experiments using pertussis toxin suggest that the recognition of the myelin molecules by receptors on the cell surface leads to the activation of one or more G-proteins in both cell types. The G-protein activation in turn leads to a [Ca]i increase which, at least in oligodendrocytes, plays a causal role in the rapid cytoskeletal reorganization that undoubtedly underlies the growth cone collapse in both cell types. However, a direct G-protein mediated cytoskeletal reorganization cannot be ruled out in either cell type. These results suggest that neurons and oligodendrocytes have similar cellular mechanisms for transducing environmental signals into meaningful growth cone behaviors. This makes it tempting to speculate about the potential similarities between target recognition by the neuronal growth cone and axon recognition by the oligo-GC.

It has become well accepted that collapse of the neuronal growth cone probably plays a role in limiting spinal cord regeneration. A role for similar inhibitory mechanisms in limiting remyelination can now be proposed. Antisera or immunoglobulin directed against spinal cord homogenates (29, 31) and specific antibodies (15) promote remyelination in an animal model of MS. This suggests that there are factors which inhibit remyelination and that antibodies to these factors neutralize their inhibitory effects. CNS-myelin has inhibitory effects on oligodendrocytes *in vitro* (17). These results suggest that contact with myelin or myelin debris, *in vivo*, actively and specifically inhibits oligodendrocyte motility and migration into demyelinated areas and thereby might limit remyelination.

Since myelin is a product of mature oligodendrocytes, it is possible that intercellular contact between oligodendrocytes might have inhibitory effects similar to contact with myelin. *In vitro*, oligodendrocytes express all the major components of myelin in the absence of contact with an axon (37). *In vitro*, contact between adjacent oligodendrocytes causes collapse of the oligo-GC similar to that induced by contact with myelin (16). *In vivo*, this type of interaction could affect the process of remyelination by oligodendrocytes. Oligodendrocytes might already express the inhibitory myelin molecule(s) when they begin to repopulate demyelinated areas in MS. The inhibitory effects of the contact between oligodendrocytes might prevent oligodendrocytes from migrating very far into demyelinated areas and might also prevent remyelination of axons from being initiated.

In vivo, there are undoubtedly a myriad of molecules and influences in the local environment of an injury or demyelinated region that might be inhibitory to regenerating neuronal growth cones and remyelinating oligodendrocytes. It is the neuronal growth cone that integrates these environmental factors and transduces them into meaningful behaviors such as pathway and target recognition. By analogy, it is the oligo-GC that integrates the environmental factors and transduces them into meaningful behaviors such as myelination or remyelination. The similarity of the inhibitory response of LC-GCs to the inhibitory response of the oligo-GC supports the idea that many inhibitory influences might exert their effects through common cellular mechanisms. If this is true, it might be easier to manipulate the cellular mechanisms and enhance regeneration and remyelination than it is to negate all of the inhibitory influences. Therefore, knowing the cellular mechanisms that underlie neuronal and oligodendroglial growth cone behaviors provides a rational basis for developing therapies to enhance regeneration, functional reinnervation, and remyelination of the injured or diseased central nervous system.

5 Acknowledgments.

I thank Robert M. Gould and Denise Dehnbostel for comments on the manuscript. This work was supported by: Young Investigator Fellowships from the Marine Biological Laboratory at Woods Hole; Faculty Research Awards from the UNT Health Science Center at Fort Worth; The American Paralysis Association.

6 References.

1. Bandtlow, C.E., Schmidt, M.F., Hassinger, T.D., Schwab, M.E. and Kater, S.B. (1993) Role of intracellular calcium in NI-35-evoked collapse of neuronal growth cones. Science 259:80-83
2. Bandtlow, C.E., Zachleder, T. and Schwab, M.E. (1990) Oligodendrocytes arrest neurite growth by contact inhibition. J. Neurosci. 10:3837-3848
3. Bozyczko, D., deFarra, F. and McMorris, F.A. (1990) Production of immortalized cells of the oligodendroglial lineage. Ann. NY Acad. Sci. 605:350-353
4. Caroni, P. and Schwab, M.E. (1988) Two membrane protein fractions from rat central myelin with inhibitory properties for neurite growth and fibroblast spreading. J. Cell Biol. 106:1281-8
5. Caroni, P. and Schwab, M.E. (1988) Antibody against myelin-associated inhibitor of neurite growth neutralizes nonpermissive substrate properties of CNS white matter. Neuron 1:85-96
6. Dal Canto, M.C. and Lipton, H.L. (1977) Multiple Sclerosis: animal model. Theiler's virus infection in mice. Am. J. Pathol. 88:497-500
7. Fok-Seang, J. and Miller, R.H. (1992) Astrocyte precursors in neonatal rat spinal cord cultures. J. Neurosci. 12:2751-2764
8. Forssberg, H. and Grillner, S. (1973) The locomotion of the acute spinal cat injected with clonidine i.v.. Brain Res. 50:184-186
9. Grynkiewicz, G., Poenie, M. and Tsien, R.Y. (1985) A new generation of calcium indicators with greatly improved fluorescent properties. J. Biol. Chem. 260:2440-3450
10. Igarashi, M., Strittmatter, S.M., Vartanian, T. and Fishman, M.C. (1993) Mediation by G proteins of signals that cause collapse of growth cones. Science 259:77-80
11. Jankowska, E., Jukes, M.G.M., Lund, S. and Lundberg, A. (1967): The effect of DOPA on the spinal cord; 5. Reciprocal organization of pathways transmitting excitatory action to alpha motoneurons of flexors and extensors. Acta Physiol. Scand. 70:369-388
12. Kater, S.B. and Guthrie, P.B. (1990) Neuronal growth cone as an integrator of complex environmental information. Cold Spring Harbor Symposia on Quantitative Biology. 55:359-70
13. Kater, S.B. and Mills, L.R. (1991) Regulation of growth cone behavior by calcium. J. Neurosci. 11:891-899
14. Keynes, R.J. and Cook, G.M.W. (1992) Repellent cues in axon guidance. Current Opinion in Neurobiology 2:55-59
15. Masuko, S., Nakajima, Y., Nakajima, S. and Yamaguchi, K. (1986) Noradrenergic neurons from the Locus Coeruleus in dissociated cell culture; Culture methods, morphology, and electrophysiology. J. Neurosci. 6:3229-3241

16. Miller, D.J., Sanborn, K.S., Katzmann, J.A. and Rodriguez, M. (1994) Monoclonal autoantibodies promote central nervous system repair in an animal model of multiple sclerosis. J. Neurosci. 14:6230-6238
17. Moorman, S.J. (1995) The inhibition that results from contact between two oligodendrocytes in vitro can be blocked by pertussis toxin. Glia (submitted)
18. Moorman, S.J. and Hume, R.I. (1994): Contact with myelin in-vitro evokes a release of calcium from internal stores in neonatal rat oligodendrocytes. Glia 10:202-210
19. Moorman, S.J. and Hume, R.I. (1993) Omega-conotoxin prevents myelin-evoked growth cone collapse in neonatal-rat locus coeruleus neurons in vitro. J. Neurosci. 13:4727-4736
20. Moorman, S.J. and Whalen, L.R. (1993) A model system to determine the effects of specific neurotransmitters on segmental reflexes in the spinal cord of the rat. J. Neurosci. Meth. 46:73-81
21. Norton, W.T. (1976) Isolation of myelin from nerve tissue. In Methods in Enzymology, vol. 31, S. Fleischer and L. Packer (eds.), Academic Press, NY. pp. 435-444
22. Pfeiffer, S.E., Warrington, A.E. and Bansal, R. (1993) The oligodendrocyte and its many cellular processes. Trends Cell. Biol. 3:191-197.
23. Prayoonwiwat, M.D. and Rodriguez, M. (1993) The potential for oligodendrocyte proliferation during demyelinating disease. J. Neuropathol. Exp. Neurol. 52:55-63
24. Prineas JW and Connell F (1979) Remyelination in multiple sclerosis. Ann. Neurol. 5:22-31
25. Raff, M.C., Mirsky, R., Fields, K.L., Lisak, R.P., Dorfman, S.H., Silberberg, D.H., Gregson, N.A., Liebowitz, S. and Kennedy, M.C. (1978) Galactocerebroside is a specific cell surface antigenic marker for oligodendrocytes in culture. Nature 274:813-816
26. Raine, C.S. (1983) Multiple sclerosis and chronic relapsing EAE: Comparative ultrastructural neuropathology. In: Hallpike J, Adams CWM, Tourtellotte WW (eds): Multiple sclerosis: the patient, the disease, the treatment. London, Chapman and Hall, pp 413-460
27. Raine, C.S., Scheinberg, L. and Waltz, J.M. (1981) Multiple sclerosis: Oligodendrocyte survival and proliferation in an active established lesion. Lab. Invest. 45:534-546
28. Raine, C.S. and Traugott, U. (1985) Remyelination in chronic relapsing experimental allergic encephalomyelitis and multiple sclerosis. In: Adachi M, Hirano, A., Aronson, S. (eds.) The pathology of the myelinated axon, Tokyo, Igaku-shoin Press pp 229-275
29. Rodriguez, M. (1992) Central nervous system demyelination and remyelination in multiple sclerosis and viral models of disease. J. Neuroimmuno. 40:255-264

30. Rodriguez, M., Lennon, V., Benveniste, E.N. and Merrill, J.E. (1987) Remyelination by oligodendrocytes stimulated by antiserum to spinal cord. J. Neuropathol. Exp. Neurol. 46:84-95
31. Rodriguez, M., Oleszak, E. and Leibowitz, J. (1987b) Theiler's murine encephalomyelitis virus: a model of demyelination and viral persistence. CRC Crit. Rev. Immunol. 7:325-365
32. Rodriguez, M., Pierce, M.L. and Thiemann, R.L. (1991) Immunoglobulins stimulate central nervous system remyelination: Electron microscopic and morphometric analysis of proliferating cells. Lab. Invest. 64:358-370
33. Schnell, L. and Schwab, M.E. (1990) Axonal regeneration in the rat spinal cord produced by an antibody against myelin-associated neurite growth inhibitors. Nature 343:269-272
34. Schwab, M.E. (1990) Myelin-associated inhibitors of neurite growth and regeneration in the CNS. TINS 13:452-
35. Shine, H.D., Readhead, C., Popko, B., Hood, L., and Sidman, R.L. (1992) Morphometric analysis of normal, mutant, and transgenic CNS: correlation of myelin basic protein expression to myelinogenesis. J. Neurochem. 58:342-349
36. Sommer, I. and Schachner, M. (1981) Monoclonal antibodies (O1 and O4) to oligodendrocyte cell surfaces. Dev.Biol. 83:311-327.
37. Suzuki, M. and Raisman, G. (1994) Multifocal pattern of postnatal development of the macroglial framework of the rat fimbria. Glia 12:294-308
38. Wood, P., Moya, F., Eldridge, C., Owens, G., Ranscht, B., Schachner, M., Bunge, M. and Bunge, R.P. (1990) Studies of the initiation of myelination by Schwann cells. Annals New York Acad. Sci. 605:1-14
39. Wood, P. and Bunge, R.P. (1984) The biology of the oligodendrocyte. In: Oligodendroglia; Advances in Neurochemistry vol. 5, W.T. Norton, ed. Plenum Press, NY & London. pp 1-40

Aspects of Remyelination by Endogenous and Transplanted Glial Cells in Regenerating Goldfish Visual System

S. N. Nona

Developmental Neurobiology Laboratory, Dept. Optometry &Vision Sciences, UMIST, Manchester M60 1QD, UK, Phone 0161 200 3877, Fax 0161 236 0409

Abstract. Studies in this laboratory have established that following a crush to goldfish optic nerve, Schwann cells of unknown origin invade the lesion, forming a band of peripheral-type myelin neatly demarcated from the new central-type myelin formed distal to the lesion (7,8). Furthermore, it is clear that such a phenomenon does not take place outside the optic nerve; indeed, neither the lesioned optic tract nor the lesioned spinal cord becomes colonised by Schwann cells, and across both these lesions there is a complete restoration of CNS environment (8,9). In this article, I shall ask what this regional pattern means: is Schwann cell invasion dependent only on there being a local source for these cells, or do astrocyte populations from different regions of CNS differ in their permissiveness to Schwann cell colonisation of damaged CNS tissue?

Keywords. Regeneration, CNS, astrocytes, Schwann cells, myelination, immunochemistry

1. Introduction

An important area of research is to understand why mammalian CNS axons fail to regrow after injury, while functional recovery is invariably the case in injured fish CNS. Failure of mammalian axons to regrow has been attributed, in part, to the hostile nature of the local glial environment, namely astrocytes (5,15) and oligodendrocytes (12), which when replaced with a peripheral nerve environment containing Schwann cells results in regeneration of CNS axons (1,11). It is not clear, however, why the same glial elements, which are considered inhibitory to axonal regrowth in mammals, do not impede regeneration in fish CNS (10,14).

Molecular Signaling and Regulation in Glial Cells
A Key to Remyelination and Functional Repair

2. Repair in optic nerve

Until very recently, it had been assumed that regeneration in fish visual system is accompanied by restoration of CNS environment. It seemed to us, therefore, very appropriate to re-examine glial repair in this system more thoroughly, paying particular attention to the type of repair at the site of damage. Such a study has been made possible by the recent availability of several antibodies to fish astrocytes and myelin proteins, the use of which has demonstrated, for the first time, that a damage to goldfish optic nerve does not lead to the recovery of astrocytes or oligodendrocytes at the site of lesion. Instead, it becomes colonised by Schwann cells which subsequently myelinate the regenerated axons. This is a consistent finding which is observed whenever the astrocytic *glia limitans* is compromised, and the extent of Schwann cell colonisation is proportional to the size of CNS damage (see also ref. 2). It is, thus, possible to produce a single band of S-100$^+$ Schwann cells across the nerve (Fig. 1a), two such peripheral-type domains separated by an intact CNS region (Fig. 1c), or an extensive region of Schwann cells extending from just behind the optic nerve head to near the chiasm - a gap of more than 1.8 mm (Fig. 1b). Furthermore, species specific antibodies to fish GFAP (6) and central myelin protein 36K (4) convincingly show that neither astrocytes (Fig. 2a) nor central myelin (Fig. 2b) is present in the region occupied by Schwann cells.

Lack of molecular markers for fish Schwann cell precursors has made identification of their source in the injured optic nerve very elusive. To date, it has not been possible to recognise these Schwann cells before 6 weeks post injury, when a small number of S-100$^+$ cells first appears at the lesion site, and this event neatly coincides with the commencement of remyelination. In time, the number of S-100$^+$ cells increases exponentially as a result of cell proliferation (Nona & Stafford, unpublished results), and these cells then become a permanent feature of the optic nerve, as noted in regenerated tissue 14 months old (7). Also, the fact that Schwann cells are noted in the injured optic nerve irrespective of the position of lesion suggests that the source for these cells is present all the way **along** the nerve. Goldfish optic nerve is made up of discrete fascicles which are enveloped by astrocytic *glia limitans*, and interleaved by connective tissue septae which are an integral part of optic nerve sheath (Fig 3). This structure, which is endowed with mesenchymal cells and capillaries, may contain cells of neural crest origin which are supplied locally as Schwann cells. However, is the neural crest the only true source for these Schwann cells?

One can not resist the temptation to further speculate that Schwann cell precursors may be present within the optic nerve fascicles. During normal mammalian development for example, retinal axons growing through the optic stalk recruit a distinct population of interfascicular neuroepithelial cells destined to become glia. According to Bronner-Fraser and Fraser (3), the potential for producing "crest-derived" cells exists **within** the early neural tube as well as its dorsal lip where the crest proper is located. This raises the question of what

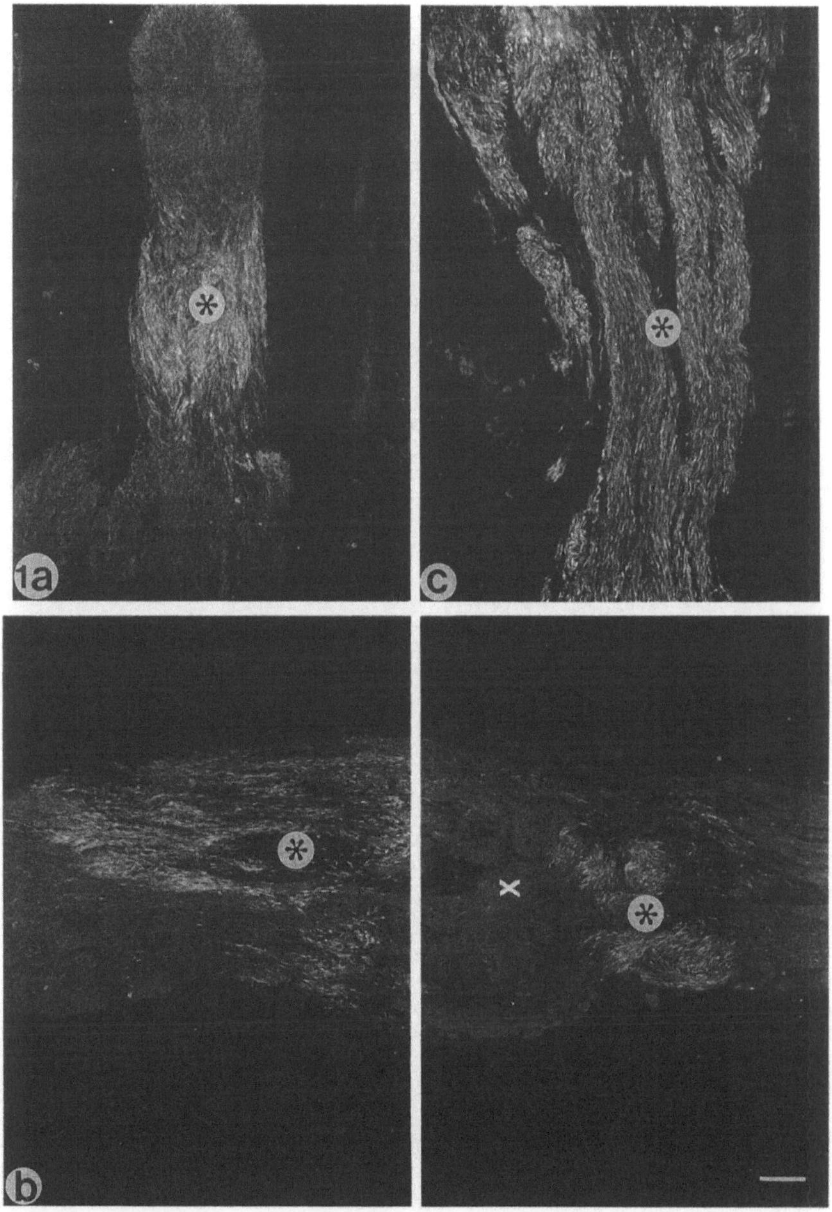

Figure 1. Goldfish optic nerves from 3 different animals 90 days post lesion, stained with anti-S-100 to highlight the Schwann cell region. (a) shows a single narrow band of S-100^+ cells(*); (c) shows a very wide band of cells (*), created by a number of adjacent crushes; and (b) shows two S-100^+ bands (*) separated by an intact CNS region (X). Bar=100μm.

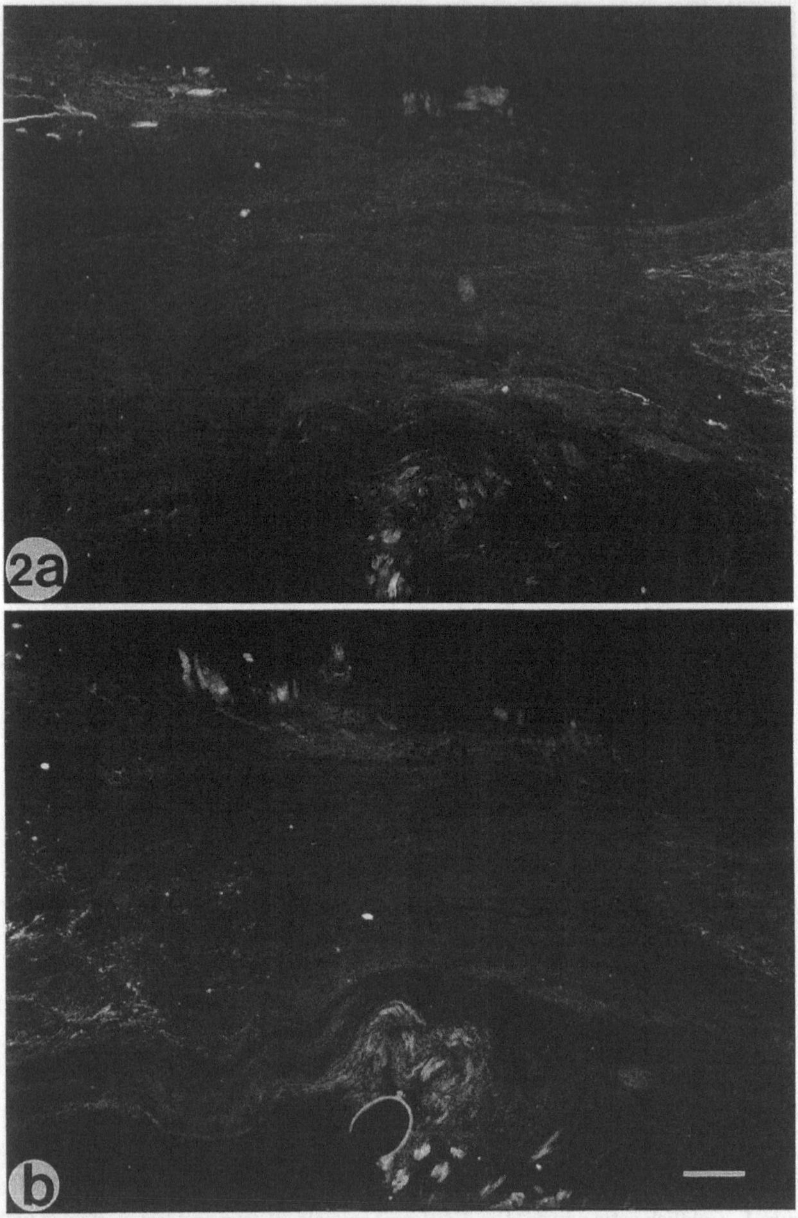

Figure 2. A wide lesion from goldfish optic nerve 90 days post crush, stained with anti-GFAP (a) and anti-36 kDa (b), showing the absence of astrocytes and central myelin, respectively, except at the margins of the lesion. Note that these are adjacent sections to that in Figure (1c) which depicts Schwann cells within the lesion. Bar=100 μm.

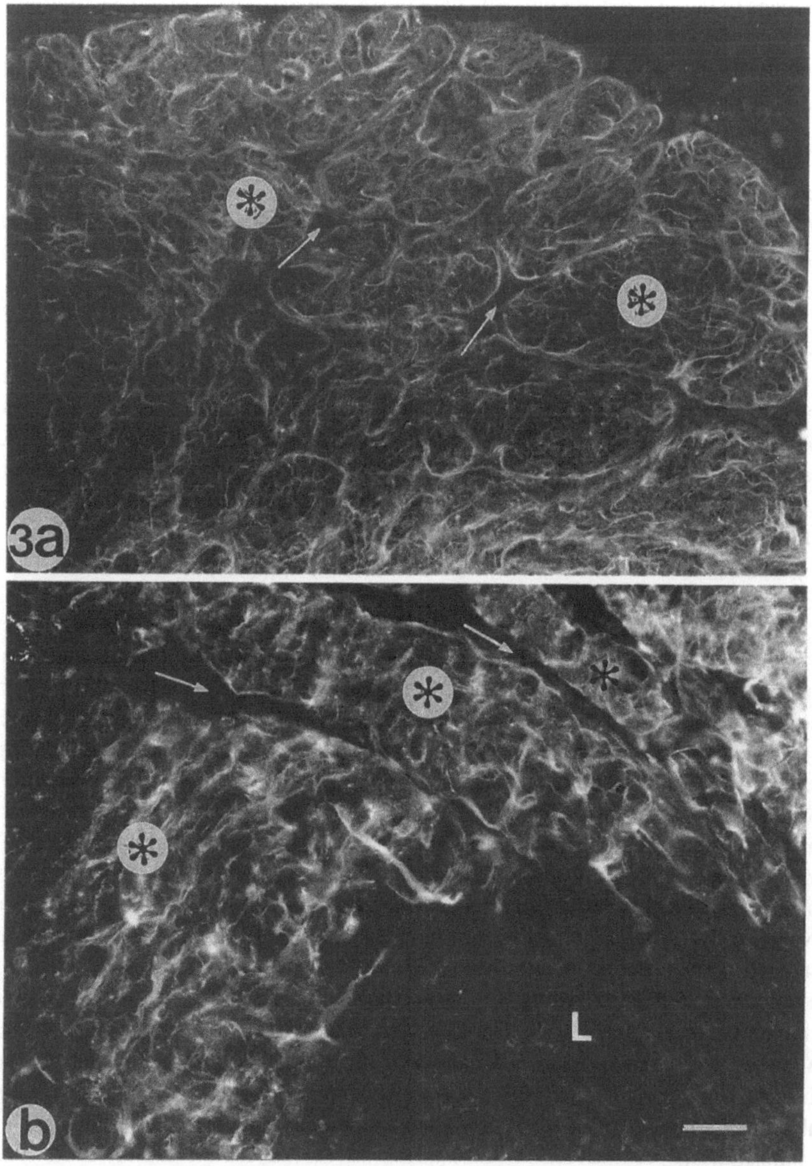

Figure 3. Cross- (a) and longitudinal (b) sections of goldfish optic nerve 10 days post crush stained with anti-GFAP to highlight the fascicular nature of the nerve (*). The nerve is interleaved with connective tissue septae (arrows). In (b), L is part of the crush area. Bar=40 μm.

happens to the neuroepithelial cells with this potential? They could be exported at the time of crest migration, of course, or, they could be (i) killed by cell death, (ii) held in suppressed form, or (iii) fate-determined into another sort of cell by signals associated with the latter differentiation of the proper astro\oligodendro-glial framework in the post-embryonic CNS. Presumably, what is different about fish is that, whatever this developmental restriction comprises, it never occurs in optic nerve where endogenous Schwann-competent cells, normally held under or suppressed come to the fore as a result of changes associated with trauma, and all our results are consistent with this. This said, it remains equally difficult to identify endogenous precursor cells as the exogenous cells.

3. Repair in optic tract

An unexpected twist to this story has been the failure to observe any Schwann cells in lesioned optic tract. Here, astrocytes from margins of the lesion bridge the gap by 30 days, and there is a complete restoration of CNS environment by 70 days (Fig. 4). It is to be emphasised that failure to note Schwann cells in lesioned tract is just as predictable as their persistent appearance in lesioned optic nerve; but why does the optic tract resist Schwann cell colonisation?

It is possible that optic tract lacks a source for Schwann cells, or that tract astrocytes, unlike astrocytes in optic nerve, somehow exclude the Schwann cells; and by forcibly introducing a fragment of peripheral nerve (from lateral line) into the tract it should be possible to distinguish between these two possibilities. Such a study is suitably conducted in "normal" regenerating tract, or in a regenerating tract in which oligodendrocyte myelination is totally abolished by X-irradiation (8). In neither case, however, do Schwann cells become established in the tract. Instead, hypertrophied astrocytes engulf the peripheral fragment (Fig. 5a) gradually expelling it through the surface of the tract - a process which is associated with a massive invasion of blood-born cells, probably macrophages (Nona & Stafford, unpublished results). By contrast transplanting a peripheral fragment into lesioned, irradiated optic nerve results in an extensive zone of myelinating Schwann cells, far exceeding that in un-irradiated nerve of comparable age (Nona & Stafford, in preparation). These results suggest that failure of Schwann cells to colonise optic tract is less likely to be due to the absence of a local source for these cells, and more likely to be due to the inhospitable nature of local environment: specifically astrocytes, as in the case of irradiated tissue.

A further piece of evidence which highlights such regional differences within goldfish visual pathway comes from the study of irradiated optic nerve, which invariably shows the presence of a small but reliable number of Schwann cells distal to the lesion, where they are normally absent (8). However, none of the Schwann cells are found beyond the boundary with the optic tract, despite the regrowth of new fibres throughout the irradiated visual pathway. These results,

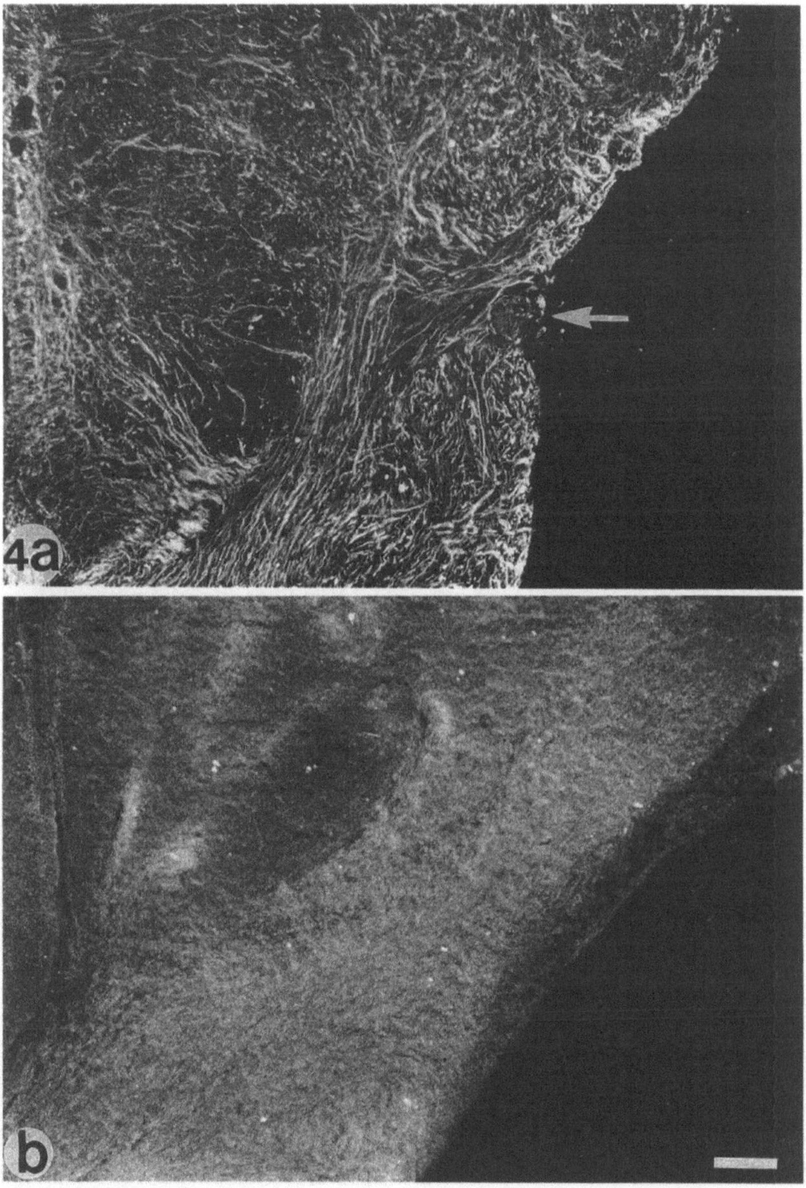

Figure 4. Optic tract 40 (a) and 70 (b) days post crush stained with anti-GFAP and anti-36 kDa, respectively, to show the recovery of astrocytes and central myelin in the regenerated tract. In (a), the site of crush is evident (arrow). Bar=100 μm.

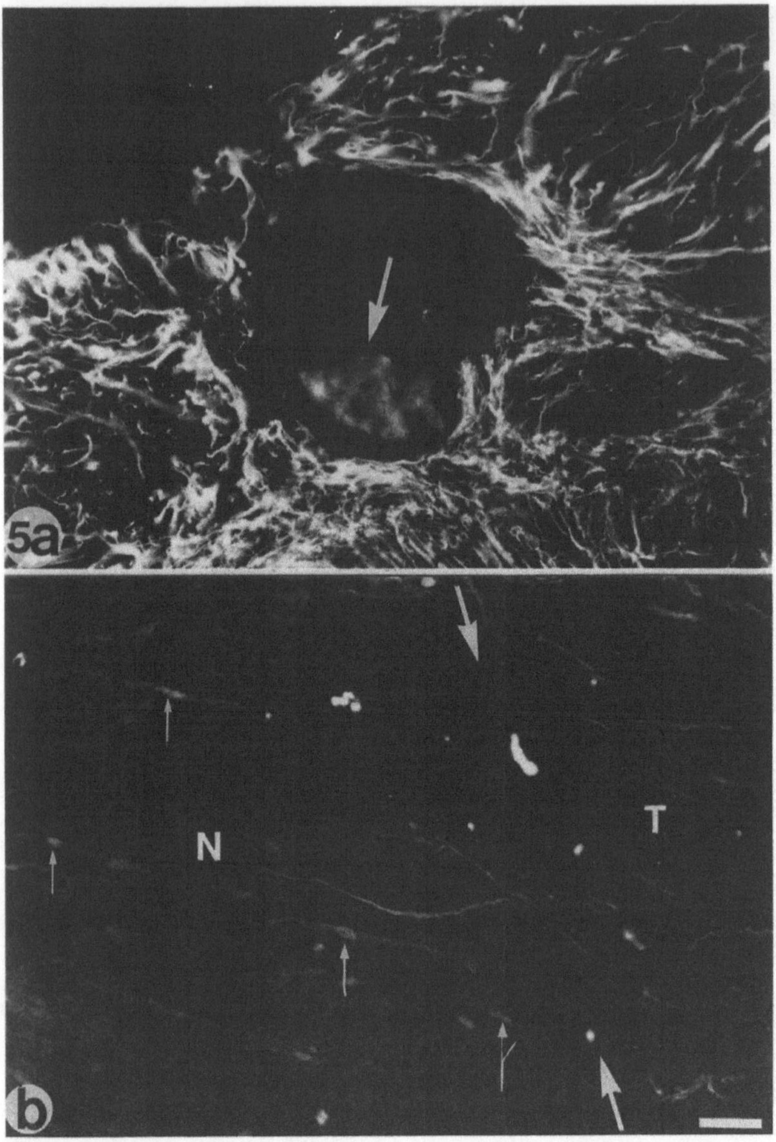

Figure 5. An 80 days old X-irradiated regenerating optic tract (a) into which had been transplanted a fragment of lateral line 1 week after lesion. Note how GFAP$^+$ hypertrophied astrocytes surround the presumed fragment which is showing autofluorescence with TRITC (arrow). (b) is optic nerve\tract boundary (large arrows) from an X-irradiated lesioned nerve 180 days old, showing that S-100$^+$ Schwann cells in the nerve (N; small arrows) are rarely found among the astrocytes of the tract (T).
Bar=40 μm.

together with those of regenerating fish spinal cord which show the failure of Schwann cells to invade the new cord (9), strongly suggest that fish brain astrocytes are less permissive to Schwann cell invasion of CNS tissue than optic nerve astrocytes. Remarkably, this regional dependence is the reverse of what has been observed in the rat where glial radiation damage admits Schwann cells to the spinal cord (13), but not to the optic nerve (Gilmore, personal communications). Identification of molecules responsible for such diverse astrocytic response will significantly improve our understanding of the strategies based on Schwann cell transplantation to repair injured CNS. Eliminating the extent of astrocytic inhibitory reaction may improve the effectiveness of the Schwann cell transplant.

Acknowledgement
I should like to thank Carole Stafford for her contribution to this work. I also thank Dr. Gunnar Jeserich for the gift of antibody to 36 kDa protein.

References

1. Berry, M., Rees. L., Hall, S. (1988) Optic axons regenerate into sciatic nerve isografts only in the presence of Schwann cells. Brain Research Bulletin 20, 223-231.
2. Blakemore, W.F. (1976) Invasion of Schwann cells into the spinal cord of the rat following local injection of lysolecithin. Neurpathology and Applied Neurobiology 2, 21-39.
3. Bronner-Fraser, M. and Fraser, S. (1989) Developmental potential of avian trunk neural crest cells in situ. Neuron 3, 755-766.
4. Jeserich, G. and Waehneldt, T.V. (1986) Characterization of antibodies against major fish CNS myelin proteins: immunoblot analysis and immunohistochemical localisation of 36kDa and IP2 proteins in trout nerve tissue. Journal of Neuroscience Research 15, 147-158.
5. Mansour, H., Asher, R., Dahl, D., Labkovsky, B., Perides, G. and Bignami, A. (1990) Permissive and non-permissive reactive astrocytes: immunofluorescence study with antibodies to the glial hyaluronate-binding protein. Journal of Neuroscience Research 25, 300-311.
6. Nona, S.N., Stafford, C.A., Shehab, S.A.S. and Cronly-Dillon, J.R. (1989) Glial fibrillary acidic protein (GFAP) from goldfish: its localisation in visual pathway. Glia 2, 189-200.
7. Nona, S.N., Duncan, A., Stafford, C.A., Maggs, A., Jeserich, G. and Cronly-Dillon, J.R. (1992) Myelination of regenerated axons in goldfish optic nerve by Schwann cells. Journal of Neurocytology 21, 391-401.
8. Nona, S.N., Stafford, C.A., Duncan, A., Cronly-Dillon, J.R. and Scholes, J. (1994) Myelin repair by Schwann cells in the regenerating goldfish

visual pathway: regional patterns revealed by X-irradiation. Journal of Neurocytology 23, 400-409.

9. Nona, S.N. and Stafford, C.A. (1995) Glial repair at the site of lesion in regenerating goldfish spinal cord: an immunohistochemical study using species-specific antibody. Journal of Neuroscience Research. In press.

10. Reier, P.J., Stensaas, L.J. and Guth, L. (1983) The astrocytic scar as an impediment to regeneration in the central nervous system. In C.C.Kao, R.P.Bunge and P.J.Reier (eds), Spinal Cord Reconstruction, Raven Press, New York, pp 163-196.

11. Richardson, P.M., McGuiness, V.M. and Aguayo, A.J. (1980) Axons fom CNS neurones regenerate into PNS grafts. Nature 284, 264-265.

12. Schwab, M. (1990) Myelin associated inhibitors of neurite growth. Experimental Neurology 109, 2-5.

13. Sims, T.J. and Gilmore, S.A. (1989) Interaction between Schwann cells and CNS axons following a delay in the normal formation of central myelin. Experimental Brain Research 75, 513-522.

14. Sivron, T. and Schwartz, M. (1994) The enigma of myelin- associated growth inhibitors in spontaneously regenerating nervous system. Trends in Neurosciences 17, 277-281.

15. Snow, D.M., Lemmon, V., Carrino, D.A., Caplan, A.I. and Silver, J. (1990) Sulfated proteoglycans in astroglial barriers inhibit neurite outgrowth in vitro. Experimental Neurology 109, 111-130.

16. Stafford, C.A., Shehab, S.A.S., Nona, S.N. and Cronly-Dillon, J.R. (1990) Expression of glial fibrillary acidic protein (GFAP) in goldfish optic nerve following injury. Glia 3, 33-42.

Remyelination of Lesions in Multiple Sclerosis

Patterns of demyelination in multiple sclerosis as defined by oligodendrocyte pathology

Wolfgang Brück[1] and Hans Lassmann[2]

1 Department of Neuropathology, University of Göttingen, Robert-Koch-Str. 40, D-37075 Göttingen, Germany

2 Research Unit for Experimental Neuropathology, Austrian Academy of Sciences, Schwarzspanierstr. 17, A-1090 Vienna, Austria

Abstract. Patterns of demyelination and oligodendrocyte pathology were studied in different types of multiple sclerosis lesions including acute, early chronic and late chronic multiple sclerosis. Oligodendrocytes within the lesions were identified by immunocytochemistry for myelin oligodendrocyte glycoprotein and by in situ hybridization for proteolipid protein messenger RNA. Cell degeneration was visualized by the detection of DNA fragmentation within the cell nuclei. All three types of multiple sclerosis tissue revealed demyelinated plaques mainly infiltrated by T lymphocytes and macrophages. Different patterns of cell destruction were found depending on the lesion type. Acute multiple sclerosis lesions showed extensive destruction of oligodendrocytes as well as of other tissue elements including astrocytes, axons and macrophages although a considerable proportion of oligodendrocytes was preserved within the lesions. Lesions obtained in the early course of the disease showed selective demyelination with marked oligodendrocyte preservation in most of the cases, resulting in a high number of remyelinating lesions. In late chronic multiple sclerosis lesions, demyelination was associated with massive oligodendrocyte destruction and loss. Remyelination was sparse in these cases and restricted to the borders of the lesions. The observed patterns of cell death and oligodendrocyte pathology suggest different pathogenetic mechanisms involving the oligodendrocyte and/or the myelin sheath as the target of the immune response. These data indicate the variability of demyelination, remyelination and oligodendrocyte preservation or damage between different multiple sclerosis patients and between different stages of the disease.

Keywords. multiple sclerosis, oligodendrocytes, demyelination, remyelination, cell death

Molecular Signaling and Regulation in Glial Cells
A Key to Remyelination and Functional Repair

1 Introduction

Multiple sclerosis (MS) is a chronic inflammatory disease of the central nervous system (CNS) which leads to selective destruction of myelin sheaths. The immunological mechanisms behind the destruction of myelin as well as the patterns of demyelination have not yet been defined; the primary target of the immune response also has not yet been identified. It is not clear whether oligodendrocytes, the myelin-forming cells of the CNS, are primarily affected in this disease or whether they are destroyed together with myelin during active demyelination (16, 18, 28).

The patterns of oligodendrocyte destruction, preservation or proliferation are still a matter of debate. There is general agreement that oligodendrocytes may be present in demyelinating lesions from patients with a fatal course of the disease who die within a short period after onset of clincal symptoms (18, 28, 32, 33, 43). Extensive remyelination may be present throughout such lesions (18, 29, 30, 33). In classical chronic MS plaques, however, oligodendrocytes seem to be lacking (28). Oligodendrocyte damage has been observed ultrastructurally in an active lesion (35).

The nature of the remyelinating cell in demyelinated MS lesions has long been discussed. In particular, it is not clear whether the remyelinating oligodendrocytes stem from preexisting cells which survived the demyelinating attack or whether they are recruited from a pool of proliferating progenitor cells. Several studies reported a loss of oligodendrocytes in active lesions followed by recruitment of newly remyelinating cells from undifferentiated progenitors (28-30). Oligodendrocyte preservation in early demyelinating lesions followed by cell destruction in the later phase of plaque formation was observed by other investigators (39). The extent of oligodendrocyte destruction also seems to vary among different patients and depend on the time of plaque formation during the course of the disease (18).

A major problem of studies on demyelination and remyelination is due to the fact that at the time of investigation only limited markers for identification of oligodendrocytes in tissue sections were available. Myelin oligodendrocyte glycoprotein (MOG) is a protein that is detectable on the surface of myelin and oligodendrocytes late during myelination (3, 20, 24). In demyelinated lesions, MOG is preserved at the surface of oligodendrocytes which survived the destruction of their myelin sheaths (22). This marker can therefore be used to identify mature oligodendrocytes in demyelinated MS lesions. Additionally, the detection of myelin protein mRNA by in situ hybridization allows the identification of oligodendrocytes which are engaged in the synthesis or maintenance of myelin (2, 5, 17). The identification of degenerating cells in tissue sections is another factor which has not yet been considered. Recently, techniques to identify degenerating cells by detecting DNA fragmentation have been developed (8, 9, 37). They have, however, not yet been applied to a larger series of MS plaques.

In the present study, patterns of oligodendrocyte pathology were studied by using the above described techniques. A comparative analysis of different types of MS tissue was performed, including lesions developing in the early or late course of chronic multiple sclerosis as well as lesions which were formed in Marburg's acute type of multiple sclerosis (23). The data obtained suggest that different pathways are involved in the pathogenesis of demyelination in multiple sclerosis.

2 Material and Methods

2.1 Categories of MS tissue

Paraffin-embedded biopsy and autopsy tissue from 28 cases was used (Table 1). The diagnosis was confirmed clinically and/or autoptically. The biopsy material used in this study was collected at the Department of Neuropathology of the University of Göttingen. The clinical history and course of 5 patients were previously described in detail (27). In these cases, brain biopsy was performed to exclude other treatable diseases such as lymphoma, glioma or toxoplasmosis. The cases were grouped in the following three categories.

Acute multiple sclerosis. Acute multiple sclerosis was defined according to the original criteria by Marburg (23). Clinically, the patients developed a severe progressive or relapsing neurological disease that led to extensive neurological deficits and death of the patients within one year after onset. Neuropathological examination revealed pronounced inflammation associated with multiple demyelinated plaques in characteristic topographical distribution throughout the brain and spinal cord.

Early chronic multiple sclerosis. Lesions of early multiple sclerosis developed during the first year after onset of the disease. Stereotactic or open biopsies were taken from these patients. The tissue usually included periplaque white matter, plaque border as well as plaque center. After establishing the diagnosis of an inflammatory demyelinating CNS process, the patients were followed for another 2-7 years (4).

Late chronic multiple sclerosis. These cases included typical chronic MS lesions with a disease duration of 1-21 years after onset. This group contained patients with relapsing and progressive multiple sclerosis (25).

In all three subgroups, active and inactive lesions, as defined below, were found. A total of 72 lesions was studied (see Table 1).

2.2 Neuropathological and immunocytochemical techniques

Paraffin sections were dewaxed and stained with hematoxylin-eosin (HE), Luxol fast blue-Periodic Schiff reagent (LFB-PAS) for myelin and Bielschowsky's silver impregnation for axons.

Immunocytochemistry was performed using an avidin-biotin or an alkaline phosphatase/anti-alkaline phosphatase technique. The following antibodies were used:

Inflammatory cells: T and B cells (CD3 and L26, Dako); macrophages (27E10, 25F9, MRP14) (1, 44, 45).

Myelin/oligodendrocytes: Myelin oligodendroglia glycoprotein (MOG) (26); 2',3'-cyclic nucleotide 3'-phosphodiesterase (CNPase); myelin basic protein (MBP); proteolipid protein (PLP) (11); HNK-1.

Astrocytes: Glial fibrillary acid potein (GFAP).

2.3 In situ hybridization

Digoxigenin-labeled riboprobes specific for proteolipid protein mRNA were used for non-radioactive in situ hybridization (5). The source and specificity of the probes, the labeling techniques and the methods for in situ hybridization were described in detail earlier (2). The specificity of the reaction was controlled by comparing hybridization with antisense and sense riboprobes. Following in situ hybridization the sections were processed for immunocytochemistry with anti-PLP antibodies as described above.

2.4 Detection of DNA fragmentation

The in situ tailing (IST) protocol was used as described in detail earlier (8, 9, 37). Tissue sections were carefully deparaffinized and air-dried before they were incubated with a reaction mixture containing terminal transferase and digoxigenin-labeled nucleotides. This procedure results in the incorporation of labeled nucleotides at sites of DNA strand breaks. Positive cells were identified by combining IST and immunocytochemistry with antibodies against MOG, CD3, GFAP and macrophages.

2.5 Morphometry

The number of cells stained per square unit of tissue by immunocytochemistry, in situ hybridization or in situ tailing was determined on serial sections according to the demyelinating activity within the lesions. The number of cells was counted in at least 10 standardized microscopic fields of 25,000 μm^2 each, defined by an ocular morphometric grid. Values given in tables and figures represent the number of cells per mm^2.

3 Results

3.1 General neuropathology

All tissues investigated contained confluent demyelinated lesions (Fig. 1a) in various stages of demyelinating activity. There were, however, remarkable differences in the morphological structures of the plaques between the different groups of MS tissue. Lesions in acute multiple sclerosis showed complete loss of myelin within the lesions. Demyelination was associated with extensive pathology of axons resulting in axon loss in the lesions. Tissue destruction was prominent. The oedematous extracellular space was mainly filled with macrophages; some glial cells or few axons were also seen.Lesions in early chronic multiple sclerosis, in contrast, revealed a quite different histological pattern. Selective demyelination was associated with almost completely preserved axonal structures (Fig. 1c). Demyelinated nerve fibers were surrounded by a dense GFAP-positive glial scar tissue (Fig. 1d). More than half of the lesions were composed of axons encompassed by thin myelin sheaths indicative of remyelination. Such phenomena of remyelination were observed throughout the entire lesions and were not restricted to the plaque edge. Demyelination was also the most prominent feature of lesions in late chronic multiple sclerosis. Axon loss within the plaques was variable and the remaining axons were embedded in a dense glial scar tissue. Remyelination, when present, was only seen at the plaque margins. Inflammation was present in all tissue specimes examined (Fig. 1b). The inflammatory infiltrate consisted of lymphocytes and macrophages which were mainly located in perivascular cuffs. Actively demyelinating lesions showed a massive infiltration with macrophages and, to a lesser extent, with lymphocytes.

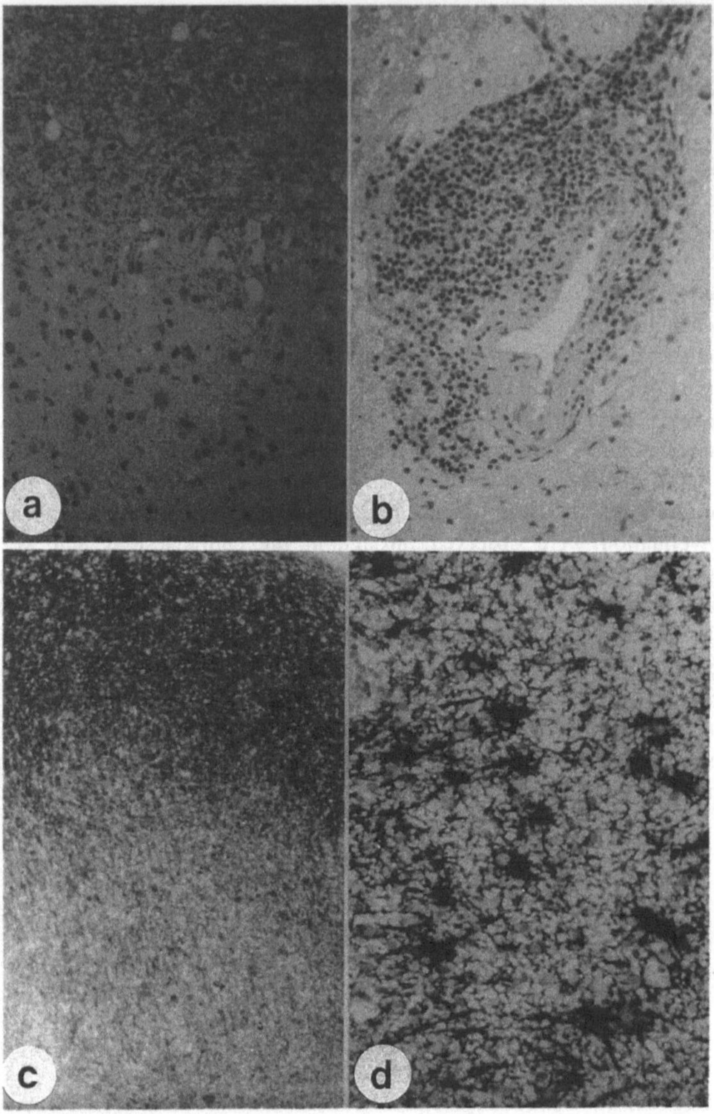

Fig. 1. Neuropathological changes in multiple sclerosis lesions. **a.** There are confluent demyelinated plaques with sharp borders to the periplaque white matter (LFB-PAS). **b.** Perivascular inflammation is present within the plaque (HE). **c.** Axons are generally preserved in early MS plaques (Bielschowsky). **d.** Astrocytic gliosis with numerous GFAP-positive astrocytes is prominent (Immunocytochemistry for GFAP).

3.2 Definition of demyelinating activity

The following stages of demyelinating activity, as defined by the presence of myelin degradation products within the macrophage cytoplasm, were distinguished (see Table 2):

Early active lesions. These lesions were generally located at the plaque border between demyelinated plaques and the periplaque white matter. There were many macrophages within these lesions seen either in the demyelinated tissue or between fibers that were still ensheathed by myelin. The macrophages contained myelin degradation products which were stained with LFB and were immunoreactive with antibodies against all major myelin proteins (Fig. 2a-b) including the minor components such as MOG or CNPase.

Late active lesions. These lesions were similar to those described above. Myelin degradation was more advanced with macrophages revealing immunoreactivity for major myelin proteins such as MBP or PLP, but not for MOG or CNPase.

Inactive, demyelinated lesions. These lesions were infiltrated by macrophages containing either empty vacuoles or PAS-positive degradation products. Many T lymphocytes were seen within these completely demyelinated plaques.

Early remyelinating lesions. Numerous lymphocytes and macrophages were also found in these lesions. Macrophages mainly contained empty vacuoles or PAS-positive degradation products, their cytoplasm sometimes revealing immunoreactivity for myelin proteins. Small clusters of axons within the plaques were surrounded by very thin myelin sheaths indicative of remyelination.

Late remyelinating lesion (Shadow plaque). In these lesions, axons are surrounded by thin myelin sheaths which resulted in a reduced intensity of myelin staining when compared with the periplaque white matter. Only few macrophages were found within these lesions. Axon density was generally lower than in the surrounding white matter. There was prominent astrocytic gliosis in all plaques.

Table 1. Summary of clinical and neuropathological findings

Case no.	Age	Sex	Disease duration[1]	Type[2]	Lesions studied active	inactive
1	21	F	2 w	Acute	1	1
2	24	F	4.5 m	Balo	0	2
3	51	F	7 m	Acute	4	0
4	47	F	3.5 m	Acute	2	1
5	29	M	1.5 m	Acute	1	0
6	46	F	12 d	Acute	1	0
7	26	F	3 m	Acute	0	1
8	23	F	4 m	Acute	0	2
9	68	F	1.5 m	Myelitis	1	0
10	25	M	1 m[3]	E	0	4
11	21	F	7.5 m[3]	E	0	4
12	21	F	11 d[3]	E	1	2
13	28	M	7 m[3]	E	6	1
14	20	F	2 m[3]	E	0	4
15	15	M	1 m[3]	E	0	1
16	21	M	2.5 m[3]	E	3	0
17	34	F	13 y	LC/SP	3	2
18	28	F	1 y	LC/SP	1	3
19	28	F	4 y	LC/Rel	0	2
20	33	F	10 y	LC/SP	0	2
21	45	F	20 y	LC/Rel	0	1
22	42	F	11 y	LC/Rel	1	0
23	30	M	6 y	LC/SP	1	1
24	46	F	7 y	LC/Rel	0	1
25	37	F	2 y	LC/PP	6	0
26	40	F	1.5 y	LC/SP	2	1
27	53	F	21 y	LC/SP	0	1
28	39	F	3 y	LC/SP	0	1

[1] d = days; w = weeks; m = months; y = years.

[2] Acute = acute MS; E = early MS; LC = late chronic MS; Rel = relapsing remitting; PP = primary progressive; SP = secondary progressive.

[3] Interval between first symptoms and biopsy.

For details see (4, 25).

Table 2. Stages of demyelinating activity defined by myelin degradation products within macrophages

	Early active	Late active	De-myelinated	Early re-myelinated	Late re-myelinated
MOG, CNPase	+	-	-	±	-
MBP, PLP	+	+	-	±	-
LFB	+	+	-	±	-
PAS	-	-	+	+	+
Vacuoles	-	-	+	+	+

3.3 Quantitative determination of oligodendrocytes

A range of immunocytochemical markers was applied to the sections for identifying oligodendrocytes within demyelinating plaques. Only a minor portion of oligodendrocytes showed immunoreactivity with antibodies against MBP, PLP or HNK-1. Cells positive for these antibodies were mostly found in remyelinating lesions. CNPase reactivity was found in a higher portion of cells. The highest numbers of oligodendrocytes were found by immunocytochemistry with anti-MOG antibodies (Fig. 2a, c-d). Proteolipid protein mRNA was found in all oligodendrocytes in the periplaque white matter as well as within the plaques (Fig. 2e-g). PLP mRNA-positive cells were especially found in remyelinating lesions (Fig. 2g). Oligodendrocyte quantification was thus performed in sections stained by immunocytochemistry for MOG or by in situ hybridization for PLP mRNA.

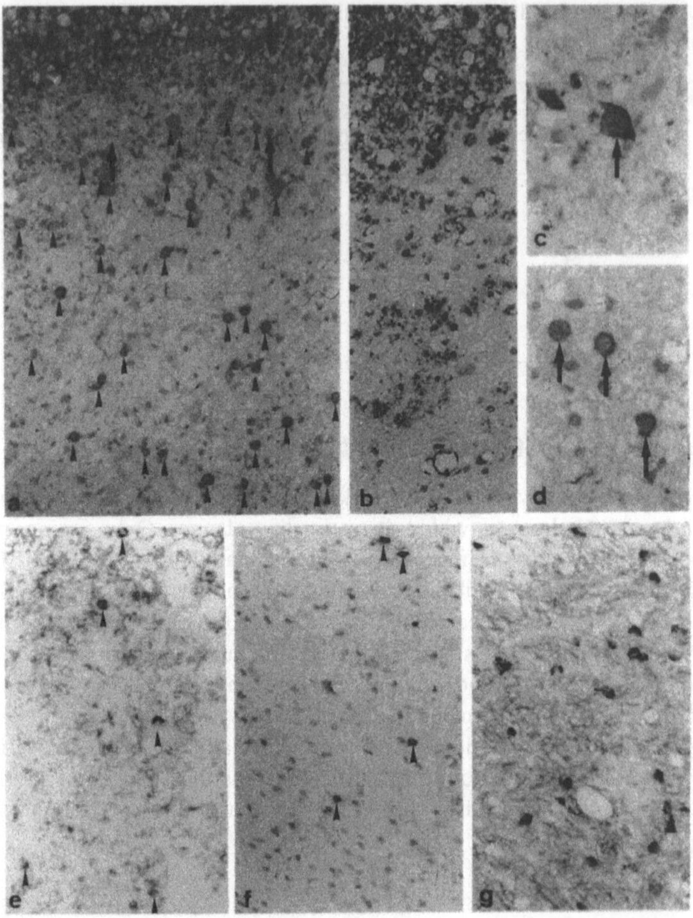

Fig. 2. Detection of oligodendrocytes by immunocytochemistry for MOG and in situ hybridization for PLP mRNA in early MS lesions. **a.** (case 13) Border of an actively demyelinating plaque. Macrophages contain MOG-reactive debris (large arrows); numerous MOG-positive oligodendrocytes (arrowheads) are in the actively demyelinating zone and in the adjacent plaque tissue. **b.** Same lesion as a. Immunocytochemistry for PLP reveals numerous macrophages with PLP-reactive debris. **c and d.** Same lesion as a. Large, activated (c) and small, quiescent (d) oligodendrocytes are MOG-positive. **e-g.** In situ hybridization for PLP mRNA and immunocytochemistry for PLP protein. **e.** (case 12) Border of an actively demyelinating plaque with macrophages containing PLP-reactive debris; some oligodendrocytes contain PLP mRNA (arrowheads). **f.** (case 11) Center of a demyelinated plaque. Few oligodendrocytes are PLP mRNA-positive (arrowheads). **g.** (case 10) Remyelinated lesion with numerous PLP mRNA-containing oligodendrocytes. Reprinted from (4).

The number of MOG-positive oligodendrocytes was generally reduced in the plaques when compared with the normal periplaque white matter (see Table 3). The number of oligodendrocytes, however, showed only minor variances in different lesions of single patients in which quantification gave similar results (4, 25). In contrast, there were significant variations in the numbers of oligodendrocytes between different multiple sclerosis cases. In early multiple sclerosis lesions, oligodendrocytes were more or less preserved within the plaques independent of the demyelinating activity of the lesions (4). There was only a minor reduction of oligodendrocytes in the majority of these cases (see Table 3). In late chronic MS lesions, however, there was a prominent and almost complete loss of oligodendrocytes (25) (see Table 3). Oligodendrocyte loss was also prominent in acute MS lesions, although a significant number was preserved even in destructive lesions (see Table 3). The number of oligodendrocytes was similarly reduced in actively demyelinating lesions compared with inactive ones in a given patient. Comparable oligodendrocyte densities were even seen in the actively demyelinating border and the inactive center of the same plaque (4).

Table 3. Quantification of MOG-positive oligodendrocytes in MS lesions

	Acute MS	Early MS	Late chronic MS
Normal white matter	375±40	326±13	370±24
Plaque	45±13	207±48*	45±21

*p < 0.01 against acute and late chronic MS lesions

The in situ hybridization for PLP mRNA gave similar numbers of oligodendrocytes when compared with the data obtained by immunocytochemistry for MOG (4). However, there were some differences when comparing the subgroups of MS cases. In the majority of actively demyelinating lesions and in lesions of acute and early chronic MS, there were more oligodendrocytes stained by immunocytochemistry for MOG than by in situ hybridization for PLP mRNA. Inactive lesions, especially in late chronic MS, in contrast, revealed more cells expressing PLP mRNA than MOG protein.

Figure. 3. Oligodendrocyte density (cells/mm^2) in the plaques of acute, early and late chronic MS. The plaques were grouped according to their oligodendrocyte density; most of the plaques in acute MS revealed oligodendrocyte densities of 50-11 cells/mm^2; early MS lesions showed nearly normal oligodendrocyte density, whereas in late chronic MS lesions the majority of cases had <10 oligodendrocytes/mm^2.

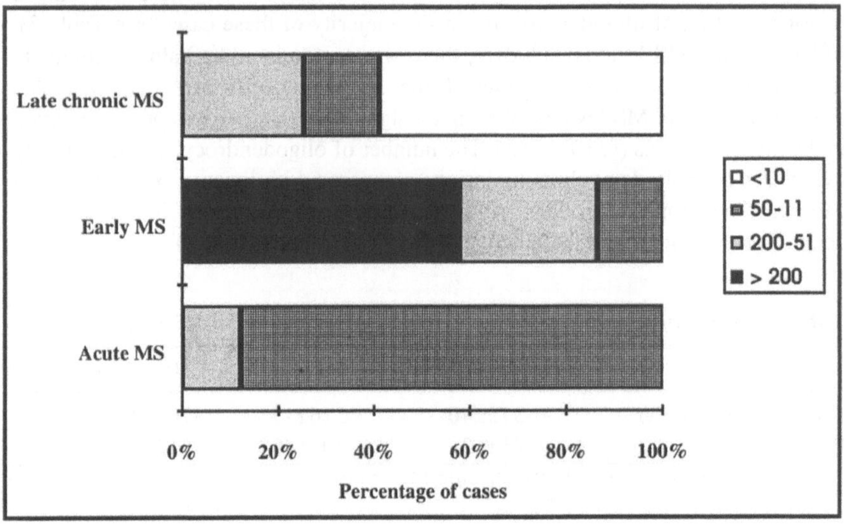

3.4 Identification of degenerating cells in multiple sclerosis lesions

The detection of DNA fragmentation in tissue sections was used to identify degenerating cells. DNA fragmentation in multiple sclerosis lesions was mainly restricted to cells located in the actively demyelinating zone at the plaque border. In acute MS lesions, degenerating cells were found throughout the entire plaque (Fig. 4a). Actively demyelinating lesions of late chronic MS, in contrast, showed DNA fragmentation only at the borders of the plaques (Fig. 4d and f). Only few degenerating cells were found in lesions of early chronic MS. The degenerating cells were mainly identified as T cells or oligodendrocytes by immunocytochemistry (Fig. 4e and g). Macrophages or astrocytes showed DNA fragmentation only in acute multiple sclerosis (Fig. 4b and c).

3.5 Patterns of oligodendrocyte pathology in actively demyelinating plaques

Four representative cases of the different MS categories were used to study the topography of oligodendrocyte distribution and cell destruction in actively demyelinating plaques and the periplaque white matter. This investigation revealed four essentially different patterns of demyelination in MS (Fig 5).

Primary destruction of myelin sheaths with very little secondary damage to oligodendrocytes. This pattern was predominantly found in cases of early chronic multiple sclerosis. Oligodendrocytes were only moderately reduced in the actively demyelinating zone or in the plaque center when compared with the surrounding white matter. Only few cells revealed DNA fragmentation.

Primary destruction of myelin sheaths with secondary oligodendrocyte loss. This pattern was found in one case of late chronic MS. Degenerating oligodendrocytes were mainly located in the actively demyelinating zone at the plaque border (Fig. 4d and h). The number of oligodendrocytes within the plaque was only moderately reduced.

Primary destruction of oligodendrocytes with secondary myelin loss. This pattern was found in one case of chronic progressive multiple sclerosis. Oligodendrocytes within the plaque and in the actively demyelinating border were almost completely lost. The overwhelming majority of degenerating oligodendrocytes with DNA fragmentation was found in a small rim in the white matter adjacent to the zone of active demyelination (Fig. 4f and i).

Acute inflammatory necrosis. In the case of severe acute multiple sclerosis, oligodendrocytes, myelin sheaths, astrocytes and axons were destroyed throughout the plaque. Degenerating cells were seen in the plaque, the periplaque white matter and even in the white matter at a considerable distance from the lesion (Fig. 4a-c).

Fig. 5. Topographical distribution of cells with DNA fragmentation in four representative MS cases. The figures given are cells/mm^2 for oligodendrocytes (upper figures) and degenerating cells (lower figures). O = oligodendrocytes; T = T-lymphocytes; A = astrocytes; M = macrophages.

	Early MS	Late chronic MS	Late chronic MS	Acute MS
Normal WM	352 0	470 0	323 3	331 9
Periplaque WM	313 0	445 2	336 57	332 56
Actively demyelinating zone	243 0	370 68	19 14	32 25
Plaque centre	259 1	149 22	0 7	54 40
Degenerating cell type	T	O, T	O, T	O, T, A, M

Fig 4. Patterns of cell death in actively demyelinating MS lesions. The sections were reacted according to the in situ tailing protocol (black reaction prduct) and then subjected to immunocytochemistry (red reaction product). **a-c.** Acute multiple sclerosis. **a.** Numerous cells show DNA fragmentation; double immunocytochemistry for PLP. The degenerating cells are identified as (**b**) macrophages (double immunocytochemistry for MRP14) or (**c**) astrocytes (double staining for GFAP). **d.** Late chronic multiple sclerosis. DNA fragmentation (arrows) is restricted to a small margin in the actively demyelinating zone at the plaque border (double labeling for PLP). **e.** Same lesion as d. Oligodendrocyte identified by immunocytochemistry for MOG shows DNA fragmentation. **f.** Late chronic multiple sclerosis. DNA fragmentation (black nuclei) is found in a small rim of the 'normal' periplaque white matter just adjacent to the plaque (double staining for PLP). **g.** Same lesion as d. T lymphocytes, identified by immunocytochemistry for CD3, show DNA fragmentation. **h.** Detail from d. Degenerating cells with DNA fragmentation are located at the plaque border in the zone of active myelin destruction. **i.** Detail from f. Degenerating cells showing DNA fragmentation are located in the periplaque white matter adjacent to the plaque border.

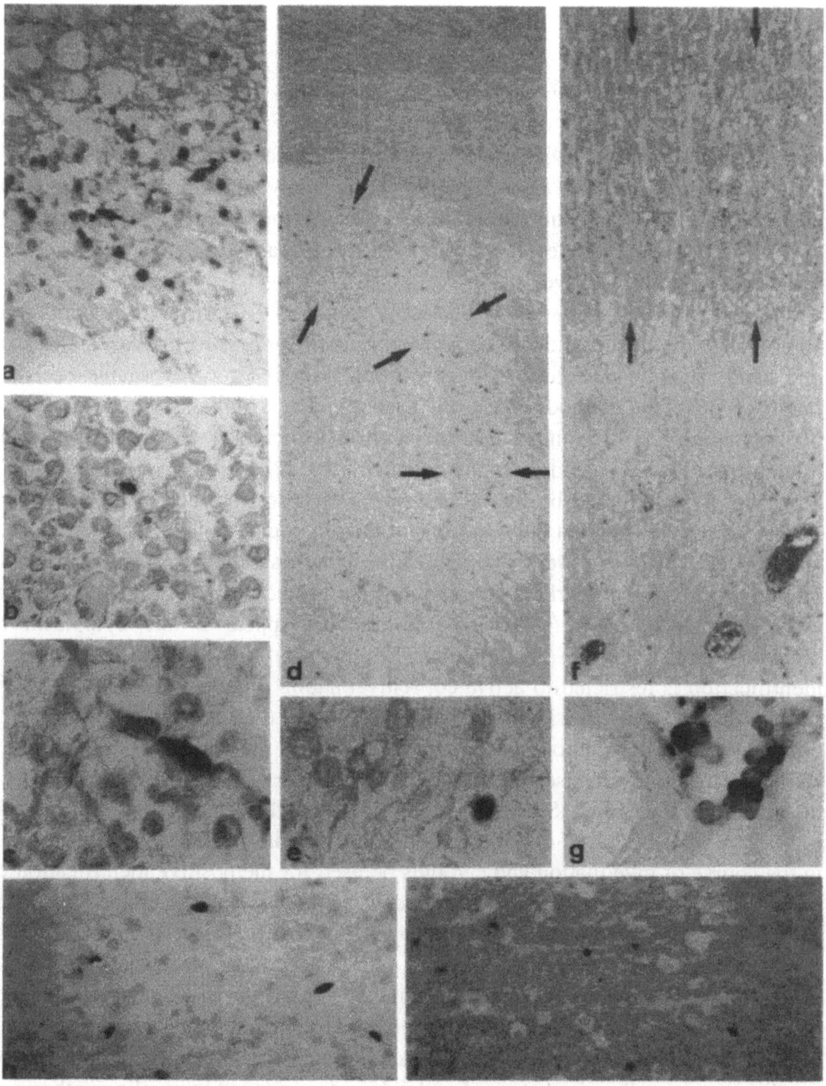
a
b
c
d
e
f
g
h
i

4. Discussion

The pathogenesis of demyelination in multiple sclerosis is unequivocally connected with the fate of oligodendrocytes in actively demyelinating lesions. The present study provides evidence for variable oligodendrocyte pathology in MS lesions, indicating a heterogenous pathogenesis of myelin and/or oligodendrocyte destruction. There are substantial differences in the number of preserved or destroyed oligodendrocytes in different MS cases as well as in different MS subgroups as defined above.

The present study is based on the combination of two different experimental approaches: the positive identification and quantification of oligodendrocytes on one hand, and the detection of degenerating oligodendrocytes on the other. A range of different methods has been used to identify oligodendrocytes within demyelinating CNS lesions. These included histochemistry (7, 13, 14), electron microscopy (31, 33) and immunocytochemistry (28, 29, 32). These studies, however, were impaired by the lack of specificity, for instance in histochemical studies, or by the fact that only portions of the oligodendrocyte population present within the plaques were recognized. This is the case when using immunocytochemical markers for myelin proteins such as MBP, PLP or myelin-associated glycoprotein (MAG). These proteins appear in the cytoplasm of oligodendrocytes during myelination (12, 15, 41); their usefulness for cell identification is restricted to activated, remyelinating oligodendrocytes (32). This situation led to a considerable controversy on the extent and course of oligodendrocyte destruction in MS.

The present study focuses on the use of two markers for positive identification of oligodendrocytes in MS lesions. Myelin oligodendrocyte glycoprotein (MOG) is a cell surface protein of oligodendrocytes and myelin sheaths which appears late in the course of myelination and can thus be used for the positive identification of mature oligodendrocytes (3, 24). MOG persists on the surface of oligodendrocytes even when the myelin sheath is destroyed, as, for instance, in the course of Wallerian degeneration (22). The second marker used in the present study was for proteolipid protein (PLP) mRNA which is present in the cytoplasm of oligodendrocytes during myelination and myelin maintenance (2, 5, 17). PLP mRNA is detectable in immature oligodendrocytes or their progenitors during myelination as well as in mature oligodendrocytes engaged in myelin synthesis or maintenance. Generally, similar numbers of oligodendrocytes were obtained when comparing immunocytochemistry for MOG and in situ hybridization for PLP mRNA. There were, however, differences in the number of positive cells depending on the type of MS tissue studied. Early multiple sclerosis lesions were marked by a high proportion of surviving, mature oligodendrocytes within the plaques (4). In these lesions, a predominance was found of MOG positive oligodendrocytes over PLP mRNA-expressing cells, indicating that mature oligodendrocytes survived the demyelinating attack but had not yet begun remyelination. In late chronic MS lesions, in contrast, a quite different pattern of oligodendrocyte differentiation was seen. The number of PLP mRNA-containing oligodendrocytes predominated the number of MOG positive cells in these cases, indicating a considerable recruitment of remyelinating cells from the precursor pool (25). These data on oligodendrocyte quantification are in accordance with the results obtained during the investigation of DNA fragmentation. There were no degenerating oligodendrocytes found during the early course of the disease. In late chronic MS lesions, however, there were many degenerating oligodendrocytes undergoing DNA fragmentation with a massive subsequent loss of oligodendrocytes within the plaques. A quite different pattern of plaque pathology was observed in lesions of acute multiple sclerosis which more resembled incomplete inflammatory necrosis than inflammatory demyelination. A range of cell types including oligodendrocytes, macrophages, T lymphocytes and astrocytes showed signs of cell degeneration. Despite the destructive nature of these lesions there was, however, a considerable number of oligodendrocytes preserved within the plaques.

The preservation or destruction of oligodendrocytes has immense consequences on the regenerative capacities within demyelinating plaques. Remyelination has been described as a prominent feature in multiple sclerosis plaques (18, 29, 30, 34). Even entire plaques have been shown to be remyelinated (18). The high number of remyelinating plaques especially in the early phase of the disease (4) correlates with the preservation of oligodendrocytes during active demyelination. The presence of remyelination raised the question on the source of the remyelinating cell. Several studies suggested that oligodendrocytes in early MS lesions may be recruited by proliferation of either mature oligodendrocytes or immature progenitors (32, 33). Since proliferation patterns were not dealt with in the present investigation, this possibility cannot be confirmed here. However, there seem to be different patterns of cell destruction and proliferation in different MS cases and MS subgroups; further study is necessary to clarify this in detail.

The mechanisms behind oligodendrocyte destruction in multiple sclerosis are still not well defined. A range of different bystander mechanisms may lead to oligodendrocyte damage (42). These include the secretion of cytotoxic cytokines (40), proteolytic enzymes (36), reactive oxygen species (10), other cytotoxic cell products or complement components (6, 38). More selective oligodendrocyte damage may be induced by the specific interaction with gamma/delta T lymphocytes (39) or anti-myelin antibodies as suggested in experimental studies (19-21). Different patterns of myelin and/or oligodendrocyte may thus result in completely different patterns of demyelination and remyelination. Oligodendrocytes themselves may be the primary target of the immune response, leading to primary oligodendrocyte destruction with secondary myelin damage. Another possible mechanism may be a secondary oligodendrocyte damage in the course of demyelination. Primary myelin damage with only slight oligodendrocyte damage is another pattern which contributes to the heterogenous pathogenesis in multiple sclerosis, as demonstrated in the representative cases above.

In conclusion, our studies on oligodendrocyte pathology in multiple sclerosis revealed different patterns of demyelination in different multiple sclerosis cases. These patterns were varied in single MS patients and depended on the stage of the disease. These data thus suggest fundamentally different pathogenetic pathways in the formation of demyelinated plaques in multiple sclerosis. Further studies will aim at confirming these observations to clarify the pathogenetic mechanisms of demyelination.

5. Acknowledgements

The present study was supported by grants from the Science Research Fund, Austria, project P 9294 MED, and the Gemeinnützige Hertie-Stiftung, project GHS 292/94, Germany.

6. References

1. Bhardwaj, R.S., Zotz, C., Zwadlo-Klarwasser, G., Roth, J., Goebeler, M., Mahnke, K., Falk, M., Meinardus-Hager, G. and Sorg, C. (1992) The calcium-binding proteins MRP8 and MRP14 form a membrane-associated heterodimer in a subset of monocytes/macrophages present in acute but absent in chronic inflammatory lesions. Eur J Immunol 22, 1891-1897.
2. Breitschopf, H., Suchanek, G., Gould, R.M., Colman, D.R. and Lassmann, H. (1992) In situ hybridization with digoxigenin-labeled probes: sensitive and reliable detection method applied to myelinating rat brain. Acta Neuropathol (Berl) 84, 581-587.
3. Brunner, C., Lassmann, H., Waehneldt, T.V., Matthieu, J.-M. and Linington, C. (1989) Differential ultrastructural localization of myelin basic protein, myelin/oligodendroglial glycoprotein and 2',3'-cyclic nucleotide 3'-phosphodiesterase in the CNS of adult rats. J Neurochem 52, 296-304.
4. Brück, W., Schmied, M., Suchanek, G., Brück, Y., Breitschopf, H., Poser, S., Piddlesden, S. and Lassmann, H. (1994) Oligodendrocytes in the early course of multiple sclerosis. Ann Neurol 35, 65-73.
5. Coleman, D.R., Kreibich, G., Frey, A.B. and Sabatini, D.D. (1982) Synthesis and incorporation of myelin polypeptides into CNS myelin. J Cell Biol 95, 598-608.
6. Compston, D.A.S., Morgan, B.P., Campbell, A.K., Wilkins, P., Cole, G., Thomas, N.D. and Jasani, B. (1989) Immunocytochemical localization of the terminal complement complex in multiple sclerosis. Neuropathol Appl Neurobiol 15, 307-316.
7. Friede, R.L. (1961) Enzyme histochemical studies in multiple sclerosis. Arch Neurol 5, 433-443.

8. Gold, R., Schmied, M., Giegerich, G., Breitschopf, H., Hartung, H.-P., Toyka, K.V. and Lassmann, H. (1994) Differentiation between cellular apoptosis and necrosis by the combined use of in situ tailing and nick translation techniques. Lab Invest 71, 219-225.
9. Gold, R., Schmied, M., Rothe, G., Zischler, H., Breitschopf, H., Wekerle, H. and Lassmann, H. (1993) Detection of DNA fragmentation in apoptosis: application of in situ nick translation to cell culture systems and tissue sections. J Histochem Cytochem 41, 1023-1030.
10. Griot, C., Vandevelde, M., Richard, A., Peterhans, E. and Stocker, R. (1990) Selective degeneration of oligodendrocytes mediated by reactive oxygen species. Free Rad Res Commun 11, 181-193.
11. Gunn, C., Richards, M.K. and Linington, C. (1990) The immune response to myelin proteolipid protein in the Lewis rat: identification of the immunodominant B cell epitope. J Neuroimmunol 27, 155-162.
12. Hartman, B.K., Agrawal, H.C., Agrawal, D. and Kalmbach, S. (1982) Development and maturation of central nervous system myelin: comparison of immunocytochemical localization of proteolipid protein and basic protein in myelin and oligodendrocytes. Proc Natl Acad Sci USA 79, 4217-4220.
13. Ibrahim, M.Z.M. and Adams, C.W.M. (1963) The relationship between enzyme activity and neuroglia in plaques of multiple sclerosis. J Neurol Neurosurg Psychiatry 26, 101-110.
14. Ibrahim, M.Z.M. and Adams, C.W.M. (1965) The relation between enzyme activity and neuroglia in early plaques of multiple sclerosis. J Pathol Bacteriol 90, 239-243.
15. Itoyama, Y., Sternberger, N.H., Kies, M.W., Cohen, S.R., Richardson, E.P. and Webster, H.deF. (1980) Immunocytochemical method to identify myelin basic protein in oligodendroglia and myelin sheaths of the human nervous system. Ann Neurol 7, 157-166.
16. Itoyama, Y., Sternberger, N.H., Webster, H.deF., Quarles, R.H., Cohen, S.R. and Richardson, E.P. (1980) Immunocytochemical observations on the distribution of myelin associated glycoprotein and myelin basic protein in multiple sclerosis lesions. Ann Neurol 7, 167-177.
17. Jordan, C., Friedrich, V.Jr. and Dubois Dalcq, M.E. (1989) In situ hybridization analysis of myelin gene transcripts in developing mouse spinal cord. J Neurosci 9, 248-257.
18. Lassmann, H. (1983) Comparative neuropathology of chronic experimental allergic encephalomyelitis and multiple sclerosis. Springer-Verlag, Berlin Heidelberg,
19. Lassmann, H., Brunner, C., Bradl, M. and Linington, C. (1988) Experimental allergic encephalomyelitis: the balance between encephalitogenic T lymphocytes and demyelinating antibodies determines size and structure od demyelinated lesions. Acta Neuropathol (Berl) 75, 566-576.
20. Linington, C., Bradl, M., Lassmann, H., Brunner, C. and Vass, K. (1988) Augmentation of demyelination in rat acute allergic encephalomyelitis by

circulating mouse monoclonal antibodies directed against a myelin/oligodendrocyte glycoprotein. Am J Pathol 130, 443-454.
21. Linington, C., Engelhardt, B., Kapocs, G. and Lassmann, H. (1992) Induction of persistently demyelinated lesions in the rat following the repeated adoptive transfer of encephalitogenic T cells and demyelinating antibody. J Neuroimmunol 40, 219-224.
22. Ludwin, S.K. (1990) Oligodendrocyte survival in Wallerian degeneration. Acta Neuropathol (Berl) 80, 184-191.
23. Marburg, O. (1906) Die sogenannte akute Multiple Sklerose (Encephalomyelitis periaxialis scleroticans). Jahrb Psychiat Neurol 27, 213-312.
24. Matthieu, J.-M. and Amiguet, P. (1990) Myelin/oligodendrocyte glycoprotein expression during development in normal and myelin-deficient mice. Dev Neurosci 12, 293-302.
25. Ozawa, K., Suchanek, G., Breitschopf, H., Brück, W., Budka, H., Jellinger, K. and Lassmann, H. (1994) Patterns of oligodendroglia pathology in multiple sclerosis. Brain 117, 1311-1322.
26. Piddlesden, S.J., Lassmann, H., Zimprich, F., Morgan, B.P. and Linington, C. (1993) The demyelinating potential of antibodies to myelin oligodendrcyte glycoprotein is related to their ability to fix complement. Am J Pathol 143, 555-564.
27. Poser, S., Lüer, W., Bruhn, H., Frahm, J., Brück, Y. and Felgenhauer, K. (1992) Acute demyelinating disease. Classification and non-invasive diagnosis. Acta Neurol Scand 86, 579-585.
28. Prineas, J.W. (1985) The neuropathology of multiple sclerosis. In J.C. Koetsier (ed.), Handbook of clinical neurology, Vol. 3 (47): Demyelinating diseases, Elsevier Science Publishers, Amsterdam, pp 213-257.
29. Prineas, J.W., Barnard, R.O., Kwon, E.E., Sharer, L.R. and Cho, E.S. (1993) Multiple Sclerosis: remyelination of nascent lesions. Ann Neurol 33, 137-151.
30. Prineas, J.W., Barnard, R.O., Revesz, T., Kwon, E.E., Sharer, L. and Cho, E.S. (1993) Multiple sclerosis. Pathology of recurrent lesions. Brain 116, 681-693.
31. Prineas, J.W., Kwon, E.E., Cho, E.S. and Sharer, L.R. (1984) Continual breakdown and regeneration of myelin in progressive multiple sclerosis plaques. Ann NY Acad Sci 436, 11-32.
32. Prineas, J.W., Kwon, E.E., Goldenberg, P.Z., Ilyas, A.A., Quarles, R.H., Benjamins, J.A. and Sprinkle, T.J. (1989) Multiple Sclerosis. Oligodendrocyte proliferation and differentiation in fresh lesions. Lab Invest 61, 489-503.
33. Raine, C.S., Scheinberg, L. and Waltz, J.M. (1981) Multiple sclerosis. Oligodendrocyte survival and proliferation in an active established lesion. Lab Invest 45, 534-546.
34. Raine, C.S. and Wu, E. (1993) Multiple Sclerosis: Remyelination in acute lesions. J Neuropathol Exp Neurol 52, 199-204.
35. Rodriguez, M., Scheithauer, B.W., Forbes, G. and Kelly, P.J. (1993) Oligodendrocyte injury is an early event in lesions of multiple sclerosis. Mayo Clin Proc 68, 627-636.

36. Sato, S., Quarles, R.H. and Brady, R.O. (1982) Susceptibility of the myelin-associated glycoprotein and basic protein to a neutral protease in highly purified myelin from human and rat brain. J Neurochem 39, 97-105.
37. Schmied, M., Breitschopf, H., Gold, R., Zischler, H., Rothe, G., Wekerle, H. and Lassmann, H. (1993) Apoptosis of T Lymphocytes in experimental autoimmune encephalomyelitis. Am J Pathol 143, 446-452.
38. Scolding, N.J., Morgan, B.P., Houston, A., Campbell, A.K., Linington, C. and Compston, D.A.S. (1989) Normal rat serum cytotoxicity against syngeneic oligodendrocytes. Complement activation and attack in the absence of anti-myelin antibodies. J Neurol Sci 89, 289-300.
39. Selmaj, K., Brosnan, C.F. and Raine, C.S. (1991) Colocalization of lymphocytes bearing gamma/delta T-cell receptor and heat shock protein hsp 65+ oligodendrocytes in multiple sclerosis. Proc Natl Acad Sci USA 88, 6452-6456.
40. Selmaj, K.W. and Raine, C.S. (1988) Tumor necrosis factor mediates myelin and oligodendrocyte damage in vitro. Ann Neurol 23, 339-346.
41. Sternberger, N.H., Quarles, R.H., Itoyama, Y. and Webster, H.deF. (1979) Myelin associated glycoprotein demonstrated immunocytochemically in myelin and myelin-forming cells of developing rats. Proc Natl Acad Sci USA 76, 1510-1514.
42. Wisniewski, H.M. and Bloom, B.R. (1975) Primary demyelination as a nonspecific consequence of a cell mediated immune reaction. J Exp Med 141, 346-359.
43. Yao, D.-L., Webster, H.deF., Hudson, L.D., Brenner, M., Liu, D.-S., Escobar, A.I. and Komoly, S. (1994) Concentric sclerosis (Balo): Morphometric and in situ hybridization study of lesions in six patients. Ann Neurol 35, 18-30.
44. Zwadlo, G., Bröcker, E.B., von Bassewitz, D.B., Feige, U. and Sorg, C. (1985) A monoclonal antibody to a differentiation antigen present on mature human macrophages and absent from monocytes. J Immunol 134, 1487-1492.
45. Zwadlo, G., Schlegel, R. and Sorg, C. (1986) A monoclonal antibody to a subset of human monocytes found only in the peripheral blood and inflammatory tissues. J Immunol 137, 512-518.

The adult human oligodendrocyte progenitor

Neil Scolding

University of Cambridge Neurology unit, Addenbrooke's Hospital, Hills Rd CAMBRIDGE CB2 2QQ, and MRC Cambridge Centre for Brain Repair, University Forvie Site, Robinson Way, Cambridge, CB2 2SR, United Kingdom

Abstract. The repair of demyelinated axons in patients with multiple sclerosis using growth factors and/or transplantation of glial cells to augment endogenous remyelination has become a more realistic prospect now that treatments are becoming available which may limit inflammatory processes within the CNS. However, until recently, little progress had been made in extrapolating the considerable experimental data now available concerning the rat glial progenitor - a cell which can be expanded in vitro, and which is then capable of successful remyelination upon transplantation into lesioned areas in the rodent CNS - to human oligodendrocyte biology. We have identified cells in cultures of adult human temporal lobe which are morphologically identical to rat oligodendrocyte progenitors and have an A2B5-+ve immunophenotype; they are negative for galactocerebroside (a marker for oligodendrocytes) and glial fibrillary acidic protein (an astrocyte marker), and also for nestin and vimentin. Like their rodent counterparts, human glial precursors are bipotential in vitro. They exhibit a similar astrocyte-driven proliferative response but, in common with oligodendrocyte progenitors in cultures derived from human embryonic spinal cord, do not divide when exposed to soluble growth factor signals known to be mitogenic for rat progenitors. The identification of a proliferative adult human oligodendrocyte progenitor, present also in tissue print preparations of fresh temporal lobe white matter, is the first step towards developing interventional strategies for promoting repair of demyelinated lesions in patients with multiple sclerosis.

Keywords. Multiple sclerosis, human oligodendrocyte progenitors, glial transplantation

1 Introduction

The limited but nonetheless significant remyelination occurring in multiple sclerosis rapidly follows the repopulation of acutely demyelinated lesions by oligodendrocytes (22). However, the origin of these reparative cells remains obscure. In vitro experiments indicate that mature oligodendrocytes do retain the capacity for generating new myelin (32), but successful remyelination in vivo in the rodent is best established by the transplantation of populations of neonatal

Molecular Signaling and Regulation in Glial Cells
A Key to Remyelination and Functional Repair

oligodendrocyte progenitor cells (12). Similarly, in feline models of demyelination, myelin repair appears to be established by inwardly migrating cells which have morphological properties which most closely resemble the (rodent) oligodendrocyte precursor (8). Furthermore it seems intuitively more likely that cells with a significant capacity for migration and for proliferation - which the mature oligodendrocyte lacks - are better equipped to establish repair of demyelinated, oligodendrocytopoenic areas.

Glial progenitors were originally isolated from and characterised in cultures derived from neonatal rat optic nerve (23). Four defining properties were delineated. Firstly, they are typically bipolar in shape, and secondly, they express surface gangliosides, in particular those recognised by the monoclonal antibodies A2B5 and GD3. They are bipotential in vitro, developing in response to environmental cues into either oligodendrocytes (acquiring first epitopes recognised by the monoclonal antibody 04, followed by galactocerebroside), or type 2 astrocytes which retain A2B5-related gangliosides and also express glial fibrillary acidic protein (GFAP); precursors have therefore been termed oligodendrocyte-type 2 astrocyte or O-2A progenitor cells (23). The fourth defining characteristic is their ability to proliferate, acting as stem cells inasmuch as they are capable of many cycles of division, and yield progeny which include cells apparently identical to the parent progenitor.

A number of astrocyte-derived proliferative signals for neonatal glial progenitors have been identified, including platelet-derived growth factor (PDGF), basic fibroblast growth factor (bFGF), insulin-like growth factor-1 (IGF-1), and neurotrophin-3 (NT-3) (reviewed in Ref 18; see also Ref 5); combinations of PDGF with either bFGF or NT-3 are particularly potent signals for rat progenitor division (5,6,15).

O-2A progenitors are also present in cultures derived from adult rodent CNS (9,29) and, like their neonatal counterparts, have the capacity for myelin repair in vivo (Ref 3; Sussmann, Targett, Compston and Blakemore, submitted). The adult pig appears to harbour a similar progenitor (J. Sussman, personal communication), and glial cells with the ultra-structural features of progenitors are present in the adult cat (8).

2 The human oligodendrocyte progenitor

Attempts to extrapolate these observations to human glial cell biology have proved much more difficult, partly because many of the antibodies used in rodent in vitro studies are unreliable in tissue sections, and partly because of the difficulty obtaining tissue for culture studies. In 1986, Kennedy and Fok-Seang showed that a bipolar, A2B5-positive cell could be cultured from human fetal brain, and appeared to exhibit similar properties to the rat O-2A progenitor; proliferation was not assessed (14). More recently, A2B5-positive radial glia with similarities to the rodent O-2A progenitor were cultured from embryonic (6-9 week) human spinal cord; interestingly, no proliferative response to either PDGF or bFGF was observed amongst either these cells or O4+ve pro-oligodendrocytes, in contrast to rodent progenitors (1). Similarly, Satoh and Kim (26) found that while immature oligodendrocytes (from human embryonic brain at 12-15 weeks) appeared constitutively to divide in untreated cultures, no significant additional proliferative response was evoked by the addition of PDGF, bFGF or both. Furthermore, at this stage, no staining with either A2B5 or GD3 antibodies was seen. More recently, cells with the properties of O-2A progenitors have been identified in vitro and in tissue sections prepared from human fetal cerebrum (25).

No evidence for the existence of an adult human oligodendrocyte progenitor appeared until 1992, when a substantial advance was made by the identification of glial cells in adult human tissue expressing O4-related antigens but not galactocerebroside, although they were capable of differentiating into oligodendrocytes (4). The oligodendrocyte precursor possessed multiple cellular processes, was A2B5-negative, and was not bipotential in vitro. Exposure to those growth factors active on rodent progenitors (bFGF and PDGF) did not induce adult human pre-oligodendrocyte proliferation. The O4-positive GalC-negative neonatal rodent counterpart, the pro-oligodendrocyte, is both bipotential and proliferative in vitro, and the human cell has therefore been called a "pre-oligodendrocyte" (4). A significant degree of plasticity in this cell has been clearly demonstrated (11). Furthermore, Dubois-Dalq's group, building on Richardson's observations concerning the specificity of PDGF-α-receptor mRNA hybridisation labelling (in the post-natal rodent CNS) for oligodendrocyte precursors, have more recently shown that cells in the adult human sub-ventricular zone may be stained using this technique - again, however, oligodendrocyte-lineage cells expressing PDGF-α-receptor message failed to proliferate in vitro (11; see also preceding chapter).

We have also studied cultured glial cells derived from adult human temporal lobe, removed surgically to treat drug-resitant epilepsy. Cultured cells, prepared using the method of Armstrong et al (4) were characterised after 48-72 hours in vitro. At this stage, 95-99% glial cells were round or multi-processed 04-positive cells; most were GalC-positive, but we also found 3% were O4-positive/GalC-negative pre-oligodendrocytes as previously described (Figure, a,b; see also Ref 4); Antel's group also have identified this adult human pre-oligodendrocyte (19). In addition, however, we found (in 12 consecutive temporal lobectomy preparations) a very small proportion (1.3 + 0.7%) of oligodendrocyte-lineage cells which comprised uni-, bi- or tripolar cells positive for A2B5 (Figure, c) but negative for GalC and GFAP (28); they were also negative for antibody markers of microglia (p150) and neurons (neurofilament) (not shown).

In contrast to the neonatal rodent progenitor, these cells were vimentin-negative, but unlike the adult rodent progenitor, they were O4-negative. Vimentin appears to represent a very early progenitor marker, while O4, as mentioned before, appears later during differentiation; it may therefore be that the cell we have identified is a little less immature than the (neonatal) rodent O-2A progenitor as conventionally described. In support of this is the observation that antibodies to the intermediate filament nestin, which stain neonatal rodent

O-2A cells in common with other primitive neuroectodermal precursors (10), fail to stain the adult human oligodendrocyte progenitor (not shown). In other respects, this cell closely resembles the rat O-2A progenitor in both morphology and immunophenotype (9,23,29).

After 5-7 days in vitro, cells with the phenotype of type-2 astrocytes (A2B5-positive/GFAP-positive) and (less commonly) newly-derived oligodendrocytes (A2B5-positive/GalC-positive) appeared, while the number of cells with the antigenic profile of O-2A progenitors diminished. The theoretical possibility that pre-existing astrocytes and oligodendrocytes acquired surface A2B5 during this time (accounting for the appearance of A2B5-positive astrocytes and oligodendrocytes) was excluded by treating cultures after 48 hours in vitro with A2B5 antibody and rabbit complement which removed all A2B5-positive cells. No A2B5/GFAP or A2B5/GalC co-positive cells were subsequently seen, suggesting that cells with the progenitor phenotype were indeed bipotential. Although a bias was apparent towards astrocyte development in serum-rich (10%) medium, and towards oligodendrocyte differentiation in serum-depleted medium (again mimicking the behaviour of rat progenitors; see 23), the small number of A2B5-positive cells prevented quantification of this finding. It was also found that A2B5-positive oligodendrocytes were more readily observed when progenitor-like cells developed in the presence of large numbers of astrocytes; in these circumstances, fewer type-2 astrocytes were evident (Figure, d,e).

The results of the immunolysis experiments A2B5 plus complement treatment of cultures after two days removing progenitors, which remained absent when cultures were re-examined 5-7 days later also indicate that de-differentiation of mature glia in vitro was unlikely to account for the identification of a progenitor cell in our cultures. Furthermore, the demonstration bi-, tri-, and occasionally multipolar cells with an A2B5-positive immunophenotype are present in tissue print preparations of adult human temporal lobe provides strong support for the existence in vivo of the progenitor here characterised in vitro (Figure, f,g).

3 Proliferation of adult human oligodendrocyte progenitors.

Given the apparent differences between the proliferative responses of human fetal and rodent neonatal oligodendrocyte progenitors, and between adult human and adult rodent O4-positive and GalC-negative pre-(or pro-)oligodendrocytes, it is perhaps only slightly surprising that bFGF and PDGF failed to stimulate proliferation of adult human oligodendrocyte progenitors. Similarly, incubation for periods of 2-9 days with IGF-1, NGF and NT-3, all of which have been reported to induce proliferation of rodent or pig oligodendrocyte progenitors and/or mature cells (reviewed in Ref 18; see also Refs 2,5), failed to induce DNA synthesis, as indicated by nuclear BrDU uptake, or a significant change in the number of A2B5-positive cells. Combinations of PDGF with either bFGF or NT-3, more potent signals for rat progenitor division (5,6,15), were also without effect, as were triple cocktails of IGF-1, NT-3 and ciliary neurotrophic factor (CNTF), or PDGF, bFGF and NT-3. Finally, in order to screen for previously unrecognised responses to combinations of growth factors (within the constraints imposed by limited amounts of neurosurgical specimens), cultured cells were incubated with a mixture containing bFGF, PDGF, IGF-1, CNTF, NGF, BDNF,

NT-3 and NT-4/5; again there was no suggestion of proliferation amongst A2B5-positive progenitors, pre-oligodendrocytes, or GalC-positive oligodendrocytes.

We also explored the possibility that our temporal lobe specimens might harbour an EGF-responsive stem cell as described in cultures derived from neonatal and adult rodent, and human fetal periventricular tissue (see Ref 20 for review). No proliferative response to EGF alone or in combination with other growth factors was found; furthemore, we were unable to prepare "neurospheres", aggregates of EGF-responsive stem cells, from adult human grey or white matter (Rosser, Svennson and Scolding, unpublished observation).

Proliferation of rat oligodendrocyte lineage cells can be induced by astrocyte-conditioned medium, B104 neuronal cell line-conditioned medium, and by culture on embryonic rat dorsal root ganglion (DRG) cells (13,23,30), and the effect of these treatments on adult human progenitors was also therefore tested. None stimulated human oligodendrocyte progenitor division, although human glial cells survived well on neonatal rat DRG cultures, making close contact with and ensheathing axons without producing complact myelin (C Shaw, N Scolding & DAS Compston, manuscript in preparation).

A proliferative response was, however, observed in bipolar and tripolar adult human progenitors growing on contiguous monolayers of flat astrocytes in cultures of adherent human glia maintained in serum-rich medium for 4-6 weeks (28). Whilst some top-dwelling cells had the immunophenotype either of oligodendrocytes (including newly derived A2B5-positive/GalC-positive oligodendrocytes, Fig. 1d,e; small arrow) or astrocytes (mostly type 2), other bi- and tripolar cells were A2B5-positive and GalC-negative/GFAP-negative. Rat O-2A progenitors proliferate on astrocyte monolayers (9,23), and the observation of pronounced BrDU uptake by human glial progenitors growing on astrocytes (Figure 1e; large arrow) indicated a similar proliferative response in adult human cells. Approximately 25% A2B5-positive cells were BrDU-positive after 48 hours'incubation with BrDU.

Triple and quadruple staining of A2B5-positive/BrDU-positive human glial progenitors confirmed that these cells were negative for GalC, GFAP and vimentin (not shown); the GFAP-positive astrocyte monolayer also contained BrDU co-positive cells, but these were flat, fibrous A2B5-negative cells easily distinguished from the top-dwelling processed glial cells.

The top-dwelling layer was also found to include a population of O4-positive cells, and approximately 3% of these had synthesised DNA during a 48 hour BrDU incubation, confirming the presence of a proliferating oligodendrocyte progenitor in cultures of adult human white matter. Rather than necessarily indicting pre-oligodendrocyte proliferation, these findings may be more readily explained by the acquisition of 04 by progenitors after DNA synthesis during the latter part of the 48 hour incubation with BrDU.

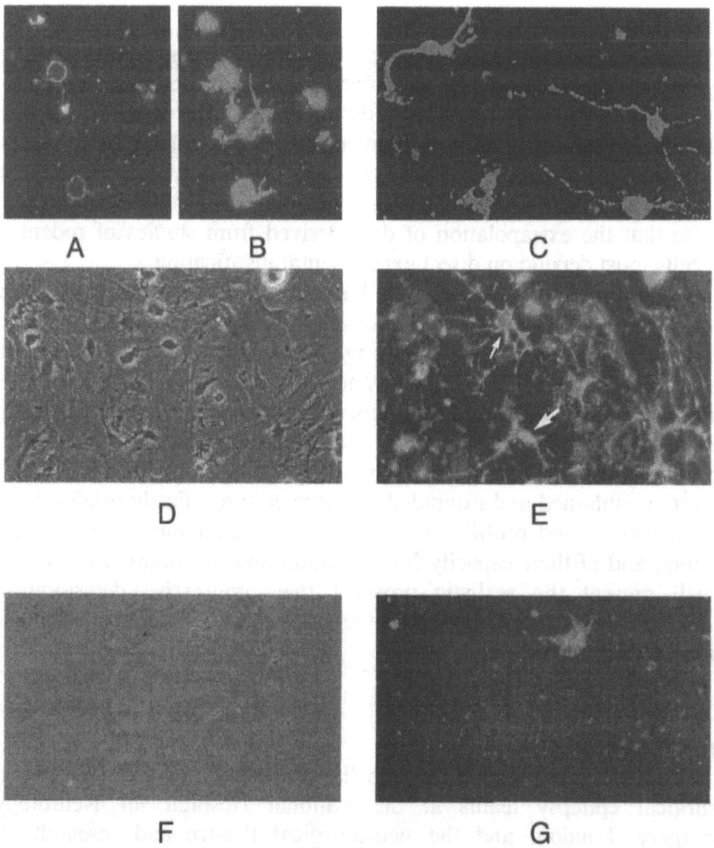

Figure1 A human glial cell with similarities to the rat O-2A progenitor.
a,b. In the first few days of culture, most of the cells replated from the initial non-adherent population were round and GalC+ve/O4+ve oligodendrocytes or GalC-ve/O4+ve pre-oligodendrocytes as previously described (Armstrong et al 1992). Here GalC is tagged with fluorescein, O4 with rhodamine; one pre-oligodendrocyte is shown. In addition, some bi- and tripolar cells were present which were A2B5 positive (c; rhodamine optics, x400) and GalC and GFAP negative (not shown). d,e. After 25-50 days in vitro, a monolayer of astrocytes had appeared in cultures of initially adherent cells, upon which a sparse layer of cells with progenitor phenotype was seen (d); after 48 hours' incubation with 10-4M BrDU, numerous cells were positive for A2B5 and had incorporated BrDU (e; large arrow; double exposure: A2B5 - rhodamine optics; BrDU - fluorescein optics, x400). Surface staining with anti-GalC antibody (applied before cell permeabilisation and also demonstrated with a fluorescein conjugate) reveals a mixed population of process-bearing BrDU-negative/A2B5-negative/GalC positive oligodendrocytes, A2B5-positive/BrDU-positive/GalC-negative progenitors (which were also GFAP-negative; not shown), and newly-derived oligodendrocytes (small arrow; A2B5-positive/BrDU-negative/GalC-positive). f,g. Fresh tissue print preparations of adult human temporal lobe white matter confirmed the presence of processed, A2B5-positive cells (phase contrast and rhodamine optics). Approximately 75% of these were GFAP-positive, the remainder GFAP-negative (not shown).

4 Discussion

These findings demonstrate that the adult human nervous system contains a glial cell type previously unrecognised in human tissue a proliferative oligodendrocyte progenitor. Whilst closely resembling the rodent O-2A progenitor cell, it is also clear that the human oligodendrocyte precursor cell has crucial differences to that of its rodent counterpart, particularly in its proliferative responses - as has previously been shown with fetal human glial progenitors, findings which emphasise that the extrapolation of data derived from studies of rodent glia to human cells must depend on direct experimental justification.

Our results also have relevance for human demyelinating disease. Recently, significant progress has been made in developing treatments designed to control and limit inflammatory damage to oligodendrocytes and myelin in multiple sclerosis (16,17). These therapeutic advances present a new opportunity to exploit an immunologically quiescent environment, and to examine the possibility that the limited endogenous capacity for remyelination might be enhanced using appropriate soluble factors, and by implanting human oligodendrocyte progenitors maintained and expanded in tissue culture. Further definition of the in vitro behaviour and proliferative responses of adult human oligodendrocyte progenitors, and of their capacity for migration and myelination in vitro and in vivo, will present the realistic prospect that progressive deterioration and permanent disability might be prevented by the development of strategies for promoting myelin repair.

Acknowledgements.

This work was made possible through the cooperation of the neurological and neurosurgical epilepsy teams at the National Hospital for Neurology and Neurosurgery, London, and the neurosurgical theatre and research staff at Addenbrooke's Hospital, Cambridge; in particular we are grateful to Professor F. Scaravilli, Dr Aljasani, and the staff in the Department of Pathology, National Hospital for Neurology and Neurosurgery, for their help in the collection of white matter. We thank Genentech for the provision of neurotrophin-3 and Amgen Inc. for the provision of IGF-1, bFGF, PDGF and CNTF. We are particularly grateful to Paula Rayner for her expert technical assistance.

References

1. Aloisi F, Giampaolo A, Russo G, Peschle C, Levi G. (1992) Developmental appearance, antigenic profile and proliferation of glial cells of the human embryonic spinal cord. Glia 5:171-181
2. Althaus HH, Kloppner S, Schmidt-Schultz T, Schwartz P. (1992): Nerve growth factor induces proliferation and enhances fiber regeneration in oligodendrocytes isolated from adult pig brain. Neurosci. Lett. 135 : 219-223.
3. Armstrong R, Friedrich VL, Holmes KV, Dubois-Dalcq M. (1990): In vitro analysis of the oligodendrocyte lineage in mice during demyelination and remyelination. J. Cell Biol. 111:1183-1195.

4. Armstrong RC, Dorn HH, Kufta CV, Friedman E, Dubois-Dalcq ME. (1992): Pre-oligodendrocytes from adult human CNS. J. Neurosci. 12 : 1538-1547.
5. Barres BA, Raff MC, Gaese F, Bartke I, Deschant G, Barde Y-A. (1994): Acrucial role for neurotrophin-3 in oligodendrocyte development. Nature 367: 371-375.
6. Bogler O, Wren D, Barnett S, Land H, Noble M. (1990): Cooperation between two growth factors promotes extended self renewal and inhibits differentiation of oligodendrocyte-type-2 astrocyte (O-2A) progenitor cells. Proc. Nat. Acad. Sci. USA 87 : 6368-6372.
7. Bottenstein, JE, Sato, GH. Growth of a rat neuroblastoma line in serum freesupplemented medium, Proc. Nat. Acad. Sci. U.S.A. 76 (1979) 514-517.
8. Carroll WM, Jennings AR. (1994): Early recruitment of oligodendrocyte progenitors in CNS demyelination. Brain 117:563-578.
9. ffrench-Constant C, Raff MC. (1986): Proliferating bipotential glial cells in the adult rat optic nerve. Nature 319 : 499-502.
10. Gallo V & Armstrong RC (1995). Developmental and growth factor-induced regulation of nestin inoligodendrocyte lineage cells. J. Neurosci. 15:394-406
11. Gogate N, Verma L, Min Zhau J, Milward E, Rusten R, O'Connor M, Kufta C, Kim J, Hudson L, Dubois-Dalcq M. (1994). Plasticity in the adult human oligodendrocyte lineage. J. Neurosci.14:4571-87.
12. Groves AK, Barnett SC, Franklin RJM, et al. (1993): Repair of demyelinatedlesions by transplantation of purified O-2A progenito r cells. Nature 362 : 453-455.
13. Hunter SF, Bottenstein JE. (1990): Growth factor responses of enriched bipotential glial progenitors. Dev. Brain Res. 54 : 235-248.
14. Kennedy PGE and Fok Seang (1986). Studies on the development, antigenic phenotype and function of human glial cells in culture. Brain 109:1261-1277
15. McKinnon RD, Matsui T, Dubois-Dalcq M, Aaronson SA. (1990): FGF modulates the PDGF-driven pathway of oligodendrocyte development. Neuron 5 : 603-614.
16. Moreau, T. Thorpe, J. Miller, D. Moseley, I. Hale, G. Waldmann, H. Clayton,D. Wing, M. Scolding, N. Compston, A. (1994) Preliminary evidence from magnetic resonance imaging for reduction in disease activity after lymphocyte depletion in multiple sclerosis. Lancet. 344 : 298-301.
17. Paty DW, Li DKB,, The IFNB Multiple Sclerosis Study Group. (1993): Interferon-beta 1b is effective in relapsing-remitting multiple sclerosis: MRI results of a multicenter, randomised, double-blind placebo-controlled trial. Neurology 43 : 662-667.
18. Pfeiffer SE, Warrington AE, Bansal R. (1993): The oligodendrocyte and its many cellular processes. Trends Cell Biol. 3 : 191-197.
19. Prabhakar S, D'Souza S, Antel JP et al (1995). Phenotypic and cell cycle properties of human oligodendrocytes in vitro. Brain Res. 672: 159-169
20. Price J. (1995) Brain stems. Current Biology 5: 232-234
21. Prineas JW, Kwon EE, Goldenberg PZ. (1989): Multiple sclerosis: oligodendrocyte proliferation and differentiation in fresh lesions. Lab. Inv. 61: 489-503.

22. Prineas JW, Barnard RO, Kwon EE, Sharer LR, Cho E-S. (1993): Multiple sclerosis: remyelination of nascent lesions. Annals Neurol. 33:137-151.
23. Raff MC, Miller RH, Noble M. (1983): A glial progenitor that develops in vitro into an astrocyte or an oligodendrocyte depending on culture medium. Nature 303 : 390-396.
24. Raine CS, Scheinberg L, Waltz JM. (1981): Multiple sclerosis: oligodendrocyte survival and proliferation in an active established lesion. Lab. Inv. 45 : 534-546.
25. Rivkin MJ, Flax J, Mozell R et al (1995). Oligodendroglial development in human fetal cerebrum. Ann. Neurol. 38:92-101
26. Satoh J and Kim SU. (1994). Proliferation and differentiation of fetal human oligodendrocytes in vitro. J. Neurosci. Res. 39:260-272
27. Scolding NJ, Zajicek JP, Wood N, Compston DAS. (1994): The pathogenesis of demyelinating disease. Prog. Neurobiol. 43 : 143-173
28. Scolding NJ, Rayner P, Sussman J, Shaw C, Compston DAS. (1995): An adult human oligodendrocyte progenitor. Neuroreport 6: 441-445
29. Wolswijk G, Noble M. (1989): Identification of an adult specific glial progenitor cell. Development 105 : 387-400).
30. Wood PM, Bunge RP. (1986a): Evidence that axons are mitogenic for oligodendrocytes isolated from adult animals. Nature 320 : 756-758.
31. Wood PM, Bunge RP. (1986b): Myelination of cultured dorsal root ganglion cells by oligodendrocytes obtained from adult rats. J. Neurol. Sci. 74 : 153-169
32. Wood PM, Bunge RP. (1991). The origin of remyelinating cells in the adult central nervous system: the role of the mature oligodendrocyte. Glia 4 : 225-232.

Strategies to Promote Central Nervous System Remyelination *In Vivo*

Kunihiko Asakura and Moses Rodriguez

Departments of Neurology and Immunology, Mayo Clinic and Mayo Foundation, Rochester, MN 55905, USA

Abstract. One important goal of multiple sclerosis research is to develop strategies to promote remyelination in the central nervous system (CNS). We previously reported that immunoglobulins directed against spinal cord homogenate promote CNS remyelination when injected into Theiler's virus-infected mice. To elucidate the mechanism of immunoglobulin-mediated CNS remyelination, we generated a monoclonal antibody (designated SCH94.03) against spinal cord homogenate, which increases remyelination four-fold when injected into SJL/J mice with Theiler's virus-induced demyelination. In live cell staining, SCH94.03 recognizes a surface antigen on well-differentiated oligodendrocytes. Double labeling experiments with other oligodendrocyte-specific antibodies indicate that this surface antigen may be a novel antigen on oligodendrocytes. In contrast with the surface labeling, SCH94.03 recognizes intracellular antigen(s) resembling cytoskeleton protein in all cells. Immunoglobulin cDNA sequence analysis reveals that SCH94.03 is encoded by immunoglobulin germline genes without somatic mutations, consistent with the conclusion that SCH94.03 is a natural autoantibody. SCH94.03 is also polyreactive toward multiple protein antigens and chemical haptens by ELISA. We propose and discuss three possible mechanisms of immunoglobulin-mediated CNS remyelination based on the results obtained from the Theiler's virus model of multiple sclerosis.

Key Words. Multiple sclerosis, demyelination, remyelination, Theiler's virus, oligodendroglia, myelin repair, natural autoantibody

1 Introduction

Multiple sclerosis (MS) is a progressive, immune-mediated demyelinating disease of the central nervous system (CNS). The etiology and pathogenesis remain uncertain, but one attractive hypothesis is that myelin degeneration is initiated by virus infection. Myelin repair in the MS lesion is limited, but has been observed at the edge of acute lesions [27,30,37]. Oligodendrocytes are abundant within some demyelinated lesions, and sometimes are present as nest of cells suggestive of proliferation [30]. Nevertheless, the reason why re-

Molecular Signaling and Regulation in Glial Cells
A Key to Remyelination and Functional Repair

myelination is incomplete remains obscure. One of the therapeutic goals for MS is to promote remyelination and to obtain complete functional recovery.

Autoimmune, toxic and infectious experimental models are available to investigate CNS demyelination and remyelination. Experimental autoimmune (allergic) encephalomyelitis (EAE) is the primary T-cell-mediated autoimmune CNS demyelination model [43]. Myelin basic protein (MBP)-reactive CD4+ T cells play an important role in this model, and as in MS, spontaneous CNS remyelination is incomplete. Localized CNS demyelination can be induced by toxins including ethidium bromide, lysolecithin, or cuprizone. Spontaneous remyelination is rapid and complete in these lesions [6,7,18,48]. Several models of virus-induced CNS demyelination are available [9]. Among these infectious agents, our laboratory has investigated demyelination induced by Theiler's murine encephalomyelitis virus (TMEV), in an effort to examine therapeutic strategies to promote CNS remyelination. Theiler's virus is a RNA virus which belongs to the family of *Picornaviridae*. Susceptible strains of mice infected intracerebrally with the Daniel's strain of TMEV develop immune-mediated CNS demyelination with relative absence of remyelination which is similar pathologically and clinically to the chronic progressive form of MS [33]. This animal model of MS provides an opportunity to investigate and evaluate various therapeutic paradigms in a quantitative manner.

Although various therapies have and are being attempted to treat MS, none of them give a completely satisfactory outcome. Most of the current therapies for MS are based on the hypothesis that MS is an autoimmune disease. Immunomodulatory therapies including corticosteroids, azathioprine, interferon-β, plasma exchange, cyclophosphamide have been reported to be effective [5]. Recently, administrations of intravenous immunoglobulin, oral myelin, copolymer-1 are being investigated as potential therapies [5], though their mechanisms of action are not clear and their clinical effectiveness remains to be proven. Using the TMEV infected model, we have investigated various therapies designed to inhibit demyelination or promote remyelination including cyclophosphamide, anti-CD4 and CD8 monoclonal antibodies [36], IL-6 [38], and immunoglobulins against spinal cord homogenates [34]. In this report, we review our recent experiments of immunoglobulin-mediated CNS remyelination using the TMEV model of MS.

2 Beneficial immune response for CNS remyelination

Immune T cells have been proposed to play a pathogenic role in MS and in various autoimmune and viral models of MS. In contrast, several studies suggest that the humoral immune system may play a role to promote remyelination in models of MS. In EAE, injections of CNS components into guinea pigs enhanced remyelination [31,42]. The mechanism of this remyelination was unclear, although it was proposed to be mediated by suppressor mechanisms. The possibility that the immune system promotes remyelination was tested in TMEV-induced demyelination. Extensive remyelination was demonstrated in chronically infected SJL/J mice after treatment with emulsions of spinal cord homogenate

(SCH) [20]. To determine the mechanism by which emulsions of SCH promoted remyelination, we raised polyclonal antisera against SCH and showed that injection of this sera enhanced CNS remyelination in mice chronically infected with TMEV [32].

2.1 Immunoglobulins against SCH promote CNS remyelination

We considered the possibility that the immunoglobulin component of the sera raised against SCH had the capacity of promoting remyelination. Therefore, we treated chronically infected mice with purified immunoglobulins (IgG) directed against SCH. This immunoglobulin treatment resulted in a dramatic increase in the number and average size of lesions that were undergoing remyelination [34,35].

To elucidate the mechanisms of immunoglobulin-mediated remyelination, we generated a panel of monoclonal antibodies (mAbs) derived from splenocytes of SJL/J mice injected with SCH in incomplete Freund's adjuvant and screened these mAbs for their binding to SCH by ELISA, protein dot blots and immunostaining of glial cells. Hybridoma cells from two wells, called 79 and 94 series, showed significant binding to SCH. For the 79 series, 14 of 49 clones and for the 94 series, 17 of 32 clones were positive for SCH. Two clones from the 94 series (SCH94.03 and 94.32) were shown to significantly promote CNS remyelination in the TMEV model. Clone SCH94.03 and 94.32 proved to be identical by immunoglobulin gene sequence analysis, therefore SCH94.03 was chosen as the consensus designation. SCH94.03 belongs to the IgMκ subclass and promotes 4-fold more CNS remyelination (Fig. 1A, Fig. 2) compared with isotype controls (Fig. 1B) when given to SJL/J mice chronically infected with TMEV [23].

3 Characterization of SCH94.03

The concept that immunoglobulin may have stimulatory effect on CNS remyelination is novel and SCH94.03 provides a unique opportunity to examine the role of immunoglobulin in demyelinating diseases. Since SCH94.03 is a monoclonal antibody, i.e., single immunoglobulin component, it has the potential to simplify the understanding of the mechanism of immunoglobulin-mediated CNS remyelination, as compared to sera or polyclonal immunoglobulins against SCH. Therefore, characterization of SCH94.03 was performed by using immunohistochemichal and molecular biological techniques.

3.1 SCH94.03 recognizes a surface antigen on well-differentiated oligodendrocytes

To investigate the cell specificity of the antigen recognized by SCH94.03, we performed immunohistochemical studies on cultured cells obtained from neonatal rat and mouse brains [3]. Among the isolated mouse and rat glial cell cultures, only oligodendrocytes showed surface staining with SCH94.03 (Fig. 3). The surface-labeled cells had long processes, characteristic of well-differentiated

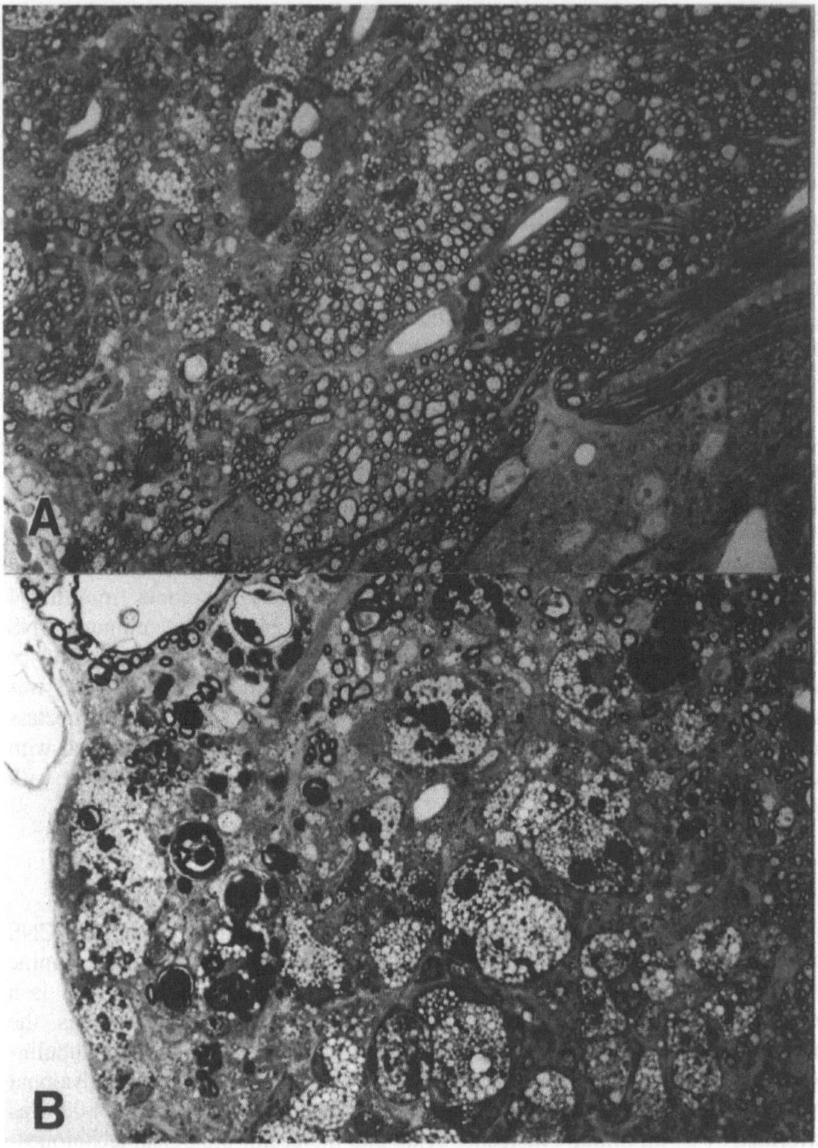

Fig. 1. CNS remyelination and demyelination in SJL/J mice infected with Theiler's virus. (A) Light micrograph of a remyelinated area in the spinal cord of chronically infected mouse treated with SCH94.03. (x400) (B) Light micrograph of a demyelinated lesion in the spinal cord of chronically infected mouse treated with isotype control IgMκ mAb. (x400) Mice were perfused with Trump's fixative (100mM phosphate buffer, pH 7.2, with 4% formaldehyde and 1.5% gluteraldehyde) and spinal cord blocks embedded with Araldite. One-micron sections were stained with paraphenylenediamine.

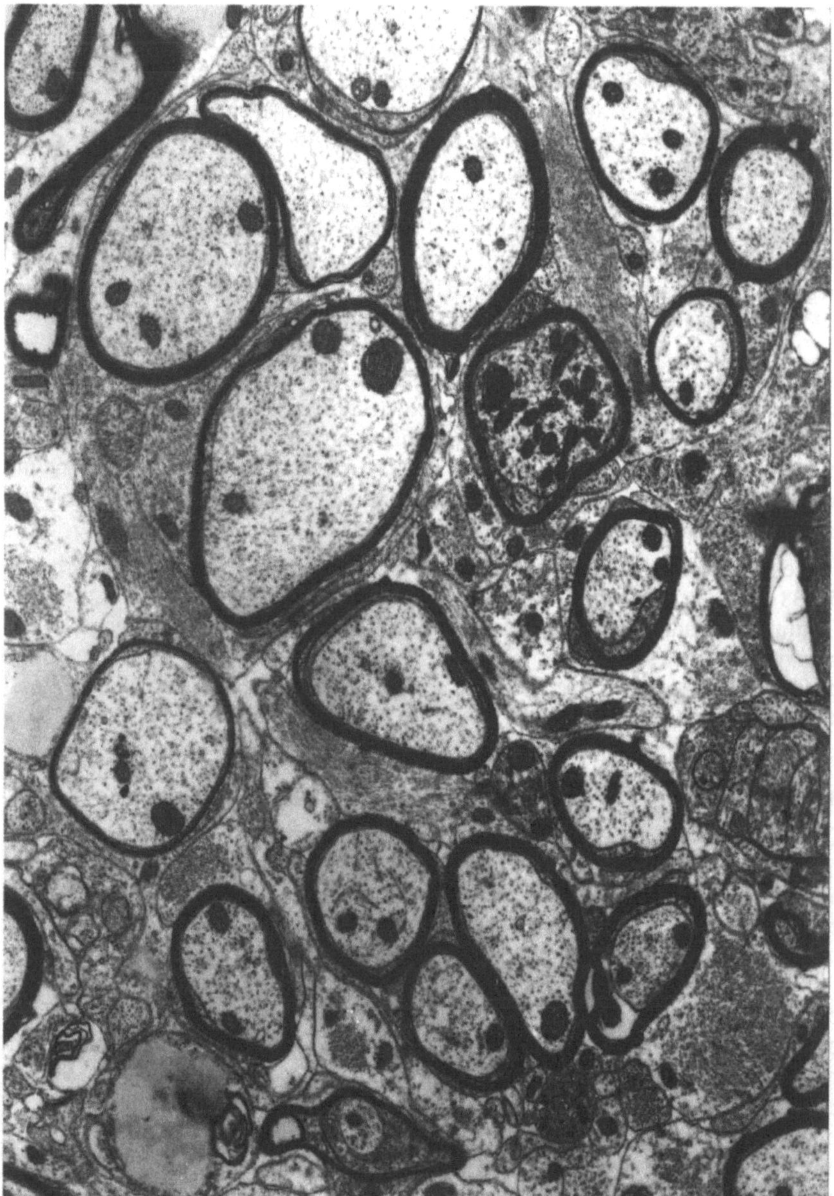

Fig. 2. Electron micrograph of spinal cord section from a chronically infected SJL/J mouse treated with SCH94.03 showing CNS remyelination. Multiple axons have abnormally thin myelin sheaths relative to axon diameter. (x9500)

oligodendrocytes. In double-labeling experiments, cells labeled on the surface with SCH94.03 were negative for glial fibrillary acidic protein (GFAP). Approximately 40 - 50% of the cells coexpressed MBP. Approximately 30 - 40% were doubly labeled with biotinylated mAb O1 which recognizes mainly galactocerebroside. Cells labeled on the surface with SCH94.03 did not express early differentiation stage-specific markers of oligodendrocytes, such as ganglioside GD3. The bipolar shaped oligodendrocyte/type-2 astrocyte (O-2A) progenitor cells ($A2B5^+$), which were blocked in differentiation by platelet-derived growth factor (PDGF) and basic fibroblast growth factor (bFGF), were negative for surface staining with SCH94.03. After removing PDGF and bFGF, $O4^+$ (proligodendroblast antigen (POA))[4] stage oligodendrocytes were negative on the surface for SCH 94.03 antigen. Surface labeling of mouse and rat oligodendrocytes with SCH94.03 was confirmed by fluorescence activated cell sorting (FACS) analysis and immunogold technique (Fig. 4A). Gold particles enhanced slightly with silver were identified on the surface of cultured cells presumed to be oligodendrocytes by electron microscopy (Fig. 5). Microglia, astrocytes, Schwann cells, myoblasts, T-lymphocytes, and B-lymphocytes were negative by either immunohistochemistry or FACS.

We also investigated surface antigen expression of various cell lines. Only rat 5.5B8 and 20.2E11 glial cell lines (kind gift from Dr. McMorris, Wistar Institute, Philadelphia, PA), which are of the oligodendrocyte lineage, were positive on the surface for SCH94.03 antigen. Other cell lines including neuroblast, fibroblast, myoblast and other oligodendrocyte lineage cell lines showed no surface reactivity.

3.2 SCH94.03 may recognize a novel surface antigen on oligodendrocytes

Oligodendrocytes and Schwann cells produce myelin in CNS and peripheral nervous system and share some unique proteins and lipids. Only a few proteins are known to be selectively expressed in oligodendrocytes but not in Schwann cells, namely, myelin/oligodendrocyte glycoprotein (MOG) [21] and myelin/oligodendrocyte-specific protein (MOSP) [11]. SCH94.03 does not stain the surface of Schwann cells. Therefore, SCH94.03 appears to recognize an oligodendrocyte-specific surface antigen. MOG and MOSP are minor components of CNS myelin. MOG appears on the surface, especially on membrane sheaths of oligodendrocytes in the late maturation stage. MOG-positive cells are always positive for MBP and 2',3'-cyclic nucleotide 3'-phosphodiesterase (CNPase) [39], whereas, approximately half of the surface-labeled oligodendrocytes with SCH94.03 do not express MBP. MOSP is constitutively expressed on the surface of oligodendrocytes. Live oligodendrocyte staining with anti-MOSP Ab shows a lacy network pattern [11] or solid pattern [12] on membrane sheaths. In contrast, oligodendrocytes labeled on the surface with SCH94.03 show strong distinct staining especially on the cell body and their proximal processes, but less on membrane sheaths. Thus, in comparison with the already-known unique surface antigens of oligodendrocytes, SCH94.03 may react with novel antigen(s) on the surface of mature oligodendrocytes.

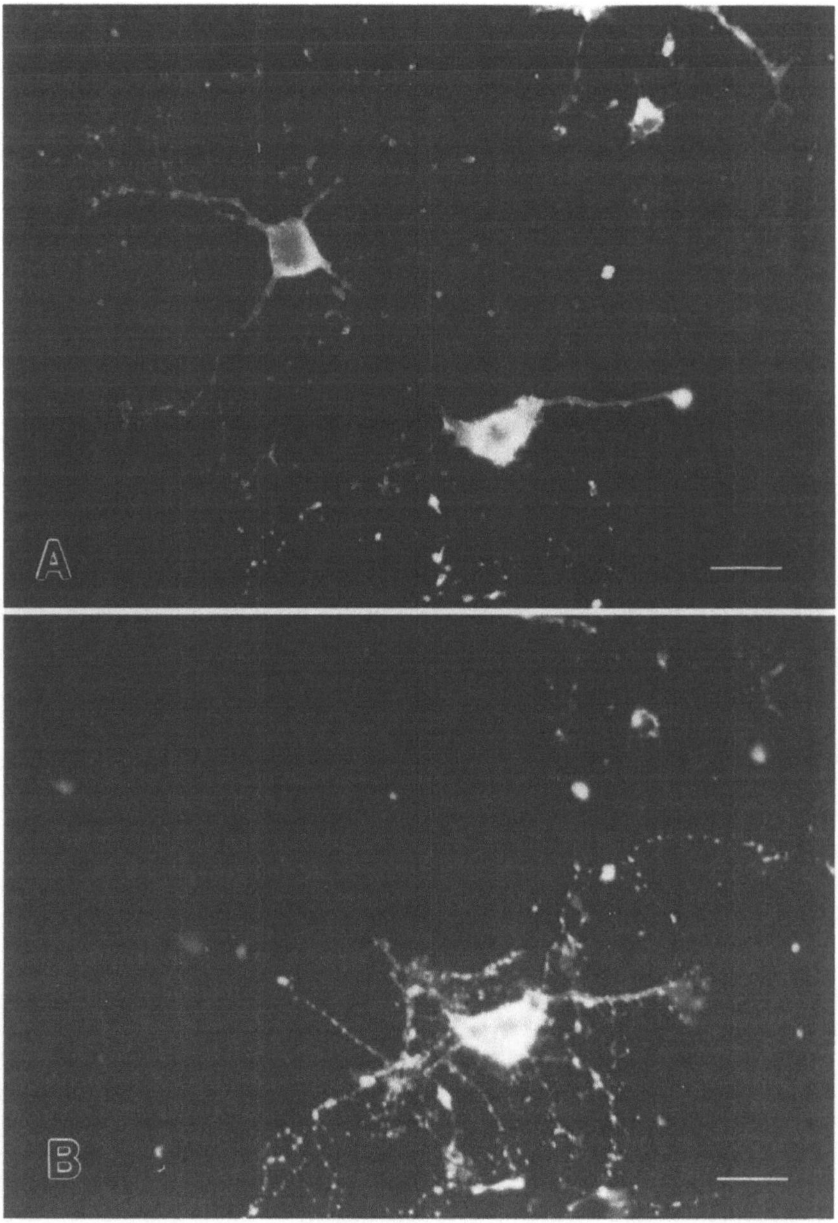

Fig. 3. Double-labeling of mouse oligodendrocytes. (A) Surface live cell staining with SCH94.03. (B) Same cells as in A stained with anti-MBP. Scale bars = 20 μm.

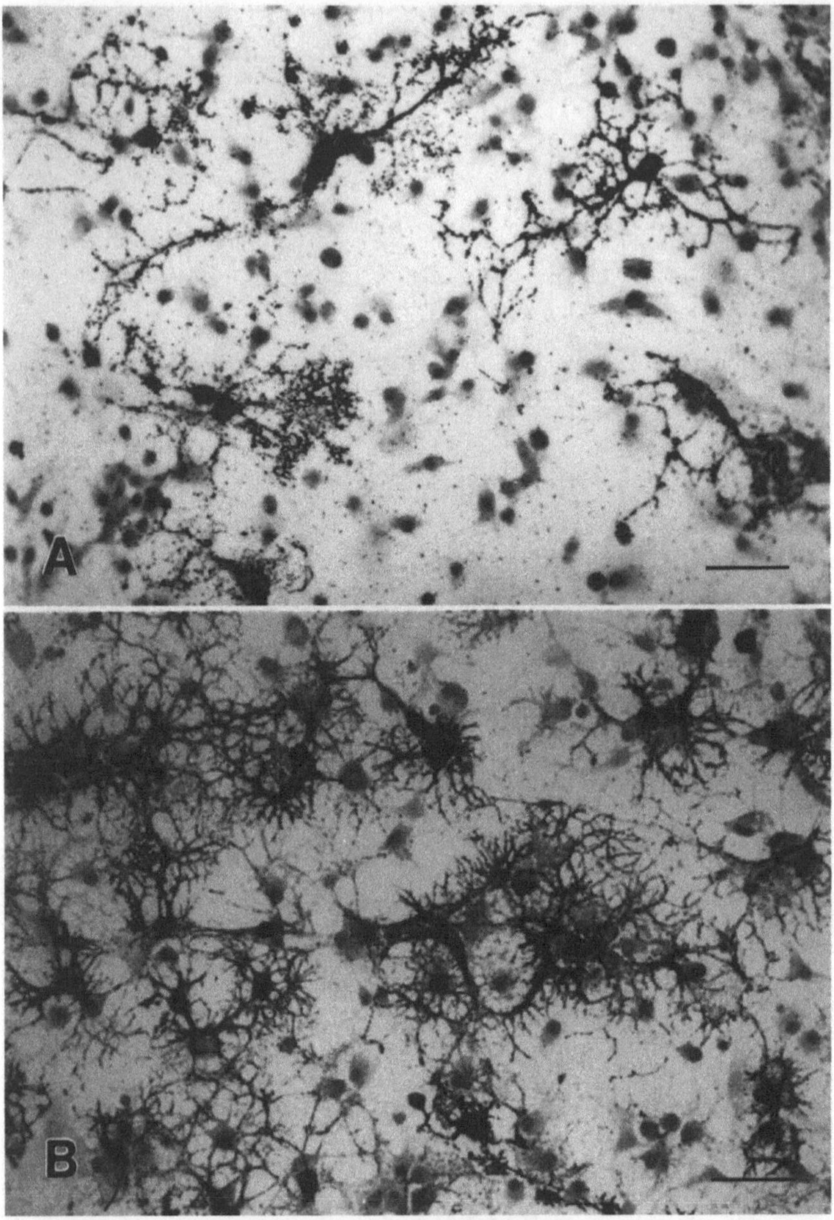

Fig. 4. Surface live staining of rat mixed glial cells with (A) SCH94.03 and (B) O1. Bound SCH94.03 and O1 mAbs were detected by immunogold followed by silver enhancement. Scale bars = 100 μm.

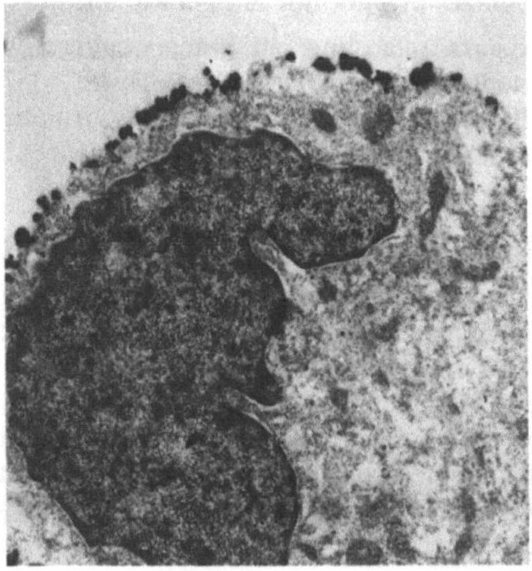

Fig. 5. Electron micrograph of rat cultured glial cell stained with SCH94.03. Bound SCH94.03 mAb was detected by immunogold followed by slight silver enhancement. (x6000)

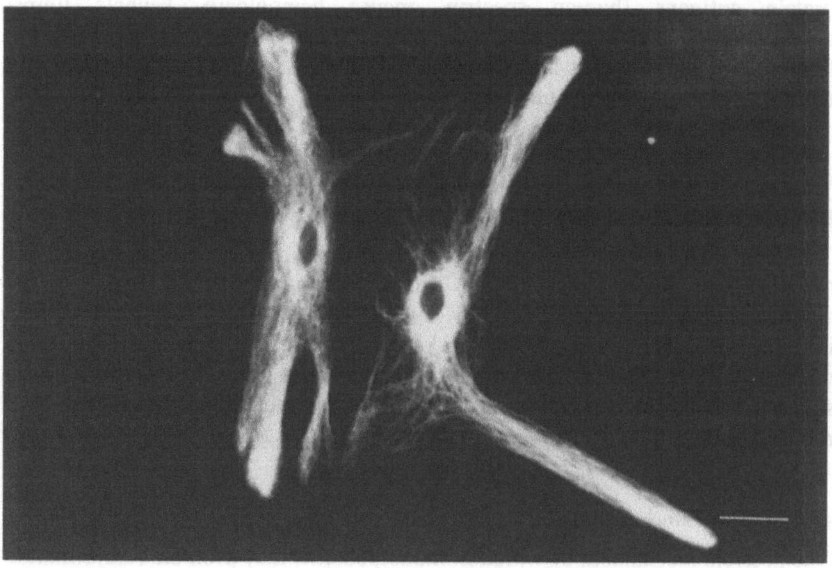

Fig. 6. Cytoplasmic staining of rat astrocytes with SCH94.03. Scale bar = 20 µm.

3.3 SCH94.03 recognizes cytoplasmic antigens in all cells

In contrast to the surface labeling, all primary-cultured cells and cell lines showed strong cytoplasmic staining with SCH94.03 [3]. The cytoplasmic staining with SCH94.03 showed a filamentous pattern which resembled cytoskeleton protein staining such as for GFAP (Fig. 6).

3.4 SCH94.03 is a natural autoantibody

Natural autoantibodies are directly encoded by immunoglobulin germline genes with no or few somatic mutations, and have been shown to be present in sera of healthy humans and rodents [10,17,19]. These natural autoantibodies are polyreactive, capable of binding to a variety of structurally unrelated antigens [41]. The physiological function of natural autoantibodies is unknown. However, by interacting with many self constituents, these natural autoantibodies and their targets may establish a vast network whereby the immune system could participate in general homeostasis.

Analysis of immunoglobulin heavy and light chain variable region cDNA sequences by the "5'-rapid amplification of cDNA ends" method confirmed that SCH94.03 uses germline immunoglobulin sequences without somatic mutations [24]. SCH94.03 is encoded by germline immunoglobulin heavy chain V23/DFL16.1/JH2 and light chain Vκ10/Jκ1 genes, without definitive somatic mutations. Solid-phase ELISA experiments showed reactivity toward several protein antigens (human spectrin, mouse hemoglobin, keyhole limpet hemocyanin) and chemical haptens ((4-hydroxy-3-nitrophenyl)acetyl, fluorescein). These results are consistent with the hypothesis that SCH94.03 is a natural or physiologic autoantibody.

4 Hypotheses of immunoglobulin-mediated remyelination gained from the study of SCH94.03

The absence of complete spontaneous remyelination in the MS lesion may be due to either the presence of inhibitory factors preventing myelin repair, or the absence of factors necessary for new myelin synthesis. The mechanism of how immunoglobulins including SCH94.03 promote remyelination in TMEV-infected mice is still unknown. The results obtained from the TMEV model have suggested to us three possible mechanisms of immunoglobulin-mediated CNS remyelination.

4. 1 Immune modulation

One possible mechanism by which immunoglobulins may promote remyelination is by their action on the immune system. This mechanism could be based on indirect immunomodulatory effects, whereby SCH94.03 inhibits pathogenic component(s) of the disease process, thereby allowing a normal physiological repair response to predominate. This hypothesis is supported by the observation that immunosuppression using cyclophosphamide and anti-CD4 or CD8 mAbs promotes remyelination in TMEV-infected mice [36]. Natural autoantibodies have also been shown to suppress the anti-acetylcholine receptor Ab response in an autoimmune model of myasthenia gravis [40] and efficiently prevent the development of diabetes in non-obese diabetic mice [1]. This supports the possibility that SCH94.03 may promote remyelination by inhibiting a pathogenic immune response.

4.2 Indirect stimulatory effect on oligodendrocytes

Immunoglobulins could react with immune or nonimmune cells in or outside of the CNS, which in turn could secrete cytokines to promote repair. Astrocytes and microglia have a wide range of critical functions for maintaining a well-balanced microenviroment in the CNS. Astrocytes and microglia, if activated, produce several cytokines important especially in the regulation of intracerebral immune responses [13]. In addition, neurons and type-1 astrocytes produce PDGF, which promotes division of O-2A progenitor cells [25]. Basic FGF acts on O-2A progenitor cells to maintain high level expression of PDGF $\alpha-$receptors and to increase the sensitivity of O-2A progenitor cells to PDGF [22]. Therefore, O-2A progenitors exposed simultaneously to PDGF and bFGF do not undergo further differentiation, but instead divide continuously as bipolar and motile progenitors [8]. In the absence of PDGF and bFGF, these cells cease to proliferate and differentiate rapidly into oligodendrocytes [8,29]. Therefore, immunoglobulins could play a role in the homeostasis of growth and differentiation factors important in the maintenance of oligodendrocytes.

4.3 Direct stimulatory effect on oligodendrocytes

Another possible and attractive mechanism to stimulate remyelination is by the direct binding of SCH94.03 or some immunoglobulin component to oligodendrocytes or their progenitor cells. Oligodendrocytes develop sequentially from precursor cells [28]. The origin of new myelin-forming cells is still debated in demyelinating diseases. One possibility is that the adult CNS contains a pool of oligodendrocyte precursor cells. Oligodendrocyte precursor cells have been shown to exist in both neonatal and adult optic nerves [44] and may play a role in the repair of demyelinated lesions. Perinatal and adult forms of oligodendrocyte precursor cells differ in many properties, but coexist during development [44,45]. The adult form of precursor cells appears to proliferate in response to growth factors [2,46] and in virus-induced demyelinating disease [15], indicating that it may give rise to new oligodendrocytes and myelin. Since surface-labeled SCH94.03-positive oligodendrocytes are well-differentiated based on their long processes or membrane sheaths, and express MBP or O1 antigen but not A2B5, GD3 and O4 antigen (POA), this indicates that SCH94.03 may act on mature oligodendrocytes but not on progenitor cells.

Alternatively, surviving oligodendrocytes in the lesions may proliferate or may dedifferentiate to initiate a new myelination program. It is possible that SCH94.03 may react to an antigen on these cells to promote remyelination. One possibility is that adult galactocerebroside positive (GC^+) oligodendrocytes can dedifferentiate, reenter the mitotic cycle, and later give rise to remyelination cells [47]. Mature oligodendrocytes isolated by a complement-mediated cell lysis method or by cytosine arabinoside treatment can be induced by bFGF to dedifferentiate and proliferate [14,16]. *In vivo*, in the spinal cord of mice chronically infected with TMEV, cells which express late oligodendrocyte differentiation markers (GC, MBP and proteolipid protein) show the greatest proliferation as identified using simultaneous immunocytochemistry and *in situ* autoradiography (^{3}H-thymidine incorporation) [26]. These experiments suggest that mature oligodendrocytes may proliferate and play an important role in remyelination following demyelinating disease. The data that SCH94.03 recognizes an antigen expressed on well-differentiated oligodendrocytes to promote remyelination is consistent with this hypothesis. Further analysis of the surface antigen on oligodendrocytes recognized by SCH94.03 may provide a clue to elucidate this remyelination potential.

References

1. Andersson, A., Forsgren, S., Söderström, S. and Holmberg, D. (1991) Monoclonal, natural antibodies prevent development of diabetes in the non-obese diabetic (NOD) mouse. J. Autoimmunity 4, 733-742.
2. Armstrong, R.C., Dorn, H.H., Kufta, C.V., Friedman, E. and Dubois-Dalcq, M.E. (1992) Pre-oligodendrocytes from adult human CNS. J. Neurosci. 12, 1538-1547.
3. Asakura, K., Miller, D.J., Murray, K., Bansal, R., Pfeiffer, S.E. and Rodriguez, M. Monoclonal autoantibody SCH94.03, which promotes CNS remyelination, recognizes an antigen on the surface of oligodendrocytes. J. Neurosci. Res. *in press*
4. Bansal, R., Stefansson, K. and Pfeiffer, S.E. (1992) Proligodendroblast antigen (POA), a developmental antigen expressed by A007/O4-positive oligodendrocyte progenitors prior to the appearance of the sulfatide and galactocerebroside. J. Neurochem. 58, 2221-2229.
5. Bansil, S., Cook, S.D. and Rohowsky-Kochan, C. (1995) Multiple sclerosis: Immune mechanism and update on current therapies. Ann. Neurol. 37 (Supplement 1), 87-101.
6. Blakemore, W.F. (1973) Demyelination of the superior peduncle in the mouse induced by cuprizone. J. Neurol. Sci. 20, 63-72.
7. Blakemore, W.F. (1973) Remyelination of the superior cerebellar peduncle in the mouse following demyelination induced by feeding cuprizone. J. Neurol. Sci. 20, 73-83.
8. Bögler, O., Wren, D., Barnett, S.C., Land, H. and Noble, M. (1990) Cooperation between two growth factors promotes extended self-renewal and inhibits differentiation of oligodendrocytes-type-2 astrocyte (O-2A) progenitor cells. Proc. Natl. Acad. Sci. USA 87, 6368-6372.

9. Dal Canto, M.C. and Rabinowitz, S.G. (1982) Experimental models of virus-induced demyelination of the central nervous system. Ann. Neurol. 11, 109-127.

10. Dighiero, G., Lymberi, P., Marié, J.-C., Rouyre, S., Butler-Browne, G.S., Whalen, R.G. and Avrameas, S. (1983) Murine hybridomas secreting natural monoclonal antibodies reacting with self antigens. J. Immunol. 131, 2267-2272.

11. Dyer, C.A., Hickey, W.F. and Geisert, E.E.J. (1991) Myelin/ associates with microtubules. J. Neurosci. Res. 28, 607-613.

12. Dyer, C.A. and Matthieu, J.M. (1994) Antibodies to myelin/ oligodendrocyte-specific protein and myelin/oligodendrocyte glycoprotein signal distinct changes in the organization of cultured oligodendroglial oligodendrocyte-specific protein: a novel surface membrane protein that membrane sheets. J. Neurochem. 62, 777-787.

13. Fontana, A., Frei, K., Bodmer, S. and Hofer, E. (1987) Immune-mediated encephalitis: On the role of antigen-presenting cells in brain tissue. Immunol. Rev. 100, 185-201.

14. Fressinaud, C., Laeng, P., Labourdette, G., Durand, J. and Vallat, J.M. (1993) The proliferation of mature oligodendrocytes in vitro is stimulated by basic fibroblast growth factor and inhibited by oligodendrocyte-type 2 astrocyte precursors. Dev. Biol. 158, 317-329.

15. Godfraind, C., Friedrich, V.L., Holmes, K.V. and Dubois-Dalcq, M. (1989) In vivo analysis of glial cell phenotypes during a viral demyelinating disease in mice. J. Cell Biol. 109, 2405-2416.

16. Grinspan, J.B., Franceschini, S.B. and Pleasure, D. (1993) Trophic effects of basic fibroblast growth factor (bFGF) on differentiated oligodendroglia: A mechanism for regeneration of the oligodendroglial lineage. J. Neurosci. Res. 36, 672-680.

17. Guilbert, B., Dighiero, G. and Avrameas, S. (1982) Naturally occurring antibodies against nine common antigens in the human sera. J. Immunol. 128, 2779-2787.

18. Hall, S.M. (1972) The effect of injections of lysophosphatidyl choline into white matter of the adult mouse spinal cord. J. Cell Sci. 13, 257-277.

19. Hartman, A.B., Mallett, C.P., Srinvasappa, J., Prabhaker, B.S., Notkins, A.L. and Smith-Gill, J. (1989) Organ reactive autoantibodies from non-immunized adult BALB/c mice are polyreactive and express non-biased VH gene usage. Mol. Immunol. 26, 359-370.

20. Lang, W., Rodriguez, M., Lennon, V.A. and Lampert, P.W. (1984) Demyelination and remyelination in murine viral encephalomyelitis. Ann. NY Acad. Sci. 436, 98-.

21. Linnington, C., Webb, M. and Woodhams, P.L. (1984) A novel myelin-associated glycoprotein defined by a mouse monoclonal antibody. J. Neuroimmunol. 6, 387-396.

22. McKinnon, R.D., Matsui, T., Dubois-Dalcq, M. and Aaronson, S.A. (1990) FGF modulates the PDGF-driven pathway of oligodendrocyte development. Neuron 5, 603-614.

23. Miller, D.J., Sanborn, K.S., Katzmann, J.A. and Rodriguez, M. (1994) Monoclonal antibodies promote central nervous system repair in animal model of multiple sclerosis. J. Neurosci.

24. Miller, D.J. and Rodriguez, M. (1995) A monoclonal autoantibody that promotes central nervous system remyelination in a model of multiple sclerosis is a natural autoantibody encoded by germline immunoglobulin genes. J. Immunol. 154, 2460-2469.

25. Noble, M., Murray, K., Stroobant, P., Waterfield, M.D. and Riddle, P. (1988) Platelet-derived growth factor promotes division and motility and inhibits premature differentiation of the oligodendrocyte/type-2 astrocyte progenitor cell. Nature 333, 560-562.

26. Prayoonwiwat, N. and Rodriguez, M. (1993) The potential for oligodendrocyte proliferation during demyelinating disease. J. Neuropath. Exp. Neurol. 52, 55-63.

27. Prineas, J.W. and Connell, F. (1978) Remyelination in multiple sclerosis. Ann. Neurol. 5, 22-31.

28. Raff, M.C., Miller, R.H. and Noble, M. (1983) A glial progenitor cell that develops in vitro into an astrocyte or an oligodendrocyte depending on culture medium. Nature 303, 390-396.

29. Raff, M.C., Lillien, L.E., Richardson, W.D., Burne, J.F. and Noble, M.D. (1988) Platelet-derived growth factor from astrocytes drives the clock that times oligodendrocyte development in culture. Nature 333, 562-565.

30. Raine, C.S., Scheinberg, L. and Waltz, J.M. (1981) Multiple sclerosis: Oligodendrocyte survival and proliferation in an active established lesions. Lab. Invest. 45, 534-546.

31. Raine, C.S., Moore, C.R. W., Hintzen, R. and Traugott, U. (1988) Induction of oligodendrocyte proliferation and remyelination after chronic demyelination: Relevance to multiple sclerosis. Lab. Invest. 59, 467-476.

32. Rodriguez, M., Lennon, V. A., Benveniste, E. N. and Merrill, J. E. (1987) Remyelination by oligodendrocytes stimulated by antisera to spinal cord. J. Neuropath. Exp. Neurol. 46, 84-95.

33. Rodriguez, M., Oleszak, E. and Leibowitz, J. (1987) Theiler's murine encephalomyelitis virus: A model of demyelination and viral persistence. CRC Crit. Rev. Immunol. 7, 325-365.

34. Rodriguez, M. and Lennon, V.A. (1990) Immunoglobulins promote remyelination in the central nervous system. Ann. Neurol. 27, 12-17.

35. Rodriguez, M. (1991) Immunoglobulins stimulate central nervous system remyelination: electron microscopic and morphometric analysis of proliferating cells. Lab. Invest. 64, 358-370.

36. Rodriguez, M. and Lindsley, M.D. (1992) Immunosuppression promotes CNS remyelination in chronic virus-induced demyelinating disease. Neurology 42, 348-357.

37. Rodriguez, M. and Scheithauer, B. (1994) Ultrastructure of multiple sclerosis. Ultrastruct. Pathol. 18, 3-13.

38. Rodriguez, M., Pavelko, K.D., McKinney, C.W. and Leibowitz, J.L. (1994) Recombinant human IL-6 suppresses demyelination in a viral model of multiple sclerosis. J. Immunol. 153, 3811-3821.

39. Scolding, N.J., Frith, S., Linnington, C., Morgan, B. P., Campbell, A.K. and Compston, D. A. S. (1989) Myelin-oligodendrocyte glycoprotein (MOG) is a surface marker of oligodendrocyte maturation. J. Neuroimmunol. 22, 169-176.

40. Sundblad, A., Hauser, S., Holmberg, D., Cazenava, P. and Coutinho, A. (1989) Suppression of antibody responses to the acetylcholine receptor by natural antibodies. Eur. J. Immunol. 19, 1425-1430.

41. Ternynck, T. and Avrameas, S. (1986) Murine natural monoclonal autoantibodies: a study of their polyspecificities and their affinities. Immunol. Rev. 94, 99-112.

42. Traugott, U., Stone, S.H. and Raine, C.S. (1982) Chronic relapsing experimental autoimmune encephalomyelitis: Treatment with combinations of myelin components promotes clinical and structural recovery. J. Neurol. Sci. 56, 65-73.

43. Williams, K.C., Ulvestad, E. and Hickey, W.F. (1994) Immunology of multiple sclerosis. Clin. Neurosci. 2, 229-245.

44. Wolswijk, G. and Noble, M. (1989) Identification of an adult-specific glial progenitor cell. Development 105, 387-400.

45. Wolswijk, G., Riddle, P.N. and Noble, M. (1990) Coexistence of perinatal and adult forms of a glial progenitor cell during development of the rat optic nerve. Development 109, 691-698.

46. Wolswijk, G. and Noble, M. (1992) Cooperation between PDGF and FGF converts slowly dividing O-2Aadult progenitor cells to rapidly dividing cells with characteristics of O-2Aperinatal progenitor cells. J. Cell Biol. 118, 889-900.

47. Wood, P.M. and Bunge, R.P. (1991) The origin of remyelinating cells in the adult central nervous system: The role of the mature oligodendrocyte. Glia 4, 225-232

48. Yajima, K. and Suzuki, K. (1979) Demyelination and remyelination in the rat central nervous system following ethidium bromide injection. Lab. Invest. 41, 385-392.

Immunomodulatory therapy of multiple sclerosis: current trials and future perspectives

P. Rieckmann

Department of Neurology, Georg-August-University of Göttingen, Robert-Koch Str. 40, D-37075 Göttingen, Germany. Tel.: 49551-396951, Fax: 49551-398405

Abstract. Significant progress has been made during the last ten years to understand the pathogenetic mechanisms underlying disease activity in patients with multiple sclerosis (MS). Some of these discoveries resulted in the development of new therapeutic strategies, which have already entered clinical or pre-clinical studies and revealed circumstantial evidence that immunoregulatory therapy can modify the course of the disease.

Some fourty different substances and regimens for immunological treatment of MS, which are under investigation in clinical trials are listed in a recent report of the Research and Medical Programs Department of the American National Multiple Sclerosis Society. Prophylactic treatment strategies can be divided into four different groups: 1) antigen specific therapy; 2) general immunosuppression; 3) immunomodulation; and 4) remyelination. Two forms of recombinant interferon-beta (IFN-b) were shown to significantly reduce the relapse rate, lesion load and gadolinium enhancement on magnetic resonance images (MRI) of the head. IFN-b_{1a}, derived from chinese hamster ovary (CHO-) cells, in addition prolonged the time to sustained progression in disability.

Copolymer 1 (COP-1), a synthetic, randomized polypetide of four amino acids, which resemble parts of myelin basic protein (MBP), revealed a significant reduction of relapses during a recently completed, multi-center clinical trial. Antigen specific therapies against MBP (oral tolerization) or strategies directed at the T cell receptor (Vb peptides) have gained considerable interest, but no conclusive clinical data are available at this time.

Induction of apoptosis in auto-reactive T cell clones, blockade of cellular trans-migration through the blood brain barrier and induction of remyelination are future perspectives for treatment development.

Introduction

Although the initiating antigen is still unknown, MS can be regarded as a T-helper type 1 (Th1)-mediated auto-immune disease of the central nervous system, which affects the myelin sheaths and finally leads to extensive demyelination with reactive astrocytosis. This T cell-mediated damage involves different steps of immunological reactions, which can be modified by various treatments: activation of lymphocytes in the lymphoid organs after antigenic challenge; expansion of

Molecular Signaling and Regulation in Glial Cells
A Key to Remyelination and Functional Repair

activated clones; migration into the circulation and across the blood brain barrier; attraction of follower lymphocytes and effector macrophages into inflammatory foci; tissue destruction, suppression of the acute response and induction of repair mechanisms.

Antigen-specific therapy

The role of myelin basic protein (MBP) as an important antigen in the induction of experimental allergic encephalomyelitis (EAE), an animal model for MS, is well established but its role during the course of MS is still not completely understood. MBP-specific T cell lines can be obtained from patients and healthy individuals and there is no obvious restriction of T cell receptor usage in either population (33). This in contrast to results obtained with encephalitogenic T cell clones in the EAE model, where certain T cell receptors (e.g. Vb 8.2) dominate the immune response (1). Based on the assumption that MBP is a relevant antigen in MS and on the results that some MS patients preferentially use certain T cell receptor Vb-chains, Vb 5.2 and Vb 6.1 receptor peptides were administered to patients in a phase 1 clinical trial. This treatment was without significant side-effects and appeared to alter the frequency of circulating MBP-specific T cells (5). No data on clinical effectiveness are currently available.

An interesting new approach of antigen specific therapy is the concept of oral tolerance, which was already described in 1911. Feeding of the antigen prior to injection of the same protein prevents a severe immune response. After promising initials trials with oral myelin fed to rats, which ameliorated and suppressed the EAE, it was demonstrated, that this treatment results in a Th2 type response in the gut-associated lymphoid tissue. Regulatory cells, which secrete suppressive cytokines, like IL-4 or TGF-b cross the blood brain barrier and enter regions of plaque formation (14, 34). In addition, oral tolerization with MBP has a suppresive effect on proteolipid protein-induced EAE, which suggest, that bystander events occur during this therapy. A pilot trial with oral bovine myelin in MS patients revealed no toxicity or serious side effects and a large multi-center trial in patients with relapsing-remitting MS is underway (35).

Copolymer-1 (COP-1) is a synthetic polypeptide prepared from four amino acids, L-alanine, L-glutamic acid, L-lysine and L-thyrosine with a molar ratio of 6.0 : 1.9 : 4.7 : 1.0 and a molecular weight of 23 kDa. It was originally shown to downregulate EAE in various animals and inhibit MBP-specific murine and human T cell lines (3). In addition, COP-1 specifically inhibits proliferation and interleukin-2 secretion by human MBP-specific T cell lines and clones in vitro (32). Recently, it was demonstrated that COP-1 blocks the binding of MHC class II products to MBP in the trimolecular complex (7). Two principal mechanisms of COP-1 activity in EAE and MS are proposed: 1) induction of antigen-specific suppressor cells and 2) interference with T cell activation by competition with MBP for the MHC class II binding site responsible for antigen presentation. A randomized, double-blind, placebo-controlled phase III study (13) demonstrated that COP-1, administered daily subcutaneously reduces the relapse rate during a two years observation period, preferentially in patients with minor disability or early disease onset. This fact may argue for epitope spreading and heterogeneous immune reactivity during the course of the disease (17). Therefore, antigen specific therapy in MS is an attractive tool specially to treat patients very early during their disease.

General immunosuppression

Immunosuppressive treatment for MS predominantly with substances, which reduce absolute leukocyte numbers and lymphocyte activity, is under intense investigation for over 20 years. In general the results of clinical studies with cyclophosphamide, azathioprine, cyclosporine and plasma exchange did not reveal significant benefit for the patients, which would outweigh toxicity and side effects (8, 37). Conclusive results were obtained with high dose, intravenous methylprednisolone pulse therapy of acute relapses (16). The severity and duration of relapses could be reduced and new clinical neurologic attacks were significantly reduced within a 2 years observation period in patients with optic neuritis after a pulse with i.v. corticosteroids (4).

Methotrexate, which is effectively used for years in chronic inflammatory diseases, like rheumatoid arthritis, was shown to have some effect on disease progression in chronic progressive MS. In a randomized, double-blind, placebo-controlled trial 52% of patients on methotrexate showed disease progression compared to 83% of those on placebo. Interestingly, by far the best results of treatment were observed on measures of upper extremity function (9). The principal mechanism of methotrexate activity is still uncertain. It has a variety of immunoregulatory effects, including inhibition of endothelial cell growth and chemotaxis; in addition, methotrexate suppresses cytokine activity and the generation of various inflammatory mediators (6).

Cladribine (2-chloro-desoxy-adenosine) is a purin analogue, which is not metabolised intracellularly. It is predominantly phosphorylated in lymphocytes and causes apoptosis of these cells. Cladribine is equally effective against proliferating and quiescent lymphocytes. In an initial placebo-controlled, cross-over study this drug was tested in 51 patients with chronic progressive MS. Average neurological scores, lesion load on MRI and concentrations of oligoclonal IgG in cerebrospinal fluid were stable or improved in patients receiving cladribine but continued to deteriorate in patients on placebo (30). The encouraging results of this trial should be interpretated with caution due to the potential toxicity of the drug and its uncommon route of application (indwelling central venous catheter).

More selective immunosuppression can be achieved with monoclonal antibodies against functional epitopes on lymphocyte subpopulations. A profound systemic depletion of lymphocytes was obtained with a humanised monoclonal antibody against CDw52 (CAMPATH-1H). This treatment of 7 MS patients in an open trial prevented new lesions on gadolinium-enhanced MRI. From these results it can be hypothezised, that lymphocyte depletion in the peripheral circulation suppresses new lesion development (21). Another trial with an antibody to CD4, an epitope present on T helper cells, is near to completion. The antibody is effective to reduce the number of circulating CD4 lymphocytes in MS patients without serious side effects (19).

Immunomodulation

The functional differences of T helper lymphocytes revealed that disease promoting and reducing cells with the same antigen specificty against MBP can be established and differentially induce EAE in susceptible animals (20, 23). These cell lines can be divided according to their cytokine production pattern: Th1 (encephalitogenic) cells produce predominantly IL-2, IFN-g, LT and TNF-a, whereas the non-encephalitogenic Th2 cells synthezise IL-4, IL-10, IL-13 (18). Cytokines as mediators of the inflammatory process during EAE and MS are known to be of pathogenetic significance for both diseases. Fluctuation of the cytokine mRNA expression pattern of blood mononuclear cells are associated with disease activity in MS patients (25). One of the major principals of immunomodulatory therapy is therefore the modification of the cytokine response after antigenic challenge. The first immunomodulatory substance to be effective in the treatment of relapsing-remitting MS is recombinant IFN-b.Two large multi-center trials demonstrated a significant reduction of relapse rates in the groups of patients treated with IFN-b in comparison to patients receiving placebo. In both studies significant less gadolinium-enhancing lesions were observed on T1-weighted images of the head (10). The glycosylated form of recombinant IFN-b (CHO-IFN-b or IFN-b_{1a}) had also an effect on disease progression. Patients receiving 6 MIU IFN-b_{1a} intramuscularly once a week had a statistically significant reduction in progression of disability compared to the placebo group (12). At present other studies with IFN-b are performed in patients with secondary progressive MS. Type 1 interferons, including IFN-b, have a variety of different effects on the immune system. Evidence is accumulating that IFN-b mainly acts a) to reverse a suppressor cell defects and b) by favouring the generation of a Th2 type cytokine response. In addition, it counteracts most of the disease promoting effects of IFN-g and TNF-a (36).

Other forms of cytokine-directed immunotherapy are currently investigated by different research groups. Tumor necrosis factor alpha (TNF-a) is considered to be of major importance in the pathogenesis of MS (11) and strategies aimed at the inhibition of this cytokine are therefore of prime interest. The major physiological antagonist of TNF-a is transforming growth factor-beta (TGF-b), which can be regarded as an important immunosuppressive cytokine. A phase 1 clinical trial in patients with primary progressive MS is currently underway (29).

The xanthine derivate pentoxifylline, a phosphodiesterase inhibitor, which increases intracellular cAMP levels, suppresses EAE (28) and induces an in vivo shift from a Th1-type to a Th2-dominated immune response (27). The drug is widely used to improve rheology in patients with microvascular disorders, has no toxicity and a well documented low side effect profile. A placebo-matched, double-blinded, cross-over clinical trial in patients with relapsing-remitting MS is currently performed at the Department of Neurology, University of Göttingen, Germany. Another phosphodiesterase inhibitor, rolipram, is equally effective in the treatment of EAE and suppresses pro-inflammatory cytokines (31).

Other strategies to inhibit TNF-a production or neutralize its effects in vivo include the administration of soluble TNF-receptors, neutralizing monoclonal antibodies against TNF or metalloproteinase inhibitors (an enzyme, which is important for post-translational modification of this cytokine) (24).

Current research on the pathogenesis of autoimmune demyelination has prompted the evaluation of new promising experimental therapies. IL-4 and IL-10 are both anti-inflammatory Th2-cytokines, which downregulate the production of Th1-cytokines. In addition, blockade of IL-12, a Th1 stimulating cytokine, can be anticipated as a promising future clinical trial. Recent evidence that the development of a Th1- or Th2-cytokine response is differentially regulated by the co-stimulatory molecules B7-1 and B7-2 opens another avenue for immunomodulatory treatment of MS (15).

Adhesion molecules are induced at the blood brain barrier by pro-inflammatory cytokines during inflammatory processes and allow the increased migration of auto-rective T cells from the circulation into the brain. Monoclonal antibodies and recombinant soluble adhesion molecules that compete with the binding of auto-reactive T cell are becoming available for initial experimental trials in MS (2, 26).

Remyelination

The extent of demyelination and reactive astrocytosis are key factors for permanent disability in patients with MS. Promoting remyelination is therefore the ultimate goal for effective treatment of MS. Several research groups pointed to the possibility of inducing myelin repair, either by using various growth factors or by attempting to transplant myelin-forming oligodendrocytes. Intravenous immunoglobulin was shown to induce remyelination in the Theiler's virus encephalomyelitis animal model and pilot study with MS patients started recently (22). Insulin-like growth factor (IGF-1) is another candidate for the promotion of remyelination in MS patients and initial trials are under consideration.

In summary, the results of recent clinical trials demonstrated that relapse rates and inflammatory lesions on brain MRIs can be reduced by recombinant IFN-b or COP-1. Basic research on the immunopathogenesis of MS has introduced new avenues for effective immunotherapy and possibly remyelination in patients with

MS. Combination of antigen specific and immunomodulatory therapy is probably the next step to control this chronic and disabling autoimmune disease of the central nervous system.

Table 1: Presumed mechanisms of action of selected drugs for immunotherapy of multiple sclerosis

Substance	Effect
Interferon-b (IFN-b)	- Antagonizes IFN-g effects and IFN-g production - Inhibits T cell proliferation - Restores impaired T cell suppressor function - Reduces cellular transmigration
Copolymer 1 (COP 1)	- Inhibits proliferation and IL-2 secretion by MBP-specific T cells - Competes with MBP for the MHC class II binding site responsible for antigen presentation - Induces antigen-specific suppressor cells
Corticosteroids	- Reduce peripheral CD4 T cells - Reduce lymphocyte responsiveness to activating stimuli - Decrease cytokine release - Inhibits IgG production - Reduces macrophage activation - Induces apoptosis in autoreactive T cells
Oral Myelin	- Induces a Th2-type cytokine response in the gut-associated lymphoid tissue.
Pentoxifylline	- Induces an in vivo shift from Th1- to a Th2-type cytokine response
Methotrexate	- Inhibits endothelial cell growth and chemotaxis - Suppresses cytokine activity and the generation of various inflammatory mediators
Cladribine	- Causes apoptosis in lymphocytes
Immunoglobulin (intravenous)	- Blocks Fc-receptor signaling - Anti-cytokine/adhesion molecule activity - Promotes remyelination

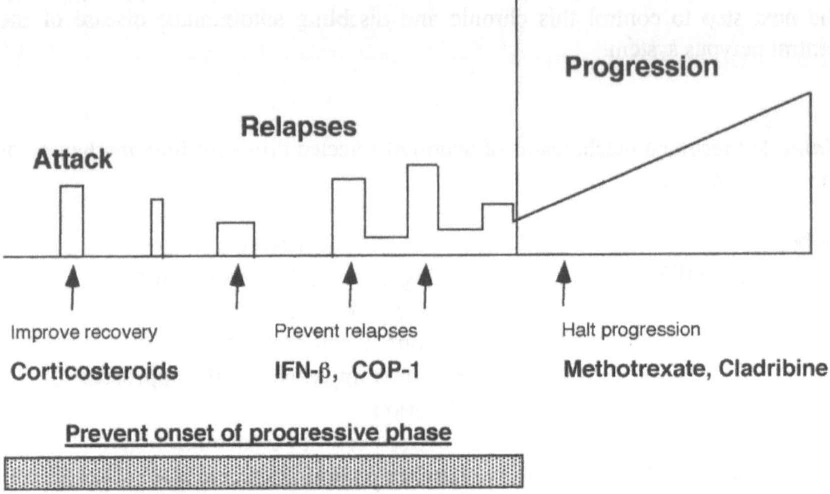

Figure1: Influence of various therapeutics on the course of multiple sclerosis

References

1. Acha-Orbea H, Mitcehll DJ, Timmermann L. (1988) Limited heterogeneity of T cell receptors from lymphocytes mediating autoimmune encephalomyelitis allows specific immune intervention. Cell 54, 263-273.
2. Archelos JJ, Jung S, Mäurer M, Schmied M, Lassmann H, Tamatani T, Miyasaka M, Toyka KV, Hartung HP. (1993) Inhibition of experimental autoimmune encephalomyelitis by an antibody to the intercellular adhesion molecule ICAM-1. Ann. Neurol. 34, 145-154.
3. Arnon R (1981). Experimental allergic encephalomyelitis-susceptibility and suppression. Immunol. Rev. 55, 5-30.
4. Beck RW, Cleary PA, Trobe JD. (1993) The effect of corticosteroids for acute optic neuritis on the subsequent development of multiple sclerosis. N. Engl. J. Med. 329, 1764-1769.
5. Chou YK, Morrison WJ, Weinberg AD, Dedrick R, Whitham R, Bourdette DN. (1994) Immunity to TCR peptides in multiple sclerosis. II. T cell recognition of Vbeta 5.2 and Vbeta 6.1 CDR 2 peptides. J. Immunol. 152, 2520-2529.
6. Cronstein BN, Naime D, Ostad E. (1993) The anti-inflammatory mechanisms of methotrexate. J. Clin. Invest. 92, 2675-2682.
7. Fridkis-Hareli M, Teitelbaum D, Gurevich E, Pecht I, Brautbar C, Kwon OJ, Brenner T, Arnon R, Sela M. (1994) Direct binding of myelin basic protein and synthetic copolymer 1 to class II major histocompatibility complex molecules on living antigen-presenting cells - specificity and promiscuity. Proc. Natl. Acad. Sci. USA 91, 4872-4876.

8. Goodin DS. (1991) The use of immunosuppressive agents in the treatment of multiple sclerosis. Neurology 41, 980-985.
9. Goodkin DE, Rudick RA, VanderBrug Medendorp S, Daughtry MM, Schwetz KM, Fischer J, van Dyke C. (1995) Low-dose (7.5 mg) oral methotrexate reduces the rate of progression in chronic progressive multiple sclerosis. Ann. Neurol. 37, 30-40.
10. Group IFNB Multiple Sclerosis Study and the University of British Columbia MS/MRI Analysis Group. (1995) Interferon beta-1b in the treatment of multiple sclerosis: Final outcome of the randomized controlled trial. Neurology 45, 1277-1285.
11. Hartung HP. (1993) Immune-mediated demyelination. Ann Neurol 33, 563-7.
12. Jacobs L, Cookfair D, Rudick R, Herndon R, Richert J, Salazar A, Group MCR. (1994) Results of a phase III trial of i.m. recombinant beta interferon as treatment for MS. J. Neuroimmunol. 54, 170 (Abstract W12.01).
13. Johnson KP, Brooks BR, Cohen JA, Ford CC, Goldstein J, Lisak RP, Myers LW, Panitch HS, Rose JW, Schiffer RB, et al. (1995) Copolymer 1 reduces relapse rate and improves disability in relapsing-remitting multiple sclerosis. Neurology 45, 1268-1276.
14. Khoury SJ, Hancock WW, Weiner HL. (1992) Oral tolerance to myelin basic protein and natural recovery from experimental autoimmune encephalomyelitis are associated with downregulation of inflammatory cytokines and differential upregulation of transforming growth factor beta, interleukin 4, and prostaglandin E expression in the brain. J Exp Med 176, 1355-64.
15. Kuchroo VK, Das MP, Brown JA, Ranger AM, Zamvil SS, Sobel RA, Weiner HL, Nabavi N, Glimcher LH. (1995) B7-1 and B7-2 costimulatory molecules activate differentially the Th1/Th2 developmental pathways: Application to autoimmune diseas therapy. Cell 80, 707-718.
16. Kupersmith MJ, Kaufman D, Paty DW, Ebers G, McFarland H, Johnson K, Reingold S, Whitaker J. (1994) Megadose corticosteroids in multiple sclerosis. Neurology 44, 1-4.
17. Lehmann PV, Forsthuber T, Miller A, Sercarz EE. (1992) Spreading of T-cell autoimmunity to cryptic determinants of an autoantigen. Nature 358, 155-157.
18. Liblau RS, Singer SM, McDevitt HO. (1995) Th1 and Th2 CD4+ T cells in the pathogenesis of organ-specific autoimmune diseases. Immunol. Today 16, 34-38.
19. Lindsey JW, Hodgkinson S, Metha R, Mitchell D, Enzmann D, Steinman L. (1994) Repeated treatment with chimeric anti-CD4 antibody in multiple sclerosis. Ann. Neurol. 36, 183-189.
20. Merrill JE, Kono DH, Clayton J, Ando DG, Hinton DR, Hofman FM. (1992) Inflammatory leukocytes and cytokines in the peptide-induced disease of experimental allergic encephalomyelitis in SJL and B10.PL mice. Proc Natl Acad Sci U S A 89, 574-8.
21. Moreau T, Thorpe J, Miller D, Moseley I, Hale G, Waldmann H, Compston A. (1994) Preliminary evidence from magnetic resonance imaging for

reduction in disease activity after lymphocyte depletion in multiple sclerosis. Lancet 1994, 44:298-301.

22. Noseworthy JH, O'Brien PC, van Engelen BGM, Rodriguez M. (1994) Intravenous immunoglobulin therapy in multiple sclerosis: progress from remyelination in the Theiler's virus model to a randomised, double-blind, placebo-controlled clinical trial. J. Neurol. Neurourg. Psychiatry 57, 11-14.
23. Powell MB, Mitchell D, Lederman J, Buckmeier J, Zamvil SS, Graham M, Ruddle NH, Steinman L. (1990) Lymphotoxin and tumor necrosis factor-alpha production by myelin basic protein-specific T cell clones correlates with encephalitogenicity. Int Immunol 2, 539-44.
24. Rieckmann P. (1995) Tumor necrosis factor and lymphotoxin: potent mediators of demyelination. In: H.P. Hartung (ed.). Inflammatory mediators in demyelinating diseases of the CNS and PNS. R.G. Landes, USA (in press).
25. Rieckmann P, Albrecht M, Kitze B, Weber T, Tumani H, Broocks A, Lüer W, Helwig A, Poser S. (1995) Tumor necrosis factor-alpha messenger RNA expression in patients with relapsing-remitting multiple sclerosis is associated with disease activity. Ann. Neurol. 37, 82-88.
26. Rieckmann P, Michel U, Albrecht M, Brück W, Wöckel L, Felgenhauer K. (1995) Soluble forms of intercellular adhesion molecule-1 (ICAM-1) block lymphocyte attachment to cerebral endothelial cells. J. Neuroimmunol. 60, 9-15.
27. Rieckmann P, Weber F, Günther A, Bitsch A, Broocks A, Kitze B, Weber T, Börner T, Poser S. (1995) Pentoxifylline, a phosphodiesterase inhibitor, induces immune deviation in patients with multiple sclerosis. J. Neuroimmunol. in press.
28. Rott O, Cash E, Fleischer B. (1993) Phosphodiesterase inhibitor pentoxifylline, a selective suppressor of T helper type 1- but not type 2-associated lymphokine production, prevents induction of experimental autoimmune encephalomyelitis in Lewis rats. Eur J Immunol 23, 1745-51.
29. Rowe PM. (1994) Clinical potential for TGF-beta. The Lancet 344, 72.
30. Sipe JC, Romine JS, Koziol JA, McMillan R, Zyroff J, Beutler E. (1994) Cladribine in the treatment of chronic progressive multiple sclerosis. Lancet 344, 9-13.
31. Sommer N, Löschmann PA, Northoff GH, Weller M, Steinbrecher A, Steinbach JP, Lichtenfels R, Meyermann R, Riethmüller A, Fontana A, et al. (1995) The antidepressant rolipram suppresses cytokine production and prevents autoimmune encephalomyelitis. Nature Med. 1, 244-248.
32. Teitelbaum D, Milo R, Arnon R, Sela M. (1992) Synthetic copolymer 1 inhibits human T cell lines specific for myelin basic protein. Proc. Natl. Acad. Sci. USA 89, 137-141.
33. Utz U, McFarland HF. (1994) The role of T cells in multiple sclerosis: implications for therapies targeting the T cell receptor. J. Neuropathol. Exp. Neurol 53, 351-358.
34. Weiner HL. Oral tolerance. (1994) Proc. Natl. Acad. Sci. USA 91, 10762-10765.

35. Weiner HL, Mackin GA, Matsui M, Orav EJ, Khoury SJ, Dawson DM, Hafler DA. (1993) Double-blind pilot trial of oral tolerization with myelin antigens in multiple sclerosis. Science 259, 1321-1324.
36. Weinstock-Guttman B, Rasohoff RM, Kinkel P, Rudick R. (1995) The interferons: biological effects, mechanisms of action, and use in multiple sclerosis. Ann. Neurol. 37, 7-15.
37. Yudkin PL, Ellison GW, Ghezzi A, Goodkin DE, Hughes RAC, McPherson K, Mertin J, Milanese C. (1991) Overview of azathioprine treatment in multiple sclerosis. Lancet 338, 1051-1055.

Cell Transplantation

Identification of new cell-surface receptors on oligodendrocyte-lineage cells

Antje Niehaus and Jacqueline Trotter

Department of Neurobiology, University of Heidelberg,
Im Neuenheimer Feld 364, 69120 Heidelberg, Germany.

Oligodendrocytes are the myelin-forming cells of the central nervous system (CNS). Cell-surface adhesion molecules regulating the behaviour of the precursor cells, in particular their interaction with neurones, are largely unknown (Doyle and Colman, 1993; Trotter, 1993). As well as their importance in development, precursor cell-specific molecules could also be target antigens for the immune system in demyelinating diseases where remyelination of demyelinated axons is inefficient. As tools to identify such molecules we have developed a range of immortalised glial cell lines by infecting dividing precursor cells with a range of different oncogene-carrying retroviruses (Jung et al., 1994, 1995; Trotter et al., 1993). These lines have retained many properties of the primary cells including antigenic and electrophysiological characteristics, normal migratory properties (Fok-Seang et al., 1995) and the ability to ensheath and interact with demyelinated axons after *in vivo* transplantation. Such cell lines are stable sources of large numbers of cells and the retention of these normal characteristics suggests they may aid in identifying molecules regulating glial development and glial-neuron interactions. To identify and characterize such cell-surface molecules, we have made monoclonal antibodies using as immunogen whole cells or glycoprotein fractions from the cell lines obtained via lectin affinity chromatography. A range of antibodies were obtained by this approach which react with surface antigens of oligodendroglial-lineage cells.

One of the antibodies obtained reacts specifically with the extracellular region of a 330kD transmembrane glycoprotein. The antigen is expressed by proliferating oligodendrocyte precursor cells and absent from myelin associated glycoprotein-positive oligodendrocytes. It shows no staining of GFAP-positive astrocytes or L1-positive neurons, as shown in Fig. 1. The antigen is expressed on the cell surface as shown in Fig. 2. Stainings of sections of P6 murine cerebellum show expression of the antigen in the forming white matter tracts. It is also expressed by Schwann cells.

Molecular Signaling and Regulation in Glial Cells
A Key to Remyelination and Functional Repair

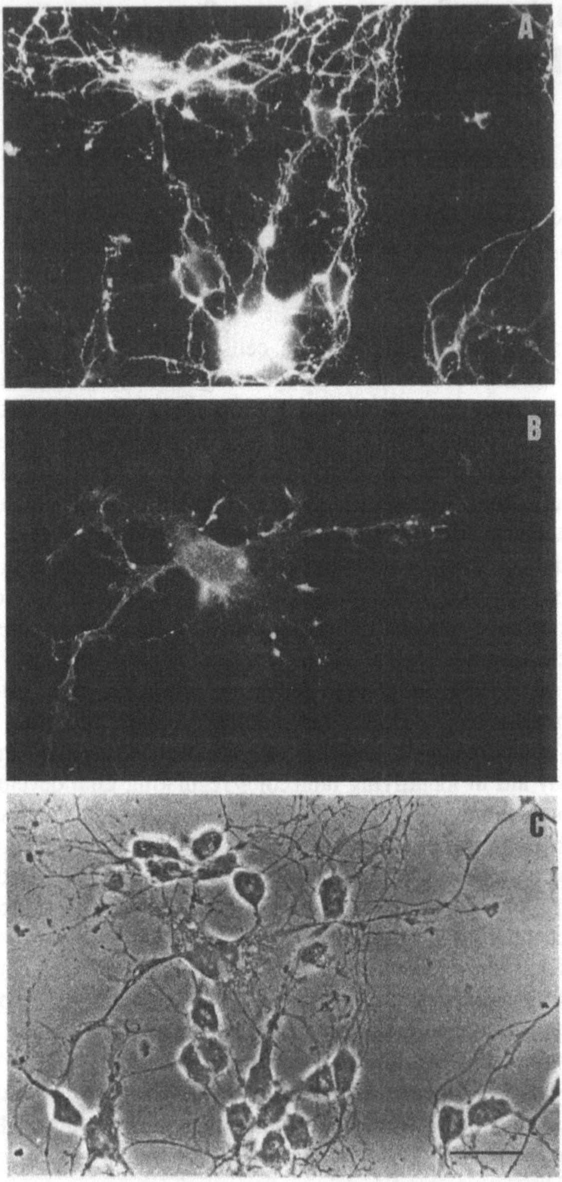

Figure 1. Immunofluorescence staining of cerebellar cultures from 6-day old mice with the antibodies AN2 and L1. The cultures were maintained in culture in Eagles basal medium containing 10% horse serum for 2 days and then fixed and labelled with polyclonal antibodies to the L1 glycoprotein (A), specific for neurons (Rathjen and Schachner, 1984) and monoclonal AN2 antibody (B). (C) shows the corresponding phase contrast picture. Bar: 17 μm.

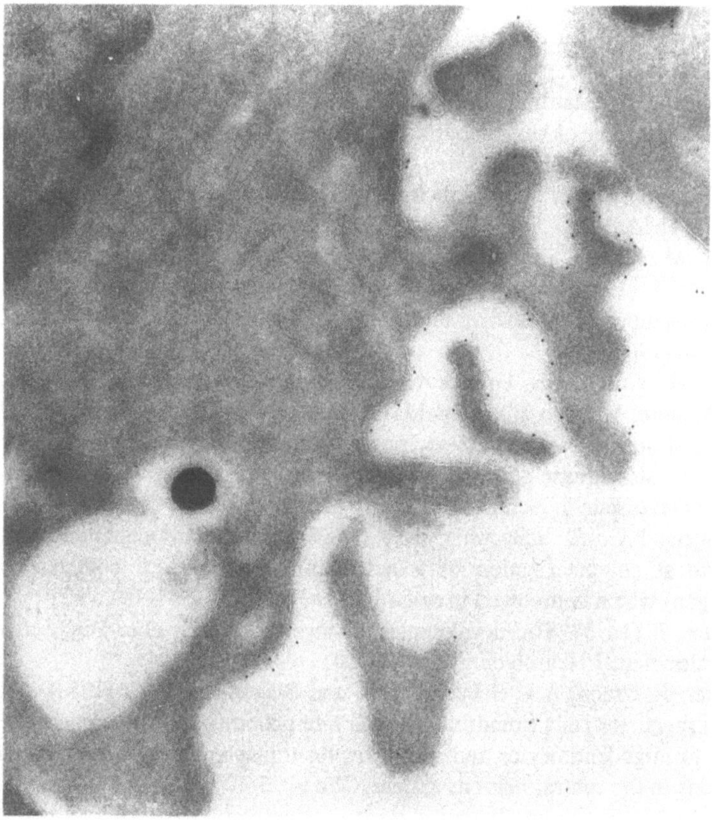

Figure 2. Localisation of the AN2 antigen on oligodendrocyte precursor cells.
Sections of the oligodendrocyte precursor cell line Oli-*neu* (Jung et al., 1995) were prepared for electron microscopy (Griffith, 1993) and labelled with polyclonal antibodies against the AN2 antigen and protein A gold. x 25 000.

Metabolic radiolabelling and enzymatic digestion experiments show that a fraction of the protein carries glycosaminoglycan chains, thus suggesting that the antigen is a part-time proteoglycan. During development of the murine CNS, expression of the protein in western blots is first seen at E10, peaks between P6 and P12 and is down-regulated in the adult brain. The function of the antigen and the developmental fate of these early antigen-positive cells is under investigation.

Acknowledgements

A.N. is a member of the graduate programme "Molecular and Cellular Neurobiology" of the University of Heidelberg; supported by the DFG Schwerpunkt "Struktur und Funktion von Glia".

References

1. Doyle, J.P. and Colman, D.R. (1993) Glial-neuron interactions and the regulation of myelin formation. Curr. Opinion in Cell Biol. 5, 779-785.
2. Fok-Seang, J., Matthews, G., ffrench-Constant, C., Trotter, J. and Fawcett, J.W. Migration of oligodendrocyte precursors on astrocytes and meningeal cells. Dev. Biol. 171, 1-15.
3. Griffith, G. (1993). Fine structure immunocytochemistry. Springer Verlag, Berlin Heidelberg.
4. Jung, M., Crang, A.J., Blakemore, W.F., Hoppe, D., Kettenmann, H. and Trotter, J. (1994) *In vitro* and *in vivo* behaviour of glial cells immortalized with a temperature sensitive mutant of the *SV40 T* oncogene. J.Neurosci. Res. 37, 182-196.
5. Jung, M., Krämer, E., Grzenkowski, M., Aguzzi, A., Khazaie, K., Tang, K., Blakemore, W., von Blankenfeld, G., Kettenmann, H. and Trotter, J. (1995) Lines of murine oligodendrocyte precursor cells immortalised via an activated *neu* tyrosine kinase show distinct degrees of interaction with axons *in vitro* and *in vivo*. Eur. J. Neurosci. 7, 1245-1265..
6. Rathjen, F.G. & Schachner, M., (1984). Immunocytochemical and biochemical characterization of a new neural cell surface component (L1-antigen) which is involved in cell adhesion. EMBO J. 3, 1-10.
7. Trotter, J. (1993). The development of myelin-forming glia. Perspectives on Developmental Neurobiology 1, 149-154.
8. Trotter, J., Crang, A.J., Schachner, M. and Blakemore, W.F. (1993). Lines of glial precursor cells immortalised with a temperature-sensitive oncogene give rise to oligodendrocytes and astrocytes on transplantation into demyelinated lesions in the central nervous system. Glia 9, 25-40.

Transplantation of Oligodendrocyte Progenitors and CG4 Cells into the Dveloping Rat Brain: Differences and Similarities

Espinosa, A., Espejo, D. and de Vellis, J.

University of California, Los Angeles, Mental Retardation Research Center, Departments of Neurobiology and Psychiatry 760 Westwood Plaza, 68-225 NPI Los Angeles, CA. 90024-1759 and Biobehavioral Science, Brain Research Institute, UCLA School of Medicine, Los Angeles, CA. 90024-1759

1 Introduction

Unlike most organs, the Central Nervous System (CNS) is very complex, containing a variety of cell types which form not a homogenous organ but rather a group of diverse components acting in concert to perform CNS functions. Once the CNS has developed during the early postnatal life of the organism, a function of each cell type is the maintenance of the newly formed structures such as synapses, myelin, and blood brain barrier. Neurons, astrocytes, and oligodendrocytes, form this well organized system and favorable extracellular milieu. The cell-cell interactions and biochemical exchanges between the various cell types contribute to the function and health of the CNS. The CNS can lose its functionality as a consequence of injury or disease. When CNS is damaged, neurons and glial cells respond in different ways and generally the result is poor regeneration. In the case of trauma, axonal regrowth is limited and restricted to a small area around the wound. Synaptic sprouting is observed, undamaged neurons form new branches and synapses in order to replace some of the lost ones. However, oligodendrocytes (the myelin producing cells in CNS) unlike Schwann cells (PNS) do not promote regrowth of the damaged axons. Furthermore, it has been recently shown that oligodendrocytes can block regeneration (24). The astrocyte, the second major glial cell type, responds to injury by forming a scar at the site of the wound and blocks axonal regrowth as well. The enormous difference between PNS and CNS regeneration capacities has led to the conclusion that the CNS and in particular glial cells respond to injury by actively blocking "repair". Besides injury, disease can also alter or impair CNS function. Many diseases affect neuronal population in a direct manner while others affect glial cells, astrocytes or oligodendrocytes, directly and consequently alter neuronal function. A condition proper to oligodendroglial cells is myelin deficiency. Myelin deficiency can result either from failure to produce the normal amount of myelin during development or from demyelination. The lack of myelin or abnormal myelin formation during CNS development is known as dysmyelination (For Review see 21). Myelin deficient mutants belong to this category (Review:20). The neurological dysmyelinating disorder Pelizaeus-Merzbacher (26) is the result of a mutation in the PLP/DM20

Molecular Signaling and Regulation in Glial Cells
A Key to Remyelination and Functional Repair

gene (19). The rodent PLP mutants Jimpy, Jimpy msd mice, myelin-deficient (md) rat present dysmyelination similar to the human condition and the disease is lethal in all cases (18,1). In the past it has been assumed that CNS dysfunction due to either injury or disease was irreversible. However, during the last two decades it has become clear that the CNS with the proper stimuli has the capacity to regenerate. In the case of remyelination it is still unclear if new myelin-forming cells can be regenerated in sufficient numbers to rescue the myelin deficient CNS. For this reason many researchers have chosen transplantation of myelin forming cells from either PNS or CNS (11,17 and many others) as the alternative of providing healthy cells to the host in order to myelinate impaired CNS (27). Recently various studies on oligodendroglial transplants into dysmyelinating mutants have been described (14,3). In our own laboratory different kind of transplant experiments have been designed with the final goal of their therapeutic application. Nevertheless the optimization of the system to achieve this goal involves extensive characterization on the behavior of the graft in the in vivo environment (10). By grafting young oligodendrocytes (5), or their progenitors (6) into the developing rat brain within the first week of postnatal life we have learned that they survive, migrate (7) and integrate within the host parenchyma upon transplantation. In the present report we summarize previous findings on the transplantation of young oligodendrocytes (5). We studied and compared after transplantation the behavior of wild type oligodendroglial progenitors (OLP) and, CG4 (central glial four) cells at a progenitor stage (16). Either one of the cell types as grafted into myelin deficient (md) and unaffected litter mates rat pups. We also used wild type rat pups for long-term transplantation experiments up to 6 months. Two implant sites were chosen, the hippocampus and the dorsal subarachnoidal space of the cerebellum.

2 Methodology Applied

2.1 Source of Grafting Cells

Oligodendroglial progenitors and young oligodendrocytes were prepared from primary glial cultures as described (2). For CG4 grafts passages 12 to 15 were used. Both cell types were maintained in a chemically defined medium named GDM (glial development medium) specifically designed for the maintenance of young oligodendrocytes and their progenitors. This medium consists of Delbecco's modified Eagle's/Ham's F-12 (1:1) enriched with 5 mg/liter insulin, 16.1 mg/liter putrescine, 50 mg/liter transferrin, 0.8/liter sodium selenite and 2.2 g/liter sodium bicarbonate. At the time of transplantation both cell types were bearing the following markers: A2B5+ 04ñ GC-, GFAP- the OLP phenotype.

2.2 Preparation and Labeling of Cells to be Grafted

CG4 cells or oligodendroglial progenitors were plated on plastic for grafting coated coverslips with either poly-d-lysine for OLP or poly ornithine for CG4 cells. Both cell types were plated and maintained in GDM 5% CO_2, 37øC and humidity to aturation. One day after plating cells were labeled with the fluorescent dye fast blue as described (5,6,7).

2.3 Preparation of Host-Pups

Wistar-Wild type rat pups or pups born to md carrier mothers were used. Unless stated otherwise pups within 3 to 5 days fter birth were used. For animals of this age the anesthetic procedure of choice was cool body temperature. For 16d of ageanimals Avertin was used as anesthetic. The same anesthetic was used for perfusion. These procedures have been recently described in detail (7,8). For cellular grafts p4 or p16 wild type rat pups were used.

2.4 Cell Suspension Preparation and Grafting Procedure

These methods have been described in (5,7). Immediately before grafting cells were rinsed twice with Hanks solution without calcium and magnesium at room temperature, then they were scraped gently and transferred to an Eppendorf tube and centrifuged 30 sec. Cells were resuspended in 10 æl of Hanks solution and placed in a Hamilton syringe to proceed with the grafting step. The single cell suspension was injected within a period of 5 min (1.4 to 2 æl) and 5 more min. were allowed for the graft to remain in its new environment and the needle was withdrawn within stepwise over 5 min to avoid backup of the cell suspension.

2.5.2 Cerebellar Grafts

OLP grafts were performed on P4 and P16 rat pups. For CG4 grafts pups were 4 days of age. The procedure consisted in the implantation of 60 x 103 cells in the subarachn"idal space without touching the cerebellar parenchyma. Analysis of section was done as described for hippocampal grafts (5,7,8).

3 Summary and Interpretation of Findings

3.1 Behavior of Grafted OLP and CG4 Cells in Hippocampus

3.1.1 Survival and Migration of OLP

The behavior of OLP grafted into the hippocampus of normal and myelin deficient rat pups confirmed our previous observations (5,6). These cells were easily recognized by the FB label, they were able to survive, migrate, and integrate within the host parenchyma (5). Wild type hosts were sacrificed at various time points from 1 week up to 6 months post transplants. The labeled cells were still visible even after 6 months post-grafted. Samples from md rats were collected at various time points between 1 and 21 days post-transplant (24 days of age on the basis of the md mutants life span). Both environments (normal or md) proved permissive for the survival and migration of FB+ OLP. At all the time points explored in both normal or md brain the majority of the FB+ cells were found at the level of the injection site easily recognized by a charcoal mark. A characteristic pattern of cell migration was observed: starting from the injection site, a gradient of FB+ cells could be observed along white matter tracks in normal animals or where white matter should be in md hosts. These cells were found as single cells or in small clusters at different distances from the injection site as shown in diagram (Fig. 1a). The largest distance was 1.4cm. Grafted cells were also seen at the level of the fourth ventricle and of the cerebellum. These cells might have traveled via CSF. Some FB+ cells were seen at the level of the cortical plate. A low power view of FB+ cells at the implant site (hippocampus) is shown in Fig.4D.

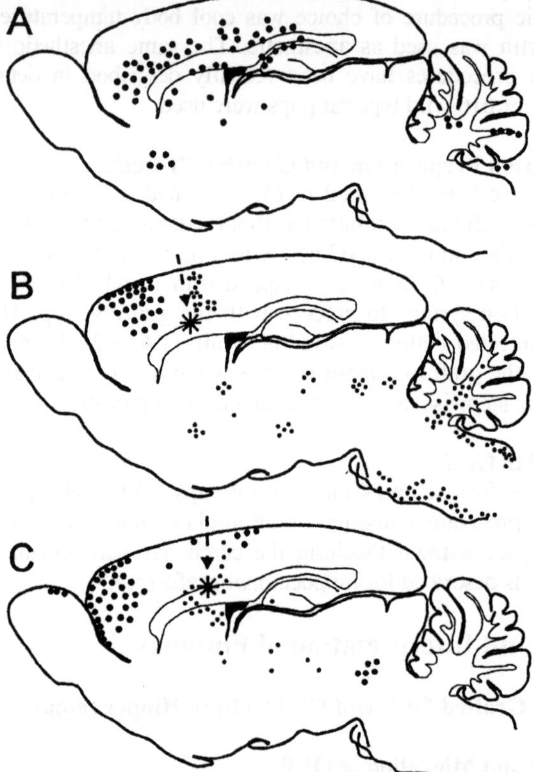

Fig. 1 Diagrammatic representation of FB+ cells 27 days post grafting. A) wild type FB+ oligodendrocytes. The majority of the cells integrated into white matter. Few cells were found in the 3rd ventricle and some others in 4th ventricle. Some of these cells entered the cerebellar parenchyma moving in a defined pattern B) and C) views of md and unaffected littermates 2 weeks post-transplant. Large dots represent FB+ CG4 cells intensely labeled. Small dots represent the progenies of the original CG4 cells grafted. These cells have a weaker FB fluorescence. The star indicates the implant site recognized by the charcoal mark.

3.1.2. Survival and Migration of CG4 Cells

A total of 56 animals from md carriers were used. Samples from hippocampal grafts were collected at various time points during the md rat life span. Some unaffected hosts were sacrificed up to 104 days post-transplant. The survival of grafted CG4 cells was comparable to that of OLP. CG4 cells migrated extensively within host parenchyma. The migratory phenomenon seemed to be more extensive in md hosts. This might be due to the absence of myelin from the host. However, in unaffected hosts the CG4 cells also migrated. A major difference was observed between wild type OLP and CG4 grafts: While in all the cases of grafted OLP the FB label remained intense and virtually unchanged even after several months post-transplant, CG4 cells displayed intense FB only at the implant site and they were visible (up to 1.6mm)from the implant site. However, the fluorescent label was much less intense than in OLP grafts and the correlation between intensity of the label and distance from the implant site was inversely proportional (Fig. 1b and c). At distances greater than 1.6cmm from the implant site the FB label was not visible anymore. However, immunoreactivity for MBP was seen along host axonal tracts indicating that the CG4 progeny reached areas such as the cerebellum and the brain stem (See Section 3.2). Our previous experience on the effect and behavior of the FB dye used on cells in cultures or co-cultures indicates that the label does not transfer from cell to cell when used under the conditions and amounts we use here. We also found that dilution of the dye occurs only through cell proliferation. However, it is only after the 3rd generation (for CG4 cells) that the dye's dilution can be noticed in either live or fixed cells, and it is not until the 6th generation of cells that the label is very weak or not visible anymore. Initially, examination of the samples under regular epyfluorescent microscope and on the basis of the FB label it seemed as if FB+ cells were present almost exclusively within or around the implant site. Further examination of the samples with a video camera revealed a characteristic pattern of FB+ cells labeled at various intensities. The cells with a very intense label were those found at the implant site. As cells moved, the dye diluted indicating that those cells were the progeny of the graft. FB+ cells were easier to find in md hosts. Wild type hosts were allowed to grow up to 6 months post-transplant. Sagittal sections from these animals allowed the visualization of FB+ cells that also expressed myelin markers (5).

3.2. Characterization of Grafted Cells by Immunocytochemistry

3.2.1 OLP Grafts: Synthesis of Myelin Components

Immunocytochemistry for oligodendrocyte (ol) markers on sagittal sections from normal and md hosts grafted with OLPs revealed the synthesis of myelin components such as GC, MBP, PLP and sulfatides by grafted cells (FB+). The expression of these markers by OLP indicated that upon migration they are able to proceed through their known developmental and maturational patterns. This phenomenon was observed in md and unaffected litter mates and the identification of grafted cells was unambiguous since a majority of the cells (>90%) displayed a very intense FB label. In normal litter mates at 31 days post-transplant the colocalization of GC or 04 with FB was almost one to one (>85% of FB+ cells were also GC+). This phenomenon was consistent at the various time points studied. In contrast in md host sections the population of grafted cells (FB+) included some cells (25 to 30%) that didn't express either GC or GFAP (5). This finding indicates that the md environment can delay or down regulate the synthesis of myelin components such as GC expression by grafted cells. In fact, we have previously reported that in the md rat mutant the oligodendroglial cell population fails to proceed throughout the developmental stages necessary for the cell to mature as a result of the PLP mutation (13). Since the neuronal population in these mutants has been reported to be normal and their metabolism unaltered (13), it is possible that immature md host oligodendroglia produce a factor(s) that interact with a subpopulation rather than with all of the wild type grafted cells delaying their maturation. Alternatively, it is possible that wild type oligodendroglia usually produces and secretes some factor(s) that are necessary for OLP to mature and that grafted cells are not producing enough of it to reach those cells that migrate far away from the implant site. It could be that in the case of md OLP, the production of this factor is either down-regulated or produced in an insufficient amount as a consequence of the PLP mutation, resulting in a maturational delay of host and grafted cells. More research is needed to approach this question about epigenetic regulation of grafted cells.

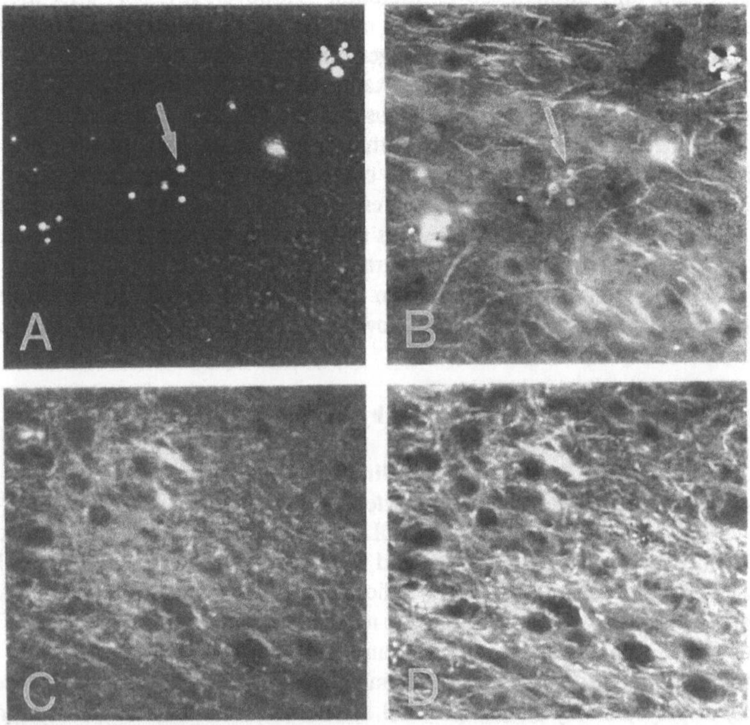

Fig. 2 Views of a sagittal section of an md host brain 15 days post-transplant. A) double exposure view in phase contrast/blue fluorescence of the implant site, CG4 cells display an intense label. B) The same site viewed after MBP immunocytochemistry. Blue cells and host fibers are intensively labeled for MBP. C) view of the same section 4mm caudal to the implant site where grafted cells are not visible under UV light however numerous host fibers are MBP+ indicating that the progenies of grafted (FB-) CG4 cells are present in this region (200x).

3.2.2. CG4 Grafts: Synthesis of Myelin Components

Double immunofluorescence on sagittal sections from hosts grafted with CG4 cells that were A2B5+, 04ñ, CG-, GFAP- at the time of implantation revealed the presence of GC, MBP and sulfatides in FB+ cells. The most surprising finding was that despite the low number of visible FB+ cells there was an extensive amount of MBP+ fibers within md hosts (Fig. 2). These fibers were found in hippocampus, cerebellum, brain stem and even olfactory bulb. On the basis of these results we re-examined all the samples with a video camera to establish a correlation between the intensity of the blue fluorescence and cell migration assessed mainly by the MBP+ immunoreactivity md host fibers. In unaffected animals the MBP+ fibers from graft and host were indistinguishable. Some of these animals were kept and observed daily. They were healthy and didn't present any differences in terms of behavior even six months post-transplant. In grafted animals CG4 cells did not form tumors, nor cause symptoms such as seizures that might indicate intracranial high pressure due to excessive cell growth.

3.3 Cells Grafted in Ectopic Locations

3.3.1 Subarachnoidal Grafts of OLP

The data from our previous experiments as well as those described here have shown that OLP are able to survive, migrate and integrate in a functional manner within host CNS. In all these studies, however, cells were grafted into host parenchyma close or within the area where grafted cells are expected to perform their function. Our next questions were: What would be the behavior of OLP grafted in ectopic locations? Would they survive? Or will the environment be adverse to them? Will they migrate and if so, will they follow a well determined pattern or, will they move at random? In order to answer these questions we performed the following series of experiments. The OLP phenotype at implantation time was the same as for hippocampal grafts: A2B5+ , 04ñ, GC- GFAP- . Hosts were either 4 or 16 days of age wild type Wistar rat pups. These experiments were extremely difficult due to the size of the cerebellum at this age (p4). For this reason many of the animals were FB-. However, those animals that were FB+ gave very interesting results. Hosts were sacrificed at various time points. Sagittal frozen sections were examined for graft location. Both groups of hosts that received grafts at P4 and P16 displayed similar results. The majority of grafted cells were seen within host parenchyma. These cells were intensively FB labeled. Twenty-five days post-transplantation OLP were found to migrate throughout molecular layer and were located along white matter tracts (Fig.3). The FB+ cells were distributed in a similar manner to that of host oligodendrocytes. These cells were mainly located among the host MBP+ myelin tracts. The distribution of FB+ cells within host cerebella was consistently reproducible. This phenomenon was also observed in P-16 hosts indicating that wild type OLPs are able to recognize and interact with the host microenvironment in a specific and organized manner. FB+ cells were able to move across the molecular and granular layers reaching the white matter. Once within white matter grafted cells pursued their migration always along host myelinated fibers as shown diagrammatically in Fig. 4.

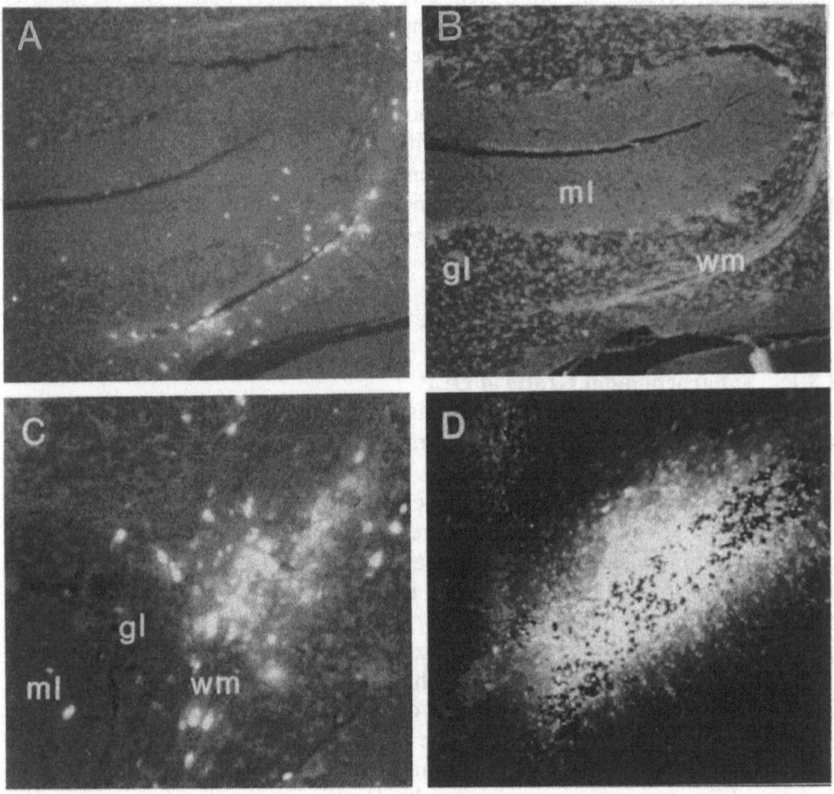

Fig. 3 Double fluorescence views of oligodendrocytes distribution within host parenchyma 29 days post-transplant. A) Low power view (100x) of a cerebellar sagittal section. FB+ cells have followed a well defined migration pattern reaching the white matter and migrating along host myelinated fibers. These fibers are MBP+ as seen in B). C) high power view (200x) of another section from same host. FB+ cells have migrated long distances within white matter. At this magnification various intensities of FB can be observed indicating that some cells have proliferated. D) low power view of Ol progenitors grafted into the hippocampus of a P-5 rat pup. This photograph shows that when cells are grafted within the host parenchyma they migrate in different directions, while those grafted in the subarchnoidal space A) migrate on their own into host parenchyma, in an organized manner, perhaps following specific signals of their micro-environment (200x).

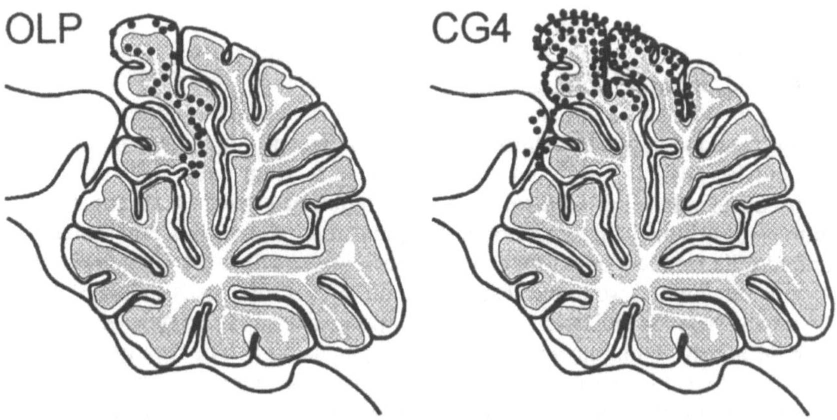

Fig. 4 Diagramatic representations of the migration patterns for CG4 cells and oligodendroglial cells grafted into the subarachnoidal space of the cerebellum of P-5 rat pups. Both cell types migrated into host parenchyma. However, CG4 cells were found in the molecular and granular layers of the cerebellum while less than 2% r eached the white matter. In contrast, oligodendrocytes followed a well-defined path through both the molecular and granular layers reaching the white matter (>90% of the cells). Once in the white matter, they moved very long distances.

3.3.2 Subarachnoidal Grafts of CG4 Cells.

We also explored the effect of host microenvironment on CG4 cell migration upon being implanted in the subarachnoidal space of the cerebellum instead of grafting them directly into host parenchyma. Procedures used were the same as for OLP. Hosts were 4 days of age rat pups born to md carrier mother. Two litters were used separately. Examination of the samples revealed major differences in behavior between CG4 cells and wild type OLP. CG4 cells were found in three different places: the majority (70%) of FB+ cells migrated long distances along meninges. Some of these cells got into cerebral cortex via frontal migration to the implant site. Other cells (23%) were found within cerebellar parenchyma in both the molecular and granular layers. However, less than 2% of grafted cells were found in the white matter as seen in Fig. 4. It has been reported that OLP migrate long distances in vivo and in vitro (10). However, migration of GC+ cells remains controversial. In the case of mouse oligodendrocytes (14) reported migration in the brain out of the "solid graft". These authors proposed either the existence of (non-characterized) trophic factors attracting grafted cells, or a passive movement of grafted cells while axons elongated. The developmental stage of those migrating cells was not identified. Rosenbluth et al (22) have reported the opposite result in spinal cord. Cells from embryonic solid graft labeled with FB did show short or no migration in the host spinal cord (12). The authors found myelin formations in the graft area, but most of the cells were observed within the implant zone. In the first case (14) the stages of migrating cells could be very informative since O2A progenitors and GC+ cells, could migrate, however it is not clear yet if there is a relationship between migration distance and the degree of maturity of the migrating cell. In the second case (12) it could be possible that even though they were from embryonic origin the environment might have enhanced the maturation of grafted OLP earlier in the brain since myelinogenesis occurs in the spinal cord. In previous experiments we grafted cells from cultures where at least 65% of the cells were GC+ at the time of implantation and we found that more than 50% of FB+ cells migrated out of the implant site (5) indicating that not only progenitors but also oligodendrocytes have the ability to migrate within young host parenchyma. The next question was if OLP and CG4 cells that have shown good survival, migration and functional integration also display the ability to integrate within host tissue in a histotypic manner. This study is of particular importance since a minimal host injury is caused during the grafting procedure leaving the parenchyma intact. This strategy is advantageous not only because a single graft site is required, but also avoids reactive gliosis usually produced by needle insertion. Consequently, the microenvironment remains unaffected, facilitating graft integration without inflammation and avoiding or reducing possible graft rejection.

4. Concluding Remarks and Prospectives.

The ultimate goal of neural transplant studies is their application as a therapy for theimpaired human CNS. Since 1985 attempts to cure conditions such as Parkinson's were done by allografts of adrenal medulla. However, graft survival was poor and benefits were transient (15). The use of fetal tissue and co-grafts of CNS/PNS cells (for NGF supply) allowed graft survival and clinical improvement. On the basis of such results, fetal cells have become the best source for human neural grafts. Other fetal grafts have been used for Huntington's disease, diabetes insipidus, and hereditary ataxia. Most of these studies included multiple implant sites and post-surgery immunosuppression (For review see 23). From such studies we have learned that developing tissues survive better than mature ones, that a certain amount of grafting material is needed to obtain CNS improvement, and that rejection is one of the most common fears of scientists working in transplantation.

Despite the fact that the number of human conditions being approached by transplantation is increasing rapidly, neural transplantation to approach Alzheimer's disease, spinal cord damage or myelin deficiency has only been developed in animal models. Both human and animal model studies are contributing to major advances in neurobiology. Animal models are useful in receiving various questions that need to be answered before any attempts can be made to approach by neural grafts conditions such as myelin deficiency in the human CNS.

From our own work and other reports in the literature we have learned that: a) single cell suspensions survive better than solid grafts; b) cells in culture can be manipulated prior to transplantation, and c) that newborn oligodendroglial progenitors are a good source of myelin forming cells that survive for a long time (at least over 6 months) migrate and mature within host parenchyma, d) and that human neural grafts are more successful when multiple grafts are performed (23). There are two reasons for this: the first one is the need to provide enough cells to overcome host impairment, and the second one might be related to graft cell migration. For example, if grafted cells display a poor migratory capacity they might be able to perform their normal function. Where they were grafted but not necessarily far away from the implant site. Therefore the outcome would be a partial success. This limitation has been solved by having multiple graft sites. However, multiple graft sites implicate multiple injuries to host tissue at the time of graft implantation that consequently will produce inflammation related post operative responses and increase the risk of rejection. In all of our studies we have observed cell migration at various time points post transplantation. We have demonstrated that OLP as well as various OLP-derived cell lines display different migration patterns, as well as different survival features within host CNS (6). Our comparative studies on the migration of wild type OLP and CG4 cells grafted in the subarachnoidal space of the cerebellum has shown that wild type OLP generally have the ability to recognize the regions where they belong. These grafted cells reached host white matter and continued migrating within it. This is a very important finding since it eliminates the need of multiple implant sites. Furthermore, it implies a histotypic integration of these cells within host parenchyma. In contrast, even though CG4 cells display all the properties needed to myelinate, they have lost the ability to selectively reach the white matter on their own. This finding indicates that cell lines, even when they are functional, might have acquired or lost other yet unknown features that might bring side effects and even perhaps cause further host dysfunction at short or long-term. More research is needed in order to understand the mechanism(s) and molecules that determine histotypic integration of wild type OLP and its control during development or upon transplantation. At the present time it is unclear if the proliferation and further maturation of grafted CG4 cells are intrinsic mechanisms or if the host environment directly regulates both phenomena.

Acknowledgment

We would like to thank Bryce Tiller for his help on the experimental work and Nancy Wainwright for the preparation of the manuscript. This work was supported by NICHD Grants HD06576 and HD04612.

References

1. Boison, D. and Stoffel, W. (1989) Myelin-deficient rat: A point mutation in exon III (A C, Thr 75 Pro) of the myelin proteolipid protein causes dysmyelination and oligodendrocyte death. EMBO J. 8, 3295-3302.

2. Cole, R., and de Vellis, J. (1989) Preparation of astrocyte and oligodendrocyte cultures from primary rat glial cultures. In Shahar, A., de Vellis, J., Vernadakis, A. and Haber, B. (eds.), A Dissection and Tissue Culture Manual of the Nervous System, Alan R. Liss, Inc., New York, pp. 121-133.
3. Duncan, I.D. (1990) Dissection of the phenotype and genotype of the X-linked myelin mutants. Ann. N.Y. Acad. Sci. 605, 110-121.
4. Espinosa de los Monteros, A., Zhang, M-S. and de Vellis, J. (1992a) Effect of exogenous transferrin in the developing rat brain. Am. Soc. Neurochem., abstract #366.
5. Espinosa de los Monteros, A., Zhang, M-S., Gordon, M., Aymie, M., and de Vellis, J. (1992b) Transplantation of cultured premyelinating oligodendrocytes into normal and myelin-deficient rat brain. Dev. Neurosci. 14, 98-104.
6. Espinosa de los Monteros, A., Bernard, R., Tiller, B., Rouget, P. and de Vellis, J. (1993a) Grafting of fast blue labeled glial cells into neonatal rat brain differential survival and migration among cell types. Int. J. Dev. Neurosci. 11, 625-639.
7. Espinosa de los Monteros, A., Zhang, M-S. and de Vellis, J. (1993b) O2A progenitor cells transplanted into the neonatal rat brain develop into oligodendrocytes but not astrocytes. PNAS 89, 50-54.
8. Espinosa de los Monteros, A., Sawaya, B.E., Guillou, M., Zakin, M.M., de Vellis, J. and Schaeffer, E. (1994) Brain-specific expression of the human transferrin gene: Similar elements govern transcription in oligodendrocytes and in a neuronal cell line. J. Biol. Chem. 269, 24504-24510.
9. Franklin, R.J.M., Bayley, S.A., Milner, R., Ffrench-Constant, C. and Blakemore, W.F. (1995) Differentiation of the O-2A progenitor cell line CG-4 into oligodendrocytes and astrocytes following transplantation into glia-deficient areas of CNS white matter. Glia 13, 39-44.
10. Gout, O., Gansmuller, R., Baumann, N. and Gumpel, M. (1988) Remyelination by transplanted oligodendrocytes of a demyelinated lesion in the spinal cord of the adult shiverer mouse. Neurosci. Lett. 89, 195-199.
11. Gumpel, M., Gout, O., Lubetzki, C., Gansmuller, A. and Baumann, N. (1989) Myelination and remyelination in the central nervous system by transplanted oligodendrocytes using the shiverer model. Devl. Neurosci. 11, 132-139.
12. Hasegawa, M. and Rosenbluth, J. (1991) Transplantation of labeled fetal spinal cord fragments into juvenile myelin-deficient rat spinal cord. Anat. Rec. 229, 138-143.
13. Kumar, S., Macklin, W.B., Gordon, M.N., Espinosa de los Monteros, A., Cole, R., Scully, S.A., de Vellis, J. (1990) Transcriptional regulation studies of myelin-associated genes in myelin-deficient mutant rats. Dev. Neurosci. 12, 316-325.
14. LaChapelle, F., Gumpel, M., Baulac, M., Jacque, C., Duc, P. and Baumann, N. (1983) Transplantation of CNS fragments into the brain of shiverer mutant mice: Extensive myelination by implanted oligodendrocytes. Dev. Neurosci. 6, 325-334.

15. Lindvall, O., Rehncrona, S., Brundin, P., Gustavii, B., Åstedt, B. , Widner, H., Lindholm, T., Björklund, A., Leenders, K.L., Rothwell, J.C., Frackowiak, R., Marsden, C.D., Johnels, B., Steg, G., Freedman, R., Hoffer, B.J., Seiger, Å., Bygdeman, M., Strömberg, I. and Olson, L. (1989) Human fetal dopamine neurons grafted into the striatum in two patients with severe Parkinson's disease: a detailed account of methodology and a 6 months follow-up. Arch. Neurol. 46, 615-631.
16. Louis, J.C., Magal, E., Muir, D., Manthorpe, M. And Varon, S. (1992) CG-4, a new bipotential glial cell line from rat brain, is capable of differentiating in vitro into either mature oligodendrocytes or type-2 astrocytes. J. Neurosci. Res. 31, 193-204.
17. Ludwin, S.K. (1987) Remyelination in demyelinating diseases of the central nervous system. CRC Critical Review in Neurobiology, Vol. 3(1), 1-28.
18. Pham-Dinh, D., Popot, J.-L., Boespflug-Tanguy, O., Landrieu, P., Boué, J., Deleuze, J.-F., Jollés, P. and Dautigny, A. (1991) Pelizaeus-Merzbacher disease: a valine to phenylalanine point mutation in a putative extracellular loop of myelin proteolipid. Proc. Natl. Acad. Sci. 17, 7562-7566
19. Pham-Dinh, D. , Boespflug-Tanguy, O., Mimault, C., Cavagna, A., Giraud, G., Leberre, G. Lemarec, B. and Dautigny, A. (1993) Pelizaeus-Merzbacher disease: a frameshift deletion/insertion event in the myelin proteolipid gene. Human Molecular Genetics, 2, 465-467.
20. Popot, J.-L, Pham-Dinh, D. and Dautigny, A. (1991) J. Membrane Biol. 120, 233-246.
21. Poster, C.M. (1968) Diseases of the myelin sheath. In J. Minckler (ed.) Pathology of the Nervous System, Vol. 1, New York, McGraw Hill, pp. 767-820.
22. Rosenbluth, J., Hasegawa, M., Shirasaki, N., Rosen, C.L. and Liu, Z. (1990) Myelin formation following transplantation of normal fetal glia into myelin-deficient rat spinal cord. J. Neurocytol. 19, 718-730.
23. Rosenfeld, J.V. (1993) Current issues in neural transplantation. Annals of the Academy of Medicine, Singapore 3, 464- 9.
24. Schwab, M.E., Kapfhammer, J.P., Bandtlow, C.E. (1993) Inhibitors of neurite growth. Annual Rev. Neurosci. 16, 565-595.
25. Skoff, R. and Knapp, P. (1990) What is the role of proteolipid protein in the biology of the normal and jimpy oligodendrocyte? Neurol. Neurobiol. 55, 307-316.
26. Seitelberger, F. (1970) Pelizaeus-Merzbacher disease. In P.J. Vinken and G.B. Bryan (eds.), Handbook of Clinical Neurology, Amsterdam, North Holland, pp 150-202.
27. Zager, E.L. and Black, P. M.cL. (1988) Neural transplantation. Surg. Neurol. 29, 350-366.

OLN-Cells: a New Oligodendroglia Cell Line with Impact for Cell Transplantation

Christiane Richter-Landsberg

Department of Biology, University of Oldenburg, D-26111 Oldenburg

Keywords. oligodendroglia, cell line, myelin proteins,

1 Introduction

The development of bipotential glial progenitor cells has been extensively studied using primary cell culture techniques (5, 8). Oligodendrocyte progenitors require multiple signals to proliferate in culture, after several cell cycles the cells stop dividing and in the absence of serum or in medium containing low concentrations of serum, the O-2A cells differentiate spontaneously into oligodendrocytes. This process is characterized by the loss of intermediate filament protein vimentin and the cell surface antigen A2B5. Concomitantly, galactocerebroside (GC) appears on the cell surfaces and numerous branched cellular processes are extended, and lateron myelin specific proteins, i.e. Wolfgram protein (CNP), myelin basic protein (MBP), myelin associated glycoprotein (MAG), and proteolipid protein (PLP) are expressed even in the absence of neurons (5).

Preparation of enriched populations of oligodendrocytes and cells of the oligodendroglia cell lineage do not generally yield sufficient quantities to analyze the mechanisms underlying glial cell specific gene expression and myelinogenesis by conventional molecular and biochemical techniques. The availability of either naturally occuring or immortalized cell lines at specific stages of the oligodendrocyte cell lineage would represent a powerful tool not only to study the cellular, biochemical and genetic factors influencing proliferation and differentiation in these cells, but also to investigate glia cell biology and repair processes of demyelinated areas in the nervous system after cell transplantation.

Several cell lines have been described which resemble glial cell progenitors, and in response to specific cell culture conditions adapt oligodendroglia properties, such as the expression of GC or MBP. These cell lines were either gained by immortalizing primary cells using oncogene transduction (1, 6, 21), or were generated spontaneously in the course of expanding O-2A cells with medium conditioned by the B 104 neuronal cell line (12). However, most of these cell lines appear to correspond to the precursor oligodendrocyte stage.

Molecular Signaling and Regulation in Glial Cells
A Key to Remyelination and Functional Repair

In this contribution, the establishment and characterization of a new permanent cell line, namely OLN-93, derived from spontaneously transformed cells in primary rat brain glial cultures is described. These cells continuously divide and constitutively express antigenic properties of mature oligodendrocytes, including GC, CNP, MBP, MAG and PLP (21).

2 Materials and Methods

2.1 Materials

All cell culture media were from Gibco/BRL (Grand Island, NY/USA). Monoclonal antibodies to GFAP and vimentin were from Boehringer (Mannheim/Germany), monoclonal antibodies against galactocerebroside (17) were kindly provided by Dr.B.Barres. Horseradish peroxidase (HRP) -conjugated and FITC-conjugated secondary antibodies were from Sigma (St.Louis,MO/ USA). Polyclonal rabbit antibodies against MAG, PLP and Wolfgram protein was a generous gift of Dr.T.V.Waehneldt (11, 25) , and polyclonal rabbit anti-MBP antiserum was from Dr. J.-M. Matthieu.

2.2 Cell culture

All cells were kept at 37^0 C and 10% $C0_2$. Growth medium was changed three times a week. Cells were viewed and photographed using an Olympus inverted microscope.

2.2.1 Establishment and Maintenance of OLN-93 Cell Line

Primary cultures of glial cells were prepared from the brains of 1- day-old Wistar rats. Cerebral hemispheres were freed from the meninges and mechanically disrupted. Single cell suspensions were transferred to culture flasks (2 brains/75 cm^2) and kept for two weeks in DMEM supplemented with 10% heat-inactivated fetal calf serum (FCS), 50 U/ml penicillin and 50µg/ml streptomycin. After two weeks, cells were removed from the flasks by mild trypsinization and replated on 10 cm dishes (1 flask per 10 dishes). During the next four weeks of culturing time, small cell foci of actively proliferating cells, spontaneously appeared in some of the culture dishes. The spontaneously formed cell clumps were well separated from the flat astrocytes and could easily be removed by mild trypsinization, using small cloning cylinders (3mm diameter). These cells were then replated on 10 cm plastic dishes, kept in DMEM/10% FCS, and were passaged once a week. The OLN-93 cell line was observed to have a stable phenotype during 2 years of culturing time. Growth properties were determined and characterization by indirect immunofluorescence and immunoblotting procedure was carried out using passage 20 to 40. For storage cells could be cryopreserved. OLN-93 cells were centrifuged (800g , 5 min), resuspended in DMEM supplemented with 20% FCS and 10% DMSO, and kept at -80^0 C or in liquid nitrogen.

2.2.2 Primary Glial Cells

Primary cultures of mixed glial cells were prepared from 1-3-day-old newborn rat brains as described above and grown in DMEM/10% FCS. For western blot analysis secondary cultures, highly enriched in oligodendrocytes or containing mainly astrocytes, were prepared from the flasks after ten days. The phase dark cells growing on the astrocytic cell layer were separated by vigorous shaking and taken off as described by McCarthy and DeVellis (14). The cells floating on top of the astrocytic cell layer were replated on poly-L-lysine-coated culture dishes (10^6 cells/ 60mm dish) and incubated for another week in serum-free DMEM to which Insulin (5 µg/ml)-Transferrin (5µg/ml)-Sodium Selenite (25 ng/ml) - supplement (Boehringer, Mannheim, Germany) was added. After one week these cultures contained a highly enriched population of differentiated oligodendrocytes with numerous cellular processes (19), and were used for immunoblot analysis. The remaining astrocytes were subcultured in uncoated 10 cm culture dishes for 2-3 weeks in DMEM supplemented with 10% FCS. Thereafter, cells were extracted in SDS-sample buffer for immunoblotting procedure (18).

2.3 Indirect Immunofluorescence Staining

For fluorescent labeling OLN-93 cells were grown on poly L-lysine-coated glass cover slips (10,000 cells per 35 mm dish). After washing with PBS, cells were fixed with 3% paraformaldehyde for 10 min. For intracellular staining paraformaldehyde-fixed cells were pretreated with 0.1% Triton X-100 (30 min). Thereafter, cells were incubated with the first antibodies for 60 min. The antibodies were used at the following dilutions: anti-galactocerebroside 1:10, anti-GFAP 1:3, anti-vimentin 1: 4, anti- MBP 1:100, anti- PLP 1:100, and anti- MAG 1:100. After washing with PBS cells were incubated with fluorescein-isothiocyanate-conjugated (FITC) goat anti-mouse IgG (1:75) or anti-rabbit IgG (1:75) for 45 min, washed with PBS and mounted. Fluorescent labeling was studied using a Zeiss epifluorescence microscope equipped with an automatic camera. Control experiments, using the secondary antibodies only, did not show unspecific staining.

2.4 Western blot analysis

Cellular monolayers were washed with PBS three times , scraped off in sample buffer containing 1% SDS and boiled for 10 min. Protein contents in the samples were determined according to Neuhoff et al. (15). For immunoblotting, total cellular extracts (10 µg protein per lane) were separated by one-dimensional SDS-PAGE using 10 % or 15% polyacrylamide gels as indicated, and transferred to nitrocellulose membranes according to Towbin et al. (22). The blots were washed and incubated with the appropriate first antibodies (anti-vimentin 1:5000; anti-GFAP 1:5000; anti-MBP 1:1000; anti-WP 1:1000; and anti-MAG 1:1000) and HRP-conjugated anti-rabbit IgG or anti-mouse IgG (1:5000), and visualized by the enhanced chemiluminescence (ECL) procedure as described by the manufacturer (Amersham, Braunschweig/Germany).

2.5. Fractionation of Rat brain myelin

Myelin was isolated from adult rat brain as described by Waehneldt et al. (24). For immunoblot analysis aliquots of 5 μg protein of the purified myelin fraction was used.

2.6 RNA extraction and PCR

Total RNA extraction was performed by the procedure of Chomczynski and Sacchi (3). One microgram of RNA was reverse transcribed, the resulting cDNA was amplified by PCR for 27 reaction cycles.

For PCR priming oligonucleotides were used matching sequences in exon 3 and exon 7 of the PLP gene, thus yielding the PLP and DM20 products. PCR products were analyzed by Southern blotting according to the manufacturer`s (Boehringer,Mannheim) protocol.

3 Results

3.1 Establishment and Growth Properties of OLN Cells

OLN-93 cells were derived from primary glial cultures of neonatal rat cerebral hemispheres. The spontaneously formed cell foci in these cultures were subcloned at low density, and the cells were kept and passaged in DMEM/10% FCS. The cell line has been grown for two years and was passaged up to 40 times. In the presence of 10% FCS cell proliferation occurred with a high rate, a doubling time of 16-18 h was determined. The growth of the cells was density dependent. At low density cells were mainly bipolar with long cellular extensions. Cells grown on poly-L-lysine coated plastic dishes in DMEM supplemented with 0.5% FCS or under serum deprivation, showed a more pronounced arborization of the cellular processes (Fig. 1). When confluency was reached, cells formed large clumps interconnected by long thin cellular processes. Under low serum conditions (0.5% FCS) and in the presence of 8-Br cAMP (1mM), cellular growth was reduced without promoting further morphological differentiation (not shown).

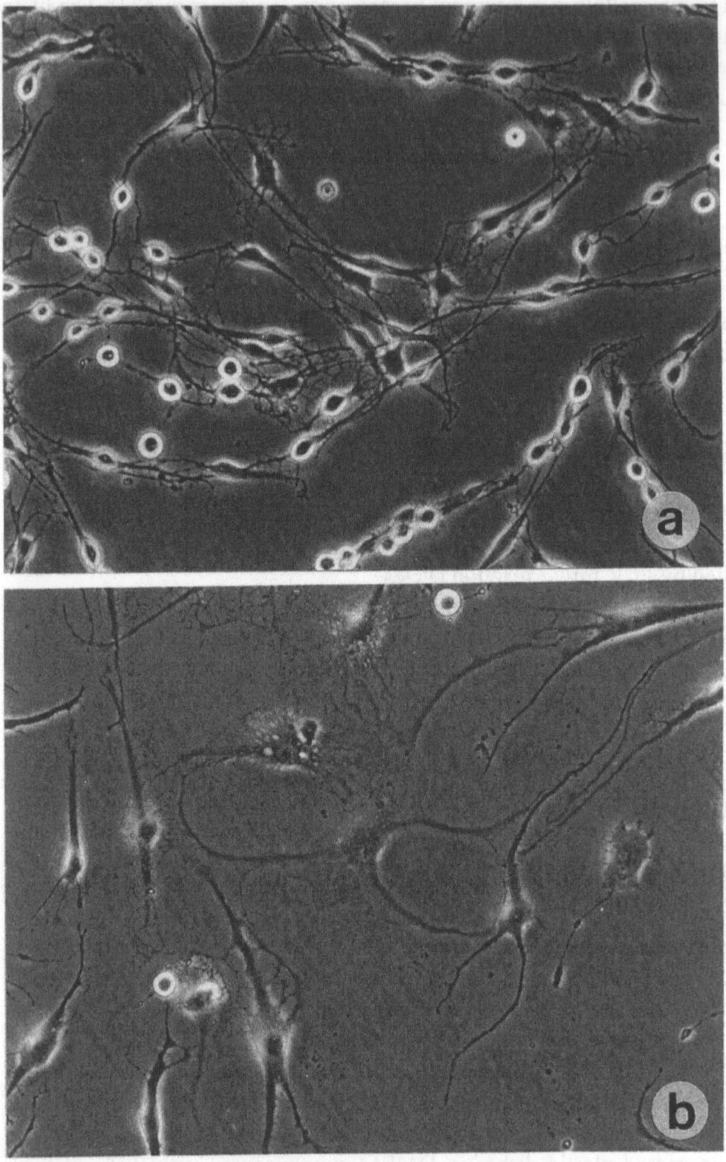

Fig. 1 Phase contrast micrographs of OLN-93 cells
cultured for 48 h in 0.5% FCS (a) or for 3 days in the absence of serum (b). Bar represents 25 µm.

3.2 Immunocytochemical Characterization

For immunocytochemistry, OLN-93 cells were grown on poly-L-lysine coated glass cover slips for 2 to 3 days. Indirect immunofluorescence staining revealed that OLN-93 cells express GC on their surfaces (Fig. 2) , and were also positively stained for MBP (Fig. 3), PLP and MAG (not shown). Staining for vimentin and the astrocytic marker GFAP was negative. The cell somata and cellular processes were heavily labelled with antibodies against the cell surface antigen N-CAM/D2 (not shown). In OLN-93 cell cultures small vesicular membrane blebs which are characteristically observed in cultures of primary oligodendroglia (4; Richter-Landsberg and Heinrich, unpublished observation) , are frequently encountered too.

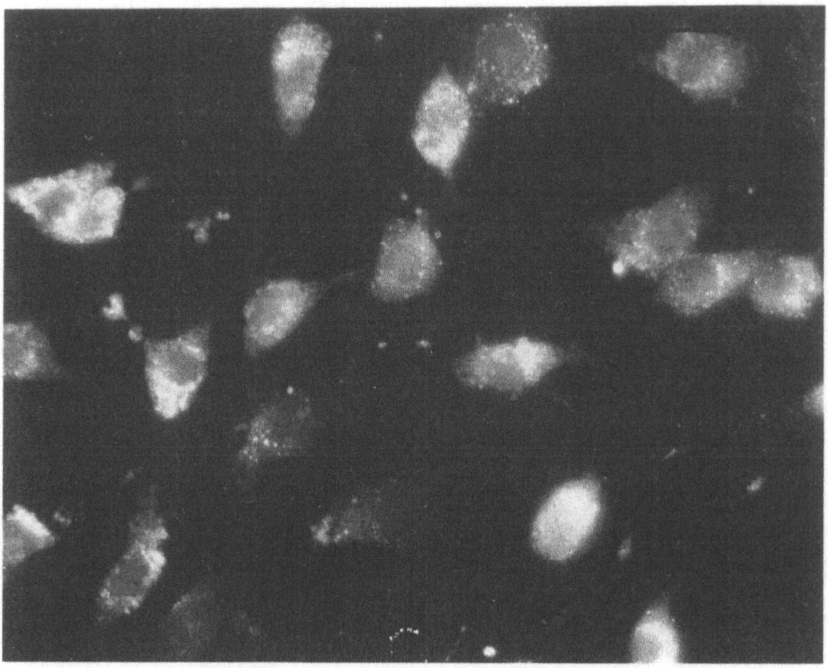

Fig. 2 Indirect immunofluorescence staining with anti-galactocerebroside antiserum. OLN-93 cells express punctuated GC-staining on their surface.

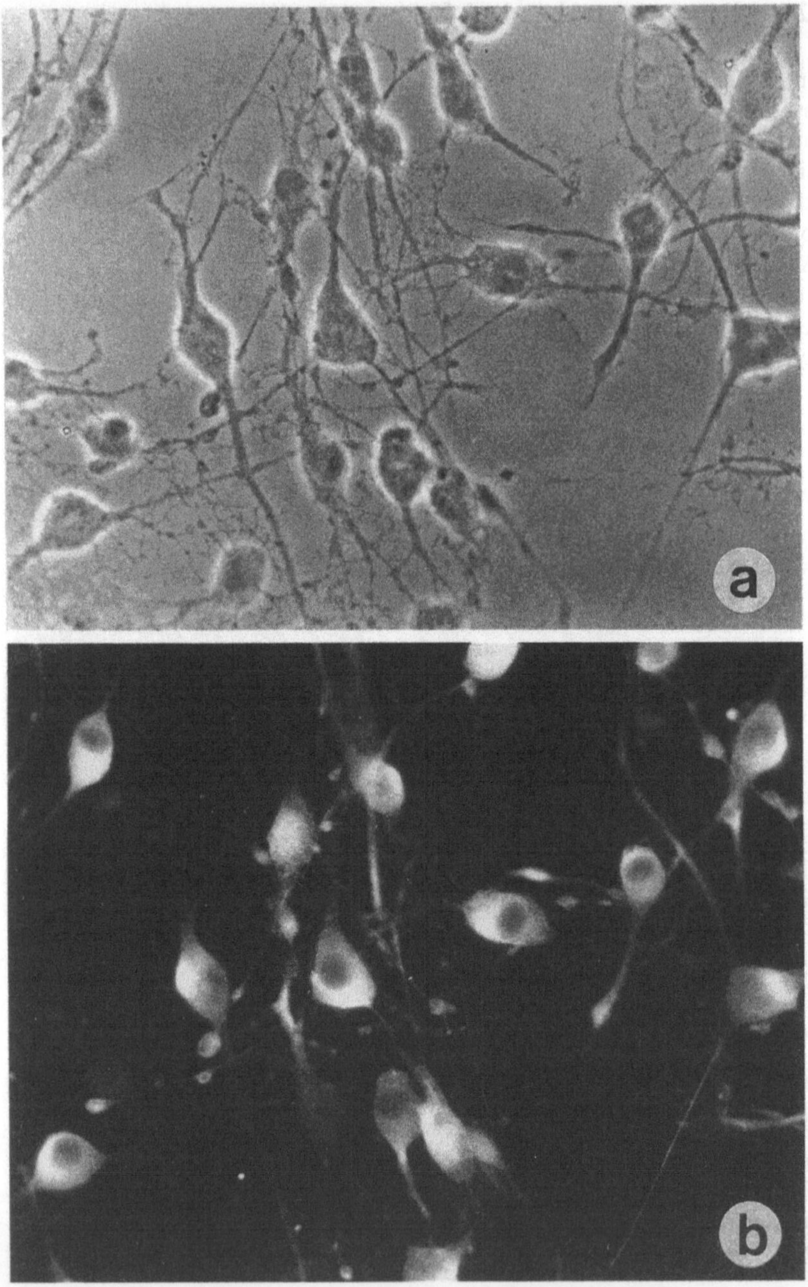

Fig. 3 Indirect immunofluorescence staining with anti-MBP antiserum.
Cells were grown for three days in medium containing 0.5% serum, in (a) the corresponding phase image to (b) is seen. Bar represents 25 μm.

3.3 Immunoblot Analysis

Cellular extracts of OLN-93 cells (OLN) were prepared and subjected to immunoblot analysis and compared to purified rat brain myelin (My). Additionally, cultures of enriched oligodendrocytes (OL) after 14 days in vitro (15 days postnatally), and cultures of astrocytes (AS) after three weeks in vitro (23 days postnatally) were prepared and analyzed.

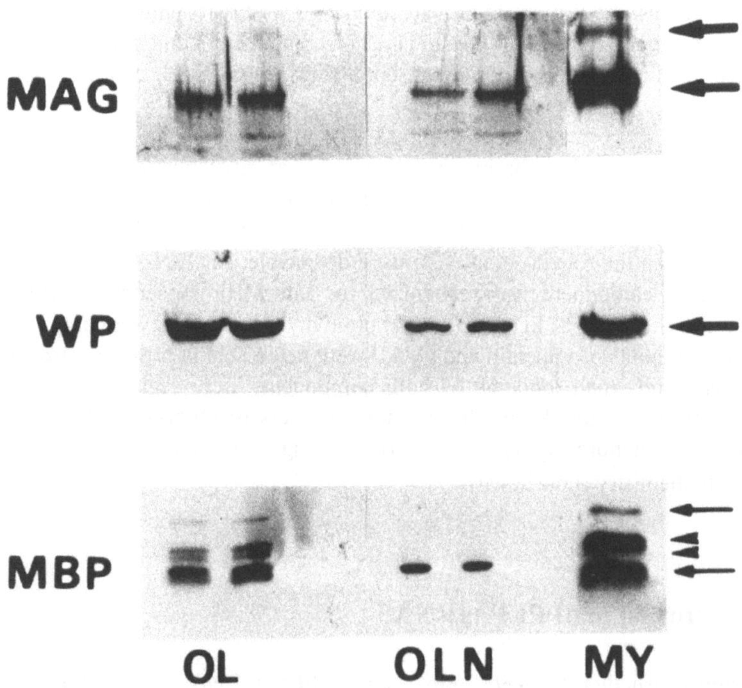

Fig. 4 Immunoblot analysis of MAG, WP and MBP in cells extracts and rat brain myelin. Cell extracts of OLN-93 cells (**OLN**) and primary cultures enriched in oligodendrocytes (**OL**), and purified rat brain myelin (**MY**) were subjected to western blot analysis, 10 µg of the cells extracts each and 5 µg of myelin were loaded on the gels.

Upper panel: arrows point to the two components recognized by anti-MAG antiserum,

Middle panel: arrow points to the 50 kDa polypeptide components recognized by anti-WP-antiserum.

Lower panel: arrows and arrow heads point to the four polypeptide components recognized by anti-MBP antiserum. Note that only one MBP isoform is expressed in OLN cells.

In rat brain myelin (Fig. 4, upper panel) two components are recognized by anti-MAG antibodies, a major broad band around 70 kDa , containing the two polypeptide chains of 67 and 72 kDa of MAG (S-MAG and L-MAG), and a high molecular weight component with an apparent molecular weight of approximately 100 kDa , representing the highly glycosylated form of MAG (16). In OLN-93 cell extracts only the smaller compound occurred, while in primary oligodendroglial cultures (OL) also the highly glycosylated form of MAG is detectable, albeit as a rather faint band (Fig. 4).

Antibodies against the Wolfgram protein complex (WP) react with a polypeptide component of approximately 50 kDa (11). This component is not only prominent in purified myelin and oligodendrocytes, but also in the OLN-93 cell line (Fig.4, middle panel).

In the immunblots of rat brain myelin and the primary oligodendroglia cell extracts (Fig. 4, lower panel), the four isoforms of myelin basic proteins with molecular weights of 21.5, 18.5, 17.0, and 14.0 kDa, respectively, which have been described in the rodent brain (10) were detectable. In OLN-93 cell extracts only a single component was recognized by anti MBP-antiserum, migrating slightly above the small 14 kDa MBP in rat brain myelin (Fig.4, lower panel).
While large amounts of vimentin and GFAP were detectable in primary astrocytic cell extracts, and small amounts of both components were seen also in cell extracts of primary oligodendroglia, the immunoblots of OLN-93 cells did not show any immunoreactivity, thus corroborating the data obtained by immunocytochemistry (not shown).

3.4 Expression of PLP mRNA

To establish whether OLN-93 cells express PLP mRNA, total RNA was extracted from cells, reverse transcribed and amplified by PCR using a set of oligonucleotide primers matching sequences in exon 3 and 7 of the PLP gene. PCR amplification products were analyzed on agarose gels and Southern blot analysis after transfer to nylon membranes using a DIG-labelled cDNA probe . Additionally, cell extracts of primary oligodendrocytes were subjected to RT-PCR analysis. Two amplification products hybridizing with the DIG-labelled PLP cDNA probe were obtained, corresponding to the PLP and DM20 protein mRNAs encoded by the PLP gene (Fig. 5 a and b). While in primary oligodendrocytes the PLP gene product was more prominent than the DM20, in OLN-93 cell extracts the DM20 mRNA represented the major species.

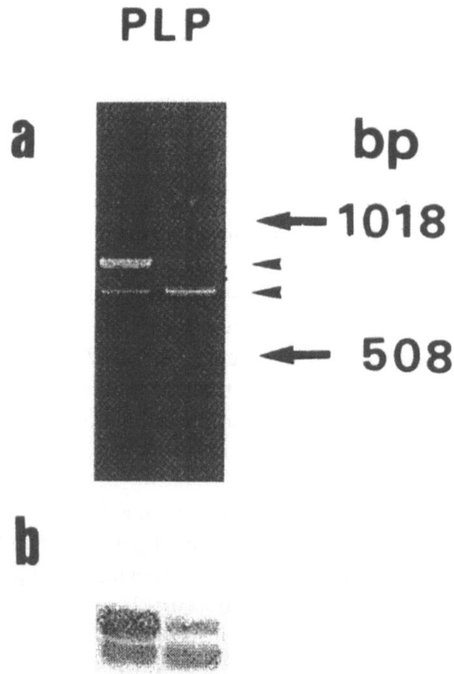

Fig. 5 Expression of PLP mRNA in OLN-93 cells
RT-PCR analysis of total RNA extracted from OLN-93 cells (right lane) or primary cultures enriched in oligodendrocytes (left lane). In (b) the Southern blot hybridization of the PCR-amplification products resolved in agarose gels (a) is seen. The two arrow heads point to the two amplification products corresponding to the PLP (upper) and DM20 (lower) products of the PLP gene.

4 Discussion

Continuous cell lines which are easy to propagate and thus are available in large amounts provide a useful means to study various aspects of cell biology. The establishment and characterization of a new permanent cell line, namely OLN-93, is described in the present report. These cells retain morphological features of O-2A- progenitors or immature oligodendrocytes, they permanently divide and resemble in their antigenic properties primary oligodendrocytes in culture. As shown by immunocytochemistry, OLN-93 cells express GC and the myelin proteins MBP, MAG, PLP and WP, but do not exhibit characteristics of astrocytes, such as the expression of GFAP or vimentin. N-CAM/D2-immunoreactivity is expressed on the surfaces of cell somata and processes. Immunoblot analysis of cellular extracts confirmed the absence of vimentin and GFAP, and the presence of MAG-, WP-, and MBP-immunoreactivity. RT-PCR analysis further demonstrated that in OLN-93 cells PLP/DM20 mRNAs are present.

In general, in vitro differentiation of O-2A progenitors into oligodendrocytes is accompanied by the coordinate expression of myelin specific components. About one week after the emergence of GC, which appears around birth, oligodendrocytes are observed to express the myelin specific proteins WP, MBP, MAG and PLP, even in the absence of neurons (4,5). In rat brain cultures MBP and MAG are expressed by oligodendrocytes at 5 to 7 days after birth, followed closely after a 1-2 day lag by PLP (4). At this time all three proteins are confined to the cell body, MAG is found in the processes approximately 10 d after birth, and MBP and PLP only after 15 days (4).

Immunoblot analysis of OLN-93 cell extracts with anti- MBP antiserum, points to the conclusion that these cells resemble young oligodendrocytes in vitro. While in adult rat brain myelin and in primary oligodendrocytes (15 d postnatally) all four MBP isoforms, normally found in rodents (10, 13) are expressed, OLN-93 cell extracts depict only one low molecular weight component (14 kDa). This polypeptide migrates slightly above the 14 kDa MBP isoform found in rat brain myelin. The altered molecular weight of the 14 kDa isoform might be the result of posttranslational modifications, such as protein phosphorylation which occurs prior to the onset of myelination (23).

The expression of the four MBP isoforms is developmentally regulated and each protein accumulates at a characteristic rate. While during mouse brain development the 14 and 18.5 kDa proteins continually accumulate slowly , the 21.5 and 17.0 kDa proteins accumulate during the time of active myelination (10-30 days after birth) and lateron disappear (10). In mixed glial cultures at early time points, the 14 kDa MBP-isoform has been described to constitute the major myelin basic protein in oligodendrocytes (2).The functional significance of the multiple MBP isoforms remains unresolved, and the new cell line might provide a useful tool, not only to study the regulation of MBP gene expression, but also the functional role of MBP isoforms in the formation and maintenance of compact myelin.

OLN-93 cells also contain the myelin- associated glycoprotein MAG, migrating as a single band with the same molecular weight as the major MAG-immunoreactive compound found in adult rat brain myelin. MAG in vivo is located in the periaxonal membranes of myelin-forming oligodendrocytes (16). Fully glycosylated MAG migrates as a broad band with an apparent molecular weight of about 100 kDa (16). In our study this compound is only seen in rat brain myelin and also faintly in cultured primary oligodendrocytes. Whether these differences reflect differences in glycosylation, which are related to developmental processes, remains to be established.

In OLN-93 cells, pronounced punctuated GC-staining is found at the somatic membranes and only few granules are seen in the leading cell processes, while PLP and MAG appeared rather diffusely, mostly in the cell somata but also faintly in the cellular extensions. MBP-staining was rather strong throughout the cytoplasm and also in the larger cell processes, while the finer arborizations were hardly detectable. Considering the localization of the myelin proteins MAG, PLP and MBP, and the results of immunblot analysis, OLN-93 cells resemble 5 to 10 day- old (postnatal time) cultured rat brain oligodendrocytes (4).

In comparison to other cell lines of the oligodendroglia cell lineage described before (e.g. 1, 6, 21), OLN-93 cells represent an oligodendroglia cell line at a later stage of differentiation, and unlike the bipotential glial cell line CG-4 (12) retain their ability to rapidly proliferate while expressing myelin specific proteins.

In conclusion, we have established a new cell line of the oligodendroglia cell type, derived from spontaneously transformed cells in primary cultures of the brains of newborn rats. These cells are rapidly dividing and express antigenic features of maturing oligodendrocytes. The absence of vimentin and GFAP, and the presence of GC, and of MBP, PLP and MAG, which are proteins of CNS myelin appearing at later stages of development, indicate that OLN-93 cells have passed the intermediate stage between the O-2A progenitor and oligodendrocyte (Fig.6). Thus, they might have lost the developmental plasticity typical for O-2A cells, without losing the ability to proliferate (9).

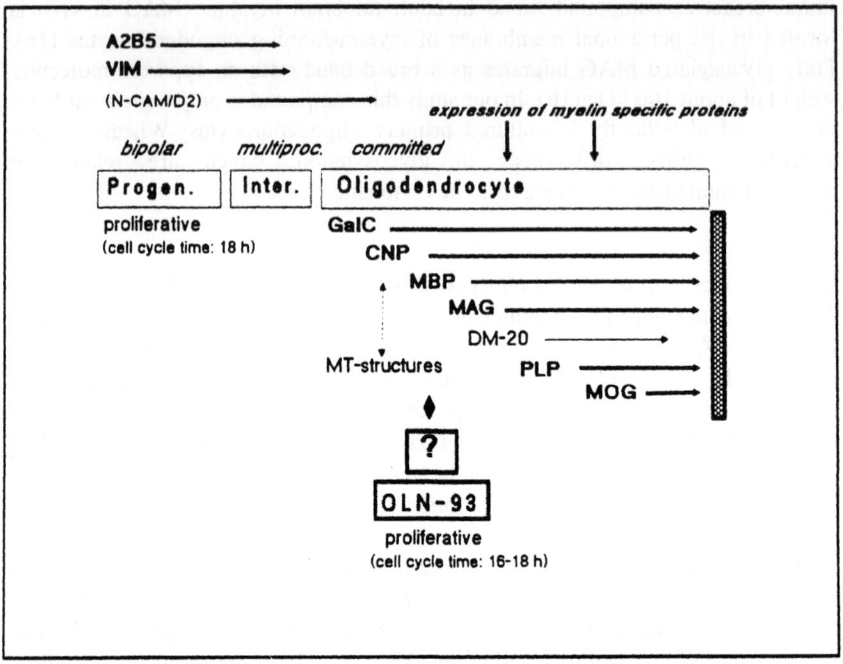

Fig.6 Oligodendrocyte differentiation from the progenitor (Progen.) to the multiprocessed intermediate stage (Inter.) . The coordinated expression of myelin specific markers is indicated by the horizontal arrows. The likely position of OLN-93 cells in this developmental scheme , according to their morphological features and antigenic profile , is indicated .

Hence, these cells will provide a useful means to investigate various important questions, such as : (1) the specific mechanisms which regulate proliferation and differentiation in oligodendroglia, (2) the regulation of myelin-specific genes and the modulation of myelin protein isoform expression, such as MBP- and PLP-isoforms, and the molecular mechanisms involved in transport processes and the incorporation into the membrane sheets; and (3) the molecular responses to and cellular interactions with other cells of the nervous system, e.g. with nerve cells and their axons. In particular, OLN-93 cells will be an interesting model system to study their regeneration potential following transplantation into normal and damaged CNS (7).

5 References

1. Almazan, G. and McKay, R. (1992) An oligodendrocyte precursor cell line from rat optic nerve. Brain Res. 579, 234-245.
2. Barbarese, E. and Pfeiffer, S.F. (1981) Developmental regulation of myelin basic protein in dispersed cultures. Proc. Natl. Acad. Sci. USA 78, 1953-1957.
3. Chomczynski, P. and Sacchi, N. (1987) Single step method of RNA isolation by acid guanidinium thiocyanate-phenol-chloroform extraction. Anal. Biochem. 162, 156-159.
4. Dubois-Dalcq, M., Behar, T., Hudson, L. and Lazzarini, R.A. (1986) Emergence of three myelin proteins in oligodendrocytes cultured without neurons. J.Cell .Biol. 102 384-392.
5. Dubois-Dalq, M. and Armstrong, R. C. (1992) The oligodendrocyte lineage during myelination and remyelination. In Martenson, R.E. (ed), Myelin: Biology and Chemistry. Boca Raton, CRC Press, pp. 81-122.
6. Foster, L.M., Phan, T., Verity, A.N. and Campagnoni, A.T. (1993) Generation and analysis of normal and shiverer temperature-sensitive immortalized cell lines exhibiting phenotypic characteristics of oligodendrocytes at several stages of differentiation. Dev. Neurosci . 15, 100-109.
7. Franklin, R.J.M. and Blakemore, W.F. (1995) Glial-cell transplantation and plasticity in the O-2A lineage-implications for CNS repair. Trends. Neurosci. 18, 151-156.
8. Gard, A.L. and Pfeiffer, S.E. (1989) Oligodendrocyte progenitors isolated directly from developing telencephalon at a specific phenotypic stage: myelinogenic potential in a defined environment. Development 106, 119-132.
9. Goldman, J. (1992) Regulation of oligodendrocyte differentiation. Trends Neurosci. 10, 359-362.
10. Kamholz, J. and Wrabetz, L. (1992) Molecular genetics of myelin basic protein. , In Martenson, R.E. (ed), Myelin: Biology and Chemistry. Boca Raton, CRC Press, pp 367-385.
11. Karin, N. J. and Waehneldt, T. V. (1985) Biosynthesis and insertion of Wolfgram protein into optic nerve membranes. Neurochem. Res. 10, 897-907.
12. Louis, J.C., Magal ,E., Muir, D., Manthorpe, M. and Varon, S. (1992) CG-4, a new bipotential glial cell line from rat brain , is capable of differentiating in vitro into either mature oligodendrocytes or type-2 astrocytes. J. Neurosci. Res. 31, 193-204.
13. Martenson, R. E. (1992) Myelin basic protein isoforms: structural and evolutionary implications of alternative exon splicing. In Martenson, R.E. (ed), Myelin: Biology and Chemistry. Boca Raton, CRC Press, pp 387-411.

14. Mc Carthy, K. D. and De Vellis , J. (1980) Preparation of separate astroglial and oligodendroglial cell cultures from rat cerebral tissue. J. Cell Biol. 85, 890-902.
15. Neuhoff, V., Phillip, K., Zimmer, H. G. and Mesecke, S. (1979) A simple, versatile , sensitive and volume-independent method for quantitative protein determination which is independent of other external influences. Hoppe Seyler`s Z Physiol. Chem. 360, 1657-1670.
16. Quarles, R. H., Colman, D. R., Salzer, J. L. and Trapp, B. D. (1992) Myelin-associated glycoprotein: structure-function relationships and involvement in neurological diseases. In Martenson, R.E. (ed), Myelin: Biology and Chemistry. Boca Raton, CRC Press, pp 413-448.
17. Ranscht, B., Clapshaw, P.A., Price, J., Noble, M. and Seifert, W. (1982) Development of oligodendrocytes and Schwann cells studied with a monoclonal antibody against galactocerebroside. Proc. Natl. Acad. Sci. USA 79, 2709- 2713.
18. Richter-Landsberg, C. and Besser, A. (1994) Effects of organotins on rat brain astrocytes in culture. J Neurochem 63, 2202-2209.
19. Richter-Landsberg, C. and Heinrich, M. (1995) S-100 immunoreactivity in rat brain glial cultures is associated with both astrocytes and oligodendrocytes. J. Neurosci. Res. 42, 657-665.
20. Richter-Landsberg,C. and Heinrich, M. (1996) OLN-93: A new permanent oligodendrocglia cell line derived from primary rat brain glial cultures. J.Neurosci. Res. submitted
21. Russo, T., Mogavero, A.R., Ammendola, R., Mesurca, M., Fiore, F., Fatatis, A., Salvatore, G. and Cimino, F. (1993) Immortalization of a cell line showing some characteristics of the oligodendrocyte phenotype. Neurosci. Lett. 159, 159- 162.
22. Towbin, H. T. M., Staehelin, T. M. and Gordon, J. (1979) Electrophoretic transfer of proteins from polyacrylamide gels to nitrocellulose sheets: procedure and some application. Proc. Natl . Acad. Sci. USA 76, 4350-4354.
23. Ulmer, J.B. (1988) The phosphorylation of myelin proteins. Progr. Neurobiol. 31, 241-259.
24. Waehneldt, T.V., Matthieu, J.M. and Neuhoff, V. (1977) Characterization of a myelin-related fraction (SN4) isolated from forebrain at two developmental stages. Brain Res. 139, 29-43.
25. Waehneldt, T. V., Malotka, J., Gunn ,C. A. and Linington, C. (1990) : Myelin proteolipid protein : cladistic tool to study vertebrate phylogeny. In Jeserich, G., Althaus ,H. H., Waehneldt, T. V. (eds) , Cellular and Molecular Biology of Myelination. Berlin Heidelberg, Springer Verl., pp361- 372.

Schwann Cell Suspension Grafts Promote Reconstruction of Transected Postcommissural Fornix in the Adult Rat

Christine Stichel, Katrin Lips, Gilbert Wunderlich and Hans Werner Müller

Molecular Neurobiology Laboratory, Department of Neurology, University of Düsseldorf, 40225 Düsseldorf, Germany

Keywords. CNS, regeneration, microtransplantation, Schwann cell

1 Introduction

Nerves of the adult mammalian CNS do not regenerate successfully after axonal injury. After an initial degeneration injured axons start to regrow, but this regrowth ceases after a few hundred micrometers without traversing the lesion site. This failure of axonal regeneration is attributed mainly to the non-permissive environment surrounding the axons. A great number of studies have shown that injured neurons are capable of regrowing to greater lengths if an appropriate environment is provided. Such a growth-promoting environment is created by peripheral nerve grafts or implanted Schwann cells (SC; for review see 10).

To date a variety of approaches have been used to introduce SC into injured CNS. In some studies SC-coated nitrocellulose-filters (7, 12), hydrophilic sponges (20) and porouse tubes made of synthetic polymers (9, 13, 17, 26) or basement membrane components (18, 19) have been implanted. Besides the fact that such prosthesis are needed in cases in which a wide gap has to be bridged (9, 11) or the geometry of the proximal and distal stump is complex (1, 22), such solid grafts itself cause extensive damage to the host tissue and their size and the need for exact positioning restrict their application to more superficial lesion areas. In an attempt to minimize these complications the microtransplantation method, that is the controlled injection of cell suspensions (8), might offer an alternative strategy to introduce the cells into the CNS tissue. Initial attempts to microinject highly purified SC into prelesioned or unlesioned adult rat brain have been successful and provided evidence that the SC integrate well (5, 21) and elicit axonal growth (6, 15, 16). With respect to clinical perspective, however, a number of important questions remain to be answered, regarding (i) the applicability and efficiency of this method also in other CNS systems, (ii) the possibility of pathway reconstruction and (iii) the long-term fate of these cells not originally located in the CNS.

The present study was designed to address some of these questions. We used a well-established lesion model, the transected postcommissural fornix of the adult rat (25). The latter was stereotactically transected at about 1.2 mm distal from the

Molecular Signaling and Regulation in Glial Cells
A Key to Remyelination and Functional Repair

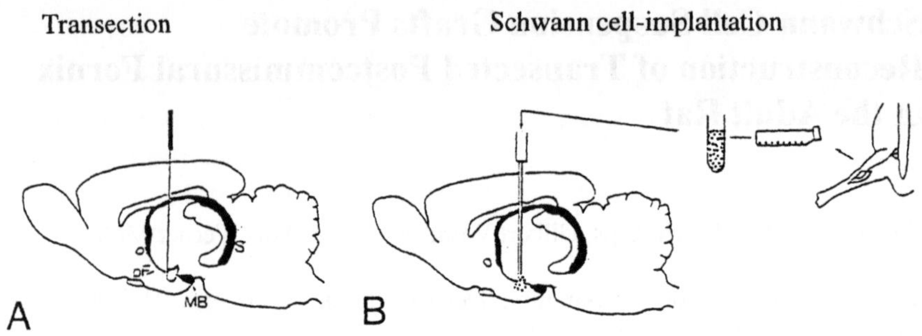

Fig. 1.1 Schematic diagrams of adult rat postcommissural fornix and surgical procedures; (A) stereotactic transection and (B) the microtransplantation of neonatal Schwann cells into the lesion site. MB, mammillary body, pF, postcommissural fornix, S, subiculum.

target region, the mammillary body (Fig. 1.1A). In previous studies we have shown that 5 days after lesion the transected fibers start to regrow but stop at the former lesion site, a region that represents the major obstacle for these regrowing fibers (23, 24). Even transplanted juvenile astrocytes were not able to guide the fibers across this barrier (25). In the present study we have analysed the influence of microtransplanted SC on the structural reconstruction of the transected fornix using immunocytochemical methods, anterograde tracing techniques and electron microscopic analysis.

2 Results and Discussion

2.1 Schwann cell cultures

Schwann cells were isolated from sciatic nerves of neonatal rats and purified in tissue culture using the method described by Brockes et al. (4). The cells were passaged up to seven times in the presence of mitogens prior implantation. The purity of the cultured SC was controlled using cell-type specific antibodies.

At the time of transplantation cultures were composed of 90-95% SC with the typical spindle-shaped bipolar morphology, an oval cell body and slender processes. These cells were strongly immunopositive for S100 and laminin (LN), weakly immunopositive for the p75 low affinity nerve growth factor receptor (LNGFR) and glial fibrillary acidic protein (GFAP) and immunonegative for the fibroblast-specific antigen Thy 1.1. This antigenic profile indicates that the SC exhibit a non-myelinating (immature) phenotype in vitro (14).

2.2 Characterisation of Schwann cell graft

Schwann cells were prelabeled with the non-toxic, vital cell tracer bisbenzimide (Hoechst 33342) and 160.000 SC (in 1.6 µl DMEM) were injected immediately after transection into the lesion site (Fig. 1.1B). The microtransplanted SC were easily visible as a dense mass of brightly fluorescent cells. The SC migrated

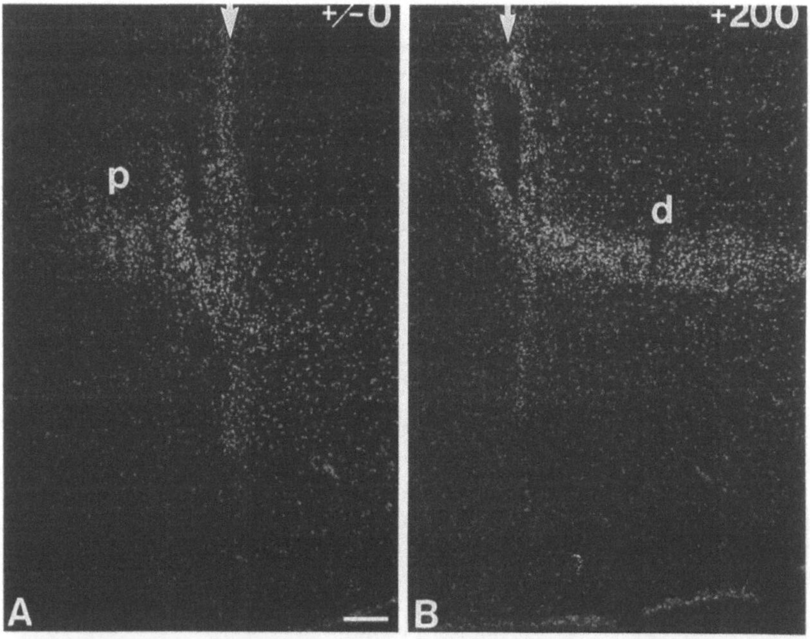

Fig. 2.1. Distribution of microtransplanted Schwann cells at 4 days after surgery.
A,B: Ten micrometer thick sections through a Hoechst-prelabeled Schwann cell transplant, demonstrating the gradient of cells radiating from the injection site (arrow) (d, distal; p, proximal). The numbers in the right left corners indicate the distance in micrometers on either side of the injection center. Scale bar, 200 μm (A, also for B).

freely in a pattern radiating from the injection center and entered the surrounding neuropil (Fig. 2.1). They travelled for considerable distances and by 4 days after transplantation individual cells could be identified up to 3 mm from the injection center. While the number of SC in the injection center progressively diminished in time, fluorescent cells were found for at least up to 8 months in this area. The migration behaviour confirms earlier studies in which extensive SC migration and similar distribution patterns have been described (2, 3, 5, 21).

Using immunocytochemical methods with antibodies against a nuclear antigen present only in proliferating cells (PCNA) we found single proliferating SC soon after transplantation. The proliferation activity decreased during the following days (Fig. 2.2) and ceased at 9-14 days postgrafting. This low and short-term proliferation activity is an important prerequisite for developing this approach to a technique with putative clinical application.

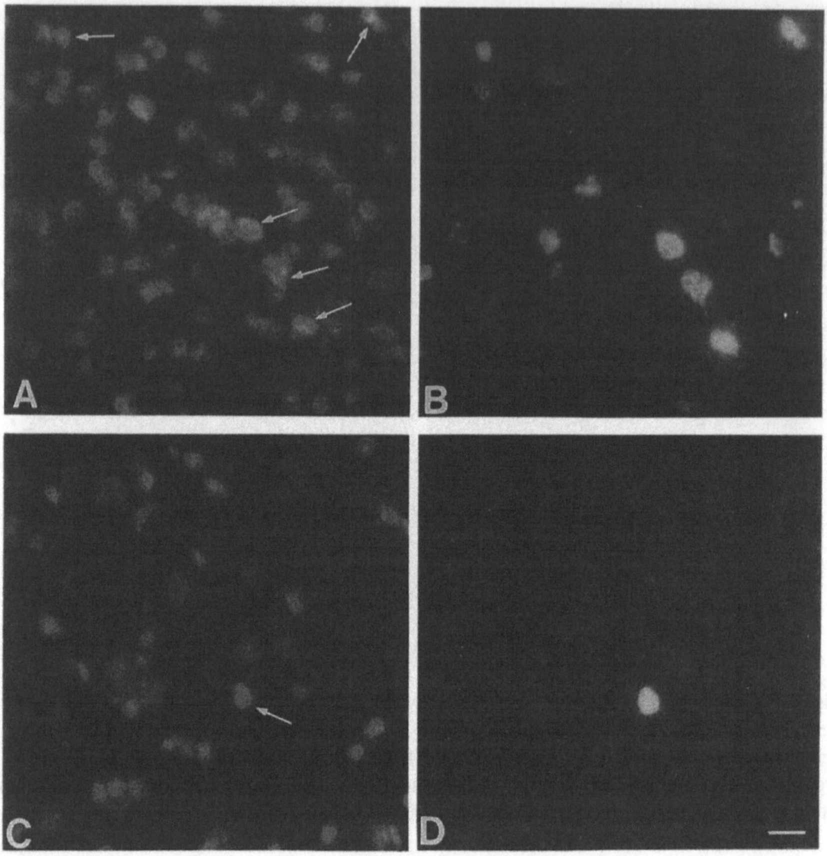

Fig. 2.2. Proliferation activity of microtransplanted Schwann cells.
PCNA-immunostaining (B, D) of Hoechst-prelabeled transplanted Schwann cells (A, C) showing mitotic activity at 4 days (A, B) and 7 days (C, D) after transplantation, respectively. Note, the small number of proliferating cells and the decrease in proliferation activity. Arrows in A and C mark double-labeled cells. Scale bar, 10 µm (D, also for A-C).

2.3 Influences of Schwann cells on de- and regeneration of transected fornix fibers

Transection of the postcommissural fornix resulted in retro- and anterograde degeneration and demyelination of axons as described previously (24, 25). The sequence of events, the time course and the pattern of regrowing fibers up to the lesion site were similar in control animals (transection only or transection plus DMEM injection) and animals with SC transplant. Thus, microtransplanted SC neither prevent degeneration and demyelination of transected axons nor accelerate the initial axonal growth or change the course of regrowing axons.

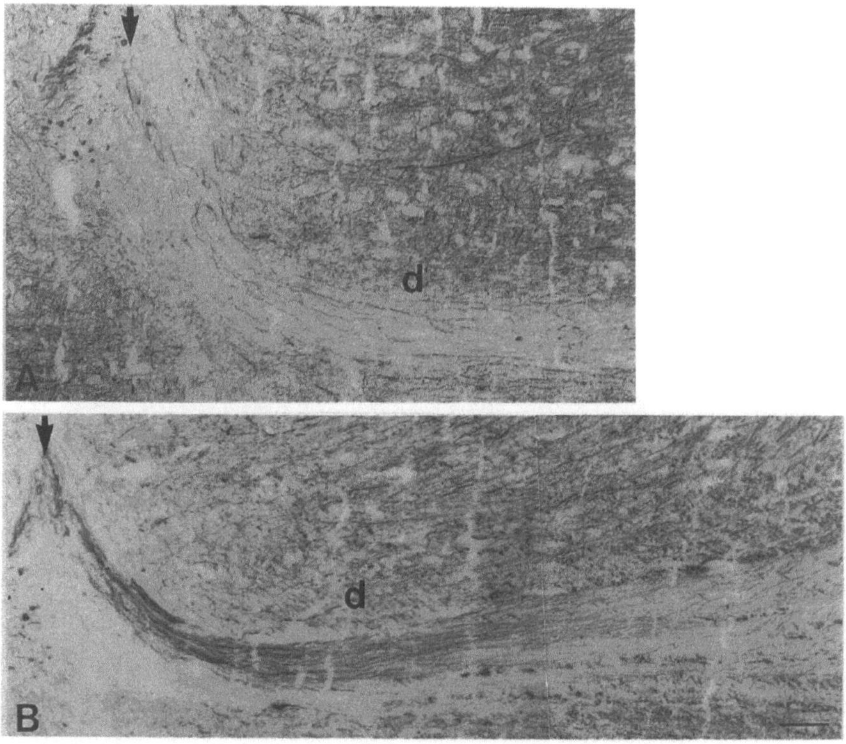

Fig. 2.3. Pattern of regrowing axons in animals with SC transplant.
Staining of neurofilament-positive axons in lesioned animals with SC transplant at 18 days (A) and 6 weeks (B) postgrafting. Regrowing fibers crossed the lesion site (arrow) and elongated within their former pathway, the distal stump (d). Note, that the number of crossing fibers increased during the first 6 weeks after injury. Scale bar, 100 μm (B, also for A).

However, in contrast to control animals (Fig. 2.3A), the regrowing fibers of animals with SC transplant crossed the former impermeable lesion site and re-entered the distal stump (Fig. 2.3B). The number of crossing fibers increased during the first 2 months after transplantation (Fig. 2.3B and C). The regenerating axons elongated about 1.2 mm up to their normal target, the mammillary body.

Anterograde tracing with horseradish peroxidase was performed to document the subicular origin of regenerating axons after SC transplantation. Following injection of WGA-HRP into the subiculum the vast majority, if not all, of the fibers elongating within the distal stump were heavily labeled, demonstrating that they emerge from the formerly transected fornix tract. Moreover, the ultra-structural analysis of these animals revealed, that the regenerating fibers

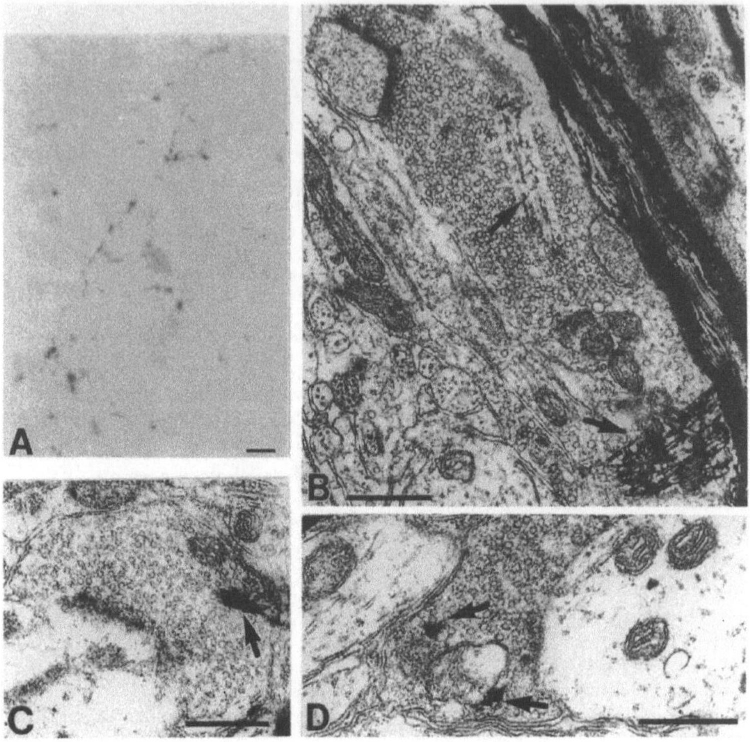

Fig. 2.4. Regenerating fibers originating from the subiculum re-establish synapses in the mammillary body.
A: WGA-HRP-filled regenerated axon within the mammillary body of an animal with SC transplant. B-D: Electron micrographs of presynaptic profiles labeled with anterogradely transported WGA-HRP. The characteristic reaction products are indicated by arrows. Scale bars, 10 μm (A), 1 μm. (B-D).

established synapses within the mammillary body (Fig. 2.4).

Within four weeks postinjury we observed remyelination of regenerating axons. The latter exhibited immunoreactivity for myelin associated glycoprotein (MAG) and myelin basic protein (MBP). The absence of the peripheral myelin protein PMP22 and the presence of the proteolipid protein (PLP), a specific central myelin protein, provided evidence that the fibers are remyelinated along their entire length by host oligodendrocytes (Fig. 2.5).

While the powerful growth-promoting influence of SC is well established, this is the first report to our knowledge that indicates that SC guide growing axons through a discrete lesion-induced nonpermissive boundary. Another surprising and important result is the amazing capacity of regrowing axons to follow their previous course up to their normal target. This precision could reflect (i) an inherent capacity of the fibers to find their appropriate target, (ii) the capability of

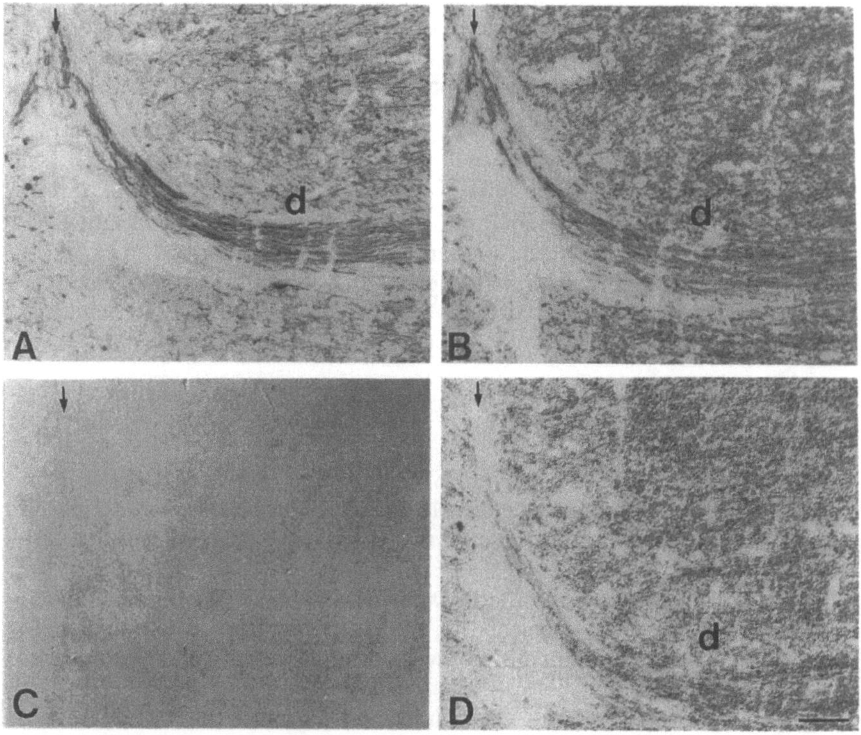

Fig. 2.5 Remyelination of regenerating fibers at 4 weeks after injury.
Immunostaining of serial sagittal sections through the lesion/implantation site (arrows) of an animal with SC transplant, using antibodies to neurofilament (NF, A), myelin associated glycoprotein (MAG, B), peripheral myelin protein 22 kDa (PMP22, C) and proteolipid protein (PLP, D). d, distal. Scale bar, 100 µm (D, also for A-C).

SC to provide guiding cues and/or (iii) the presence of boundary molecules that delineate the pathway of regenerating fornix fibers. The unexpected remyelination by host oligodendrocytes together with previous observations of primarily central myelination on SC-coated nitrocellulose filters (12) suggests that in such a SC environment oligodendrocytes dominate the competition for the remyelination of regrowing axons.

3 Concluding remarks

Our study shows that microtransplanted SC do neither prevent retrograde axonal degeneration nor demyelination but promote reconstruction of the transected fornix tract. Regenerating fibers originating in the subiculum cross the lesion site, continue to grow only within their former pathway, reinnervate their normal target and are remyelinated by host oligodendrocytes.

In conclusion, our data demonstrate that microtransplantation of SC suspensions represents a promising strategy to promote structural reconstruction of lesioned CNS pathways and provide an encouraging argument for pursuing this approach to an appropriate therapeutical strategy.

4 Acknowledgements

The authors wish to thank Prof. H.-G. Hartwig for scientific collaboration, G. Servos for excellent technical assistance, U. Vollmer and M. Lange for photographic work. This work was supported by a grant from the Deutsche Forschungsgemeinschaft (SFB 194/B5) and the Hermann and Lilly Schilling-Stiftung.

5 References

1. Aebischer, P., Guénard,V., Winn, S.R., Valentini, R.F. and Galletti, P.M. (1988) Blind ended semi-permeable guidance channels support peripheral nerve regeneration in the absence of a distal nerve stump. Brain Research 454, 179-187.
2. Baron-Van Evercooren, A., Clerin-Duhamel, E., Lapie, P., Gansmüller, A., Lachapelle, F. and Gumpel, M. (1992a). The fate of Schwann cells transplanted in the brain during development. Developmental Neuroscience 14, 73-84.
3. Baron-Van Evercooren, A., Gansmüller, A., Duhamel, E., Pascal, F. and Gumpel, M. (1992b). Repair of a myelin lesion by Schwann cells transplanted in the brain during development. Developmental Neuroscience 14, 235-242.
4. Brockes, J.P., Fields, K.L. and Raff, M.C. (1979) Studies on cultured rat Schwann cells. I. Establishment of purified populations from cultures of peripheral nerves. Brain Research 165, 105-118.
5. Brook, G.A., Lawrence, J.M. and Raisman, G. (1993) Morphology and migration of cultured Schwann cells transplanted into the fimbria and hippocampus in adult rats. Glia 9, 292-304.
6. Brook; G.A., Lawrence, J.M., Shah, B. and Raisman, G. (1994) Extrusion transplantation of Schwann cells into the adult rat thalamus induces directional host axon growth. Experimental Neurology 126, 31-43.
7. Chen, M., Harvey, A. R. and Dyson, S.E. (1991) Regrowth of lesioned retinal axons associated with the transplantation of Schwann cells to the brachial region of the rat optic tract. Restorative Neurology and Neuroscience 2, 233-248.
8. Emmett, C. J., Jaques-Berg, W. and Seeley, P.L. (1990) Micro-transplantation of neural cells into adult rat brain. Neuroscience 38, 213-222.
9. Guénard, V., Morrissey, T.K., Kleitman, N., Bunge, R.P. and Aebischer, P. (1991) Cultured syngeneic adult Schwann cells seeded in synthetic guidance channels enhance sciatic and optic nerve regeneration. Society for Neuroscience Abstracts 17, 565.

10. Guénard, V., Xu, X.M. and Bunge, M.B. (1993) The use of Schwann cell transplantation to foster central nervous system repair. Seminars in the Neuroscience 5, 401-411.
11. Harvey, A.R., Chen, M. and Dyson, S.E. (1991) Glia cells transplanted to the rat optic tract. Influence on the regrowth of retinal axons. Annals of the New York Academy of Sciences 633, 573-576.
12. Harvey, A. R., Fan, Y., Connor, A. M., Grounds, M.D. and Beilharz, M. (1993) Migration and intermixing of donor and host glia on nitrocellulose polymers implanted into cortical lesion cavities in adult mice and rats. International Journal of Developmental Neuroscience 11, 569-581.
13. Harvey, A.R., Chen, M., Plant, G.W. and Dyson, S.E. (1994) Regrowth of axons within Schwann cell-filled polycarbonate tubes implanted into the damaged optic tract and cerebral cortex of rats. Restorative Neurology and Neuroscience 6, 221-237.
14. Jessen, K. R. and Mirsky, R. (1991) Schwann cell precursors and their development. Glia 4, 185-194.
15. Martin, D., Schoenen, J., Delree, P., Leprince, P., Rogister, B. and Moonen, G. (1991) Grafts of syngeneic cultured, adult dorsal root ganglion-derived Schwann cells to the injured spinal cord of adult rats: preliminary morphological studies. Neuroscience Letters 124, 44-48.
16. Montero-Menei, C., Pouplard-Barthelaix, A., Gumpel, M. and Baron-Van Evercooren, A. (1992) Pure Schwann cell suspension grafts promote regeneration of the lesioned septo-hippocampal cholinergic pathway. Brain Research 570, 198-208.
17. Montgomery, C. T. and Robson, J.A. (1993) Implants of cultured Schwann cells support axonal growth in the central nervous system of adult rats. Experimental Neurology 122, 107-124.
18. Neuberger, T. J., Cornbrooks, C.J. and Kromer, L.F. (1992) Effects of delayed transplantation of cultured Schwann cells on axonal regeneration from central nervous system cholinergic neurons. Journal of Comperative Neurology 315, 16-33.
19. Paino, C. L., Fernandez-Valle, C., Bates, M.L. and Bunge, M.B. (1994) Regrowth of axons in lesioned adult rat spinal cord: promotion by implants of cultured Schwann cells. Journal of Neurocytology 23, 433-452.
20. Plant, G.W., Harvey, A.R. and Chirila, T.V. (1995) Axonal growth within poly(2-hydroxyethyl methycrylate) sponges infiltrated with Schwann cells and implanted into the lesioned rat optic tract. Brain Research 671, 119-130.
21. Raisman, G., Lawrence, J.M. and Brook, G.A. (1993) Schwann cells transplanted into the CNS. International Journal of Development Neuroscience 11, 651-669.
22. Shine, H. D., Harcourt, P.G. and Sidman, R.L. (1985) Cultured peripheral nervous system cells support peripheral nerve regeneration through tubes in the absence of distal nerve stump. Journal of Neuroscience Research 14, 393-401.

23. Stichel, C. C. and Müller, H.W. (1994). Relationship between injury-induced astrogliosis, laminin expression and axonal sprouting in the adult rat brain. Journal of Neurocytology 23, 615-630.
24. Stichel, C. C., Wunderlich, G., Schwab, M.E. and Müller, H.W. (1995) Clearance of myelin constituents and axonal sprouting in the transected postcommissural fornix of the adult rat. European Journal of Neuroscience 7, 401-411.
25. Wunderlich, G., Stichel, C.C., Schroeder, W.O. and Müller, H.W. (1994) Transplants of immature astrocytes promote axonal regeneration in the adult rat brain. Glia 10, 49-58.
26. Xu, X. M., Guénard, V., Kleitman, N. and Bunge, M.B. (1995) Axonal regeneration into Schwann cell-seeded guidance channels grafted ino transected adult rat spinal cord. Journal of Comparative Neurology 351, 145-160.

The biology of the transplanted oligodendrocyte progenitor

Robin J.M. Franklin

MRC Cambridge Centre for Brain Repair and Department of Clinical Veterinary Medicine, University of Cambridge, Madingley Road, Cambridge, CB3 0ES, UK.

1. Introduction

An important advance in oligodendrocyte biology during the 1980's was the development of models and techniques for studying the myelinogenic capacity of transplanted glia (8). Two basic models emerged from which our current knowledge regarding the ability of transplanted glia to myelinate host axons is largely derived. The first of these are the so-called "myelin-mutants", where a defect in one of the genes encoding central myelin proteins (such as the MBP-mutant *shiverer mouse*, or the PLP-mutant *myelin-deficient rat*) gives rise to a hypomyelinated CNS containing abundant unmyelinated or demyelinated axons available for myelination by transplanted cells. In general, the myelin mutants are short-lived, and most studies using these animals have been undertaken in neonatal or immature recipients (7,14,21). The second transplantation model involves the creation of clearly-defined, focal areas of demyelination within the white matter of adult animals by the direct injection of small volumes of gliotoxic agents such as lysolecithin or ethidium bromide. The normally-occurring, spontaneous remyelination of gliotoxin-induced areas of demyelination can be prevented by exposing the lesion area to 40 Grays of X-irradiation, thereby creating a non-repairing area of demyelination (the X-EB lesion) with which to test the behaviour of transplanted cells in the absence of competition from recruited host cells (fig.1a). The ability of transplants containing cells of the oligodendrocyte lineage to myelinate host axons was clearly demonstrated using both these models.

A second major advance during the 1980's was an improved understanding of oligodendrocyte ontogenesis based on *in vitro* approaches (18,20). The recognition of a variety of markers for specific stages of the oligodendrocyte lineage enabled the isolation of populations of cells enriched for particular phenotypes, whose contribution to transplant-mediated myelination could then be individually addressed. In an elegant and important study by Warrington, Barbarese and Pfeiffer (26) it was demonstrated that oligodendrocyte progenitors (A2B5+/O4-) cells produced more myelin over a wider area than did the more mature stages of the lineage (O4+/GalC- and GalC+ cells) when transplanted in

Molecular Signaling and Regulation in Glial Cells
A Key to Remyelination and Functional Repair

neonatal shiverer mice. As a consequence of this and other studies (13), the A2B5+ oligodendrocyte progenitor (OP) became the focus for a number of subsequent transplantation studies, some of which are reviewed in this chapter.

2. Transplantation of oligodendrocyte progenitors results in extensive remyelination of areas of persistent demyelination in the adult CNS

In order to establish the ability of OPs to remyelinate areas of demyelination in the adult CNS, OPs from neonatal rat optic nerve cultures which had been enriched to >99% purity by exposure to PDGF and bFGF, were transplanted directly into X-EB lesions (13). By 21 days following transplantation, as many as 90% of the available axons had been re-invested with central myelin sheaths, indicating that transplanted OPs were able to differentiate into myelinating oligodendrocytes within the lesion environment (fig.1b). Although the host tissue had been X-irradiated and therefore was incapable of contributing cells to the repair of the lesion, the transplant origin of the myelinating cells was confirmed by labelling the transplanted cells with the marker gene *LacZ* (fig.2). The incorporation of the *LacZ* gene did not effect the ability of the OPs to remyelinate demyelinated axons, indicating that OPs can be engineered to express a foreign protein while still retaining their function. This study also demonstrated the potential for using combinations of growth factors (in this case PDGF and bFGF) to generate large numbers of OPs *in vitro* as a prelude to transplantation, although we were also able to show that prolonged exposure resulted in a decrease in the cells' ability to undertake remyelination.

A similar degree of remyelination of X-EB lesions can be achieved using some, but not all, OP cell lines. The CG4 cell line can generate large areas of remyelination which resemble those generated by growth factor expanded populations of primary-derived OP cells, albeit with a slightly higher cell density (9). On the other hand, OP cell lines created by incorporation of the SV40 oncogene, while capable of generating myelinating oligodendrocytes, do not appear capable of extensive remyelination (2).

3. Oligodendrocyte progenitors can manifest the same bipotentiality that they show *in vitro* (although not in development), following transplantation into glia-deficient areas of adult CNS, giving rise to astrocytes as well as oligodendrocytes.

A well documented feature of the *in vitro* properties of the OP cell is that it can be induced under appropriate conditions to generate a GFAP-expressing cell, the type-2 astrocyte (20). Despite an extensive search, it has not been possible to demonstrate that this is a differentiation pathway taken by the OP cell during normal development (11,22). Nevertheless, we endeavoured to establish whether an OP cell could give rise to a type-2 astrocyte *in vivo* in pathological situations

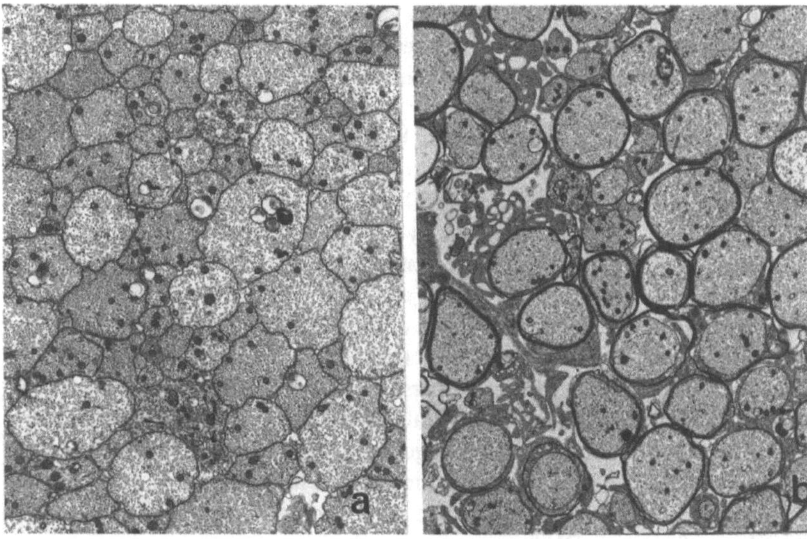

Fig.1 - (a) X-EB lesion consisting of demyelinated axons in a glia-free environment. (b) Many of the axons have been remyelinated with thin myelin sheaths 21 days after transplantation of OPs expanded *in vitro* by exposure to PDGF/bFGF (see ref. 13)

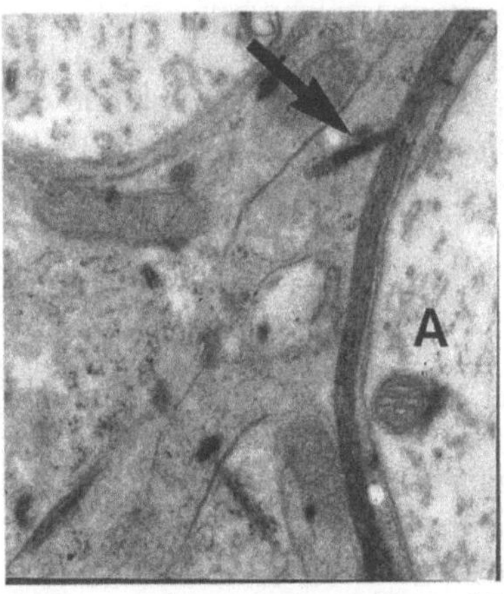

Fig.2 - The process of an OP derived-oligodendrocyte attached to a myelin sheath around axon A. The process contains an electron dense crystal of the product, generated histochemically from the *LacZ* gene product (arrow). Unstained ultrathin resin section

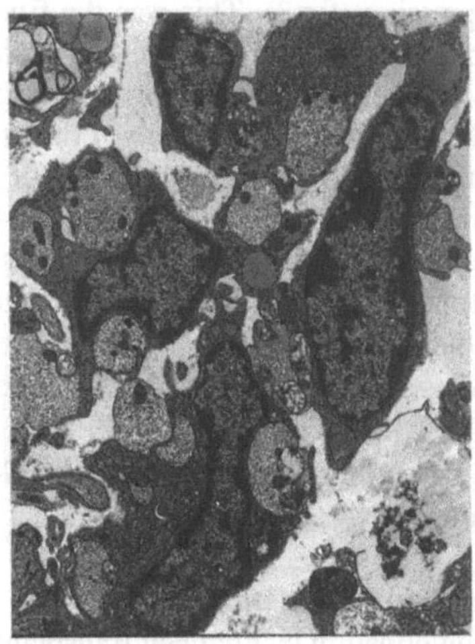

Fig.3 - Astrocytes within a X-EB lesion following transplantation of a SV40 generated OP line (see refs 2 and 10).

in which both oligodendrocytes and astrocytes are absent. Such a situation exists in the X-EB lesion (fig.1a), which therefore constitutes an environment quite different from those that occur in development.

In order to address questions of lineage by cell transplantation it is imperative that one uses populations of proven purity. Populations of progenitors which contain other phenotypes, however small the contaminant populations may be, are inadequate in that the heterogeneity of the starting population introduces uncertainty about the origin of differentiated phenotypes subsequently identified. In our experience, sufficient purity can not be obtained from primary cultures. Even PDGF/bFGF-expanded populations of OPs, although >99% pure, contained small numbers of other phenotypes. For this reason we have used OP cell lines whose clonality can be verified. The first of these OP lines was created from PDGF/bFGF-treated cultures of neonatal rat optic nerve using a temperature-sensitive mutant form of the SV40 oncogene (2). This oncogene product is active at 33°C but inactive at 39°C, and therefore inactive following transplantation into a rat host. The clonal nature of the line so generated was verified by Southern blot analysis of genomic DNA.

When a pure population of A2B5-expressing OP cells, obtained by maintaining the line at 33°C in the presence of PDGF/bFGF, was transplanted into X-EB lesions, the majority of cells remained as progenitors. However, a small proportion differentiated into myelinating oligodendrocytes or cells which were identified as astrocytes on the basis of ultrastructural features and GFAP-expression (fig.3). This was the first demonstration *in vivo* of cells akin to the type-2 astrocytes of tissue culture. That this behaviour was not a peculiarity of the SV40 line was confirmed by demonstrating a similar phenomenon using the CG4 cell line (9). The OP origin of the astrocytes within the lesion was unequivocally demonstrated by the presence of GFAP+/*LacZ*"blue"+ cells following the transplantation of *LacZ* labelled CG4's. The astrocyte generated from an OP has a distinct behaviour and morphology which distinguish it from other types of astrocyte that have been transplanted into X-EB lesions (10). Thus, in pathological situations in which there is a deficiency in both oligodendrocytes and astrocytes, OPs manifest a bipotentiality hitherto only seen in tissue culture.

4. The migratory properties of CG4 cells (an oligodendrocyte progenitor cell line) transplanted into adult CNS

Inevitably, the success that has been achieved in remyelinating persistent areas of demyelination by glial cell transplantation in animal models has fostered a belief that such an approach may be appropriate for the repair of areas of chronic demyelination in diseases such as multiple sclerosis (MS). Before such a transition can be made there are, of course, a great many additional issues that need addressing. Since MS is a multifocal condition, with areas or "plaques" of

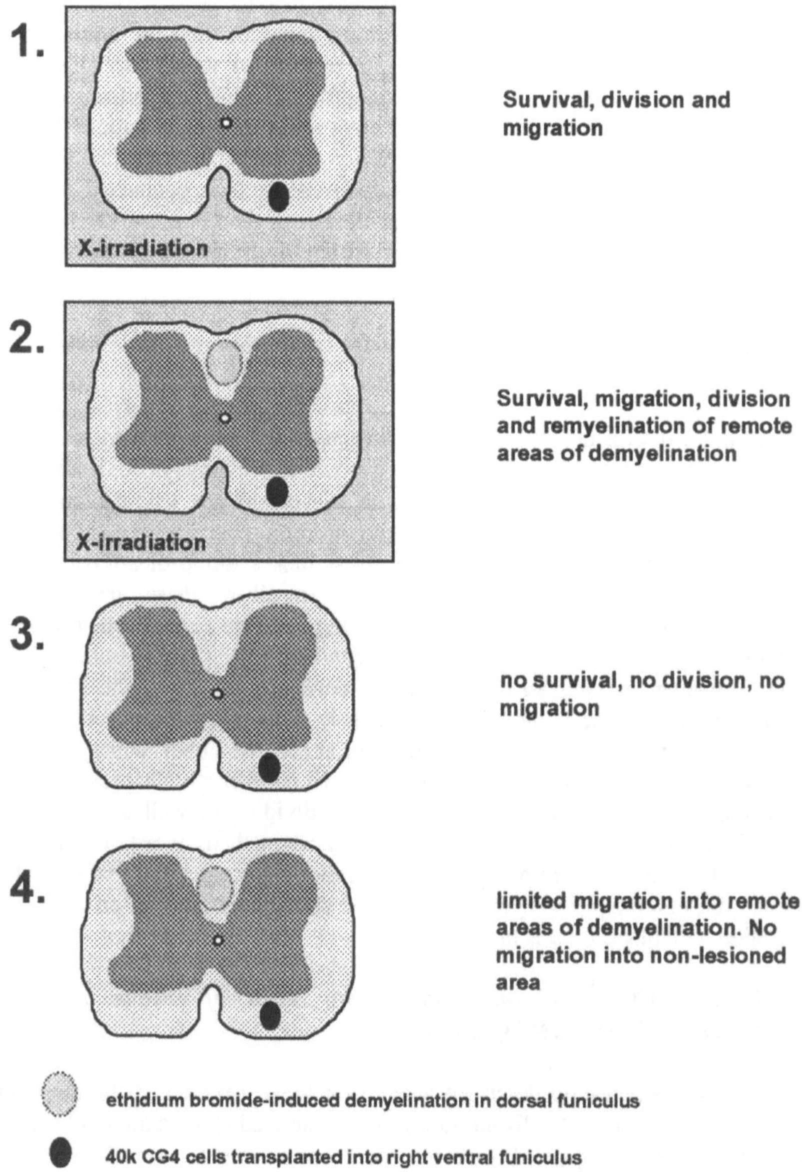

Fig.4 - Summary of CG4 migration experiments.

demyelination occurring throughout the neuraxis, an issue of some relevance is the migratory behaviour of OPs following transplantation into the adult CNS. Would OPs need to be injected directly into plaques or would the migratory behaviour of OPs *in vitro* (1,23) and in development (15,17) enable transplanted OPs to migrate through normal tissue and enter areas of demyelination? If this were to occur, then it might render unnecessary separate injections directly into each clinically relevant plaque. To address this issue we have conducted a series of experiments using the CG4 cell line, labelled with the *LacZ* marker gene to enable identification of transplanted cells within host tissue (summarised in fig.4).

4.1 Transplanted CG4 cells migrate extensively through unlesioned X-irradiated spinal cord

In an initial experiment CG4 cells were injected into the right ventral white matter of rats whose spinal cords had been exposed to 40 Grays of X-irradiation. An important aspect of migration studies is to ascertain the extent of passive spread of cells that is attributable solely to the injection procedure and should not be mistaken for active cell migration. To assess this, a group of animals were killed within 30 minutes of transplantation. In all of these animals the transplanted cells were present in a 2-2.5 cm long fusiform core of cells oriented along the longitudinal axis of the cord and confined to the right ventral white matter. However, at later survival times (7 days and 15 days) the cell had migrated away from the point of injection (fig.4). This migration occurred both radially and longitudinally, and without preference for either white or grey matter. Moreover, it was apparent by the increased number of cells present at 15 days compared with 7 days that the cells were dividing as well as migrating (fig.4). By ultrastructural and immunocytochemical methods it was possible to show that the cells were migrating in the progenitor phenotype, and that the cells migrated through the parenchymal tissue of the spinal cord and not primarily in association with blood vessels or meninges.

4.2 Migrating CG4 cells contribute to the repair of remote areas of demyelination in X-irradiated spinal cord

In light of the observation described above in 4.1, we next asked the question whether transplanted CG4 cells migrating in X-irradiated spinal cord could enter and contribute to the repair of areas of demyelination remote from the point of transplantation. To do this we repeated the experiment described above, in which CG4 cells were transplanted into the right ventral white matter, but on this occasion an area of demyelination was induced in the dorsal funiculus by injection of EB. This arrangement was chosen since the passive spread of both

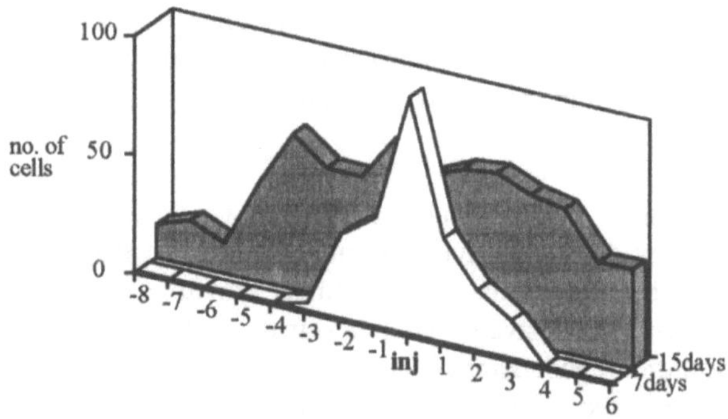

Fig.5 - Number of labelled transplanted CG4 cells in sample sections at 1 mm intervals away from the point of injection (x axis) at 7 and 15 days post-transplantation. The greater area under the "15 day" curve compared with the "7 day" curve indicates that the cells are proliferating as well as migrating in X-irradiated tissue.

cell-suspensions and EB in spinal cord white matter is mainly parallel to the longitudinal axis of the constituent fibres. By choosing two separate funiculi the danger of transplanted cells entering the lesion by passive spread was overcome, and transplanted cells could only get to the lesion by active migration. By 7 days post-transplantation cells had migrated throughout the spinal cord in a manner similar to that which occurred in unlesioned X-irradiated tissue, and labelled cells were also seen within the lesion, although none of the axons had been myelinated. However, by 15 days post-transplantation, many of the demyelinated axons within the lesion had been remyelinated. Since these experiments were undertaken in X-irradiated tissue, which has no capacity for host-derived repair, this observation indicated that migrating CG4 cells could enter areas of demyelination and contribute to their repair. While the findings of this study were encouraging in terms of repairing multifocal areas of demyelination, they carried the proviso that studies were all carried out in tissue that had been exposed to high levels of X-irradiation.

4.3 Transplanted CG4 cells fail to survive and do not migrate when transplanted into normal adult spinal cord

To establish whether the observations recorded in experiments conducted in X-irradiated spinal cord also applied to non-irradiated spinal cord, CG4 cells were transplanted into the right ventral white matter of non-irradiated spinal cord. The results of this experiment were dramatically different for those undertaken in X-irradiated spinal cord. No cells were seen to migrate away from the point of introduction. Moreover, fewer cells were seen at 15 days post-transplantation than after 7 days. At both these survival times the spread of cell was less than after 30 minutes. Indeed, cells were only seen in association with the mild degree of trauma induced in the white matter by the injection needle. This observation suggested that cells which had been introduced into normal tissue by the pressure of injection had not survived, and that survival was associated with tissue damage, albeit very mild.

Taken together, these results clearly show that while CG4 cells survive in lesioned tissue, they do not survive in normal tissue. The reason for this is not known at present but the events of development provide a possible explanation. Barres and her co-workers have shown that during post-natal development of the rat optic nerve more oligodendrocyte progenitors are generated than are required to myelinate the optic nerve (3). Approximately half the oligodendrocyte progenitors generated differentiate into myelin-forming oligodendrocytes while most of the remainder undergo apoptotic cell death. A number of molecules, including platelet derived growth factor (PDGF), insulin-like growth factor-I (IGF-I), ciliary neurotrophic factor (CNTF) and neurotrophin-3 (NT-3), are responsible for promoting survival of oligodendrocyte progenitors (3,4,5). It would appear that the limited availability of these survival factors results in the death of supernumerary progenitors since a greater number of cells survive when additional survival factor is added to the neonatal optic nerve. Thus, during the development of mature CNS white matter the appropriate number of oligodendrocyte lineage cells is achieved to some extent by limiting the availability of survival factor. By transplanting oligodendrocyte progenitor cells into adult white matter, one is adding to tissue cells that are superfluous to requirement. Such cells might therefore be expected to undergo apoptotic cell death in exactly the same manner as they would during normal development.

CG4 cells survive in X-irradiated tissue and within demyelinating lesions, regardless of whether they enter lesions by migration or direct injection. Thus, in contrast to the situation occurring in normal tissue, X-irradiated and damaged tissue support cell survival. This would indicate that survival factors are present in both these situations that are not present, or present to the same extent, in normal tissue. The factors are likely to include those factors which support oligodendrocyte progenitor survival during development and therefore include PDGF, IGF-I, CNTF and NT-3. Direct evidence for increased expression of these molecules in X-irradiated tissue, demyelinating lesions or injection needle-induced trauma is limited. However, IGF-I levels are elevated in cuprizone-induced demyelination (16), while PDGF has been implicated in the astrocyte response that accompanies stab-induced lesions of the CNS (24). Exposure of the CNS to X-irradiation has many effects which, although histologically subtle, could account for the difference in CG4 survival compared with non-irradiated tissue. For example, X-irradiation leads to activation of both astrocytes and microglia, and to changes in the permeability of the blood-brain barrier (6). The net effect of these changes presumably leads to an elevation in the expression of molecules that promote survival.

4.4 Transplanted CG4 cells can enter areas of demyelination when transplanted close to lesions but not when transplanted at a distance

When CG4 cells are transplanted into the right ventral white matter in spinal cord in which an EB-induced demyelinating lesion is present in the dorsal funiculus, labelled cells can be identified within the lesion in a proportion of animals by 18 days post-transplantation. A similar result is achieved if the position of the lesion and the location of cell transplantation is reversed, with cells transplanted into the dorsal funiculus of spinal tissue in which a EB lesion is present in the ventral white matter. The absence of cells in the surrounding unlesioned tissue is in marked contrast to the situation that occurs in X-irradiated tissue (section 4.2). If the cells are transplanted a single intervertebral space either caudal or cranial to the lesion, thereby increasing the distance over which transplanted cells need to travel to gain access to the demyelinated axons, then there is no migration away from the point of introduction and none enter the area of demyelination. It is clear from these results that only cells transplanted into the close vicinity of a lesion can enter the lesion, but will not do so if they are required to migrate through intervening normal tissue.

The observation that in some animals CG4 cells transplanted adjacent to areas of demyelination in non-irradiated tissue entered the lesion suggests that the presence of survival factors may not be confined by the boundaries of the lesion. Instead the presence of survival factors may form a diminishing gradient away from the lesion. Thus, one could imagine a penumbra of survival factors around a

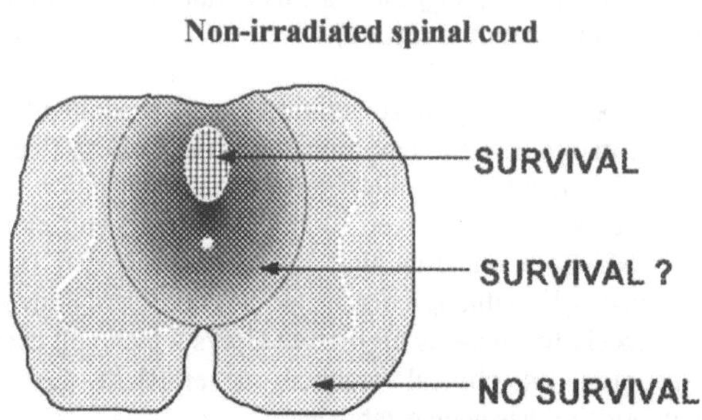

Fig. 6 - Summary of behaviour of OPs transplanted into non-irradiated adult spinal cord. When OPs are transplanted into lesions they survive and contribute to their repair. When they are transplanted into normal tissue they die. Between these two extremes is a penumbra of tissue which contains a diminishing gradients of factors, presumably similar to those in lesions and X-irradiated tissue, which permits survival and migration.

lesion within which transplanted cells may survive and gain access to the lesion (summarised in fig 6). Such a scenario might account for the observation made by other workers that remotely transplanted cells migrate into areas of demyelination (12,25), although the failure to assess properly the passive spread of cells attributable to the injection procedure or establish the extent of the lesion make these studies difficult to interpret.

4.5 OP migration in the adult CNS - Concluding remarks

The well documented migratory behaviour of oligodendrocyte progenitors *in vitro* and in development raises the possibility that one could rely on the migratory behaviour of transplanted oligodendrocyte progenitors to remyelinate multifocal areas of demyelination throughout the CNS. Moreover, the production of recruitment signals from areas of demyelination that are responsible for orchestrating the inherent repair process might also cause transplanted cells to migrate preferentially towards lesions. However, the results described above, obtained using an oligodendrocyte progenitor line, suggest that transplanted oligodendrocyte progenitors do not survive in normal tissue and will therefore not migrate within the normal tissue that separates areas of demyelination. In this case, oligodendrocyte progenitor transplantation is only likely to be successful if cells are injected either directly into, or into the close vicinity of a lesion. One must add the proviso that these studies have used a cell line to model the behaviour of oligodendrocyte progenitors obtained from primary culture and that the behaviour of the latter may differ in some respects, although preliminary observations using primary OPs suggest that this is not the case (M.T.O'Leary & W.F.Blakemore - personal communication).

The observations made following transplantation into X-irradiated tissue indicate that it is possible to manipulate the environment of the CNS so that its becomes supportive of oligodendrocyte progenitor survival and migration. More recent data has indicated that a dose of between 10 and 20 Grays is sufficient to convert a region of rat CNS from one that will not permit survival (and therefore migration) to one that will. Although one would not necessarily advocate the use of X-irradiation, especially a dose as high as 40 Grays which will eventually lead to ischaemic necrosis, in clinical practice, nevertheless the principle of environment manipulation has been established.

Acknowledgements

The author wishes to acknowledge the significant contributions made by a number of people to the work described, in particular, Drs Bill Blakemore, Sue Bayley, Sue Barnett, Andy Groves and Mark Noble. Much of the work was supported by grants from The Wellcome Trust and the Multiple Sclerosis Society of Great Britain and Northern Ireland. The author holds a Wellcome Trust Research Career Development Fellowship.

References

1. Armstrong, R.C., Harvath, L. and Dubois-Dalcq, M.E. (1990) Type-1 astrocytes and oligodendrocyte-type-2 astrocyte glial progenitors migrate toward distinct molecules. J.Neurosci.Res. 27, 400-407.
2. Barnett, S.C., Franklin, R.J.M. and Blakemore, W.F. (1993) In vitro and in vivo analysis of a rat O-2A progenitor cell line containing the temperature sensitive mutant gene of the SV40 large T antigen. Eur.J.Neurosci. 5, 1247-1260.
3. Barres, B.A., Hart, I.K., Coles, H.S.R., Burne, J.F., Voyvodic, J.T., Richardson, W.D. and Raff, M.C. (1992) Cell death and control of cell survival in the oligodendrocyte lineage. Cell 70, 31-46.
4. Barres, B.A., Raff, M.C., Gaese, F., Bartke, I., Dechant, G. and Barde, Y.-A. (1994) A crucial role for neurotrophin-3 in oligodendrocyte development. Nature 367, 371-375.
5. Barres, B.A., Schmid, R., Sendtner, M. and Raff, M.C. (1993) Muliple extracellular signals are required for long-term oligodendrocyte survival. Development 118, 283-295.
6. Chiang, C.S., McBride, W.H. and Withers, H.R. (1993) Radiation-induced astrocytic and microglial responses in mouse brain. Radiother.Oncol. 29, 60-68.
7. Duncan, I.D., Hammang, J.P., Jackson, K.F., Wood, P.M., Bunge, R.P. and Langford, L. (1988) Transplantation of oligodendrocytes and Schwann cells into the spinal cord of the myelin-deficient rat. J.Neurocytol. 17, 351-360.
9. Franklin, R.J.M. (1993) Reconstructing myelin-deficient environments in the CNS by glial cell transplantation. Sem.Neurosci. 5, 443-451.
9. Franklin, R.J.M., Bayley, S.A., Milner, R., ffrench-Constant, C. and Blakemore, W.F. (1995) Differentiation of the O-2A progenitor cell line CG-4 into oligodendrocytes and astrocytes following transplantation into glia-deficient areas of CNS white matter. Glia 13, 39-44.
10. Franklin, R.J.M. and Blakemore, W.F. (1995) Glial-cell transplantation and plasticity in the O-2A lineage -implications for CNS repair. Trends Neurosci. 18, 151-156.
11. Fulton, B.P., Burne, J.F. and Raff, M.C. (1992) Visualization of O-2A progenitor cells in developing and adult rat optic nerve by quisqualate-stimulated cobalt uptake. J.Neurosci. 12, 4816-4833.
12. Gout, O. and Dubois-Dalcq, M. (1993) Directed migration of transplanted glial cells toward a spinal cord demyelinating lesion. Int.J.Dev.Neurosci. 11, 613-623.
13. Groves, A.K., Barnett, S.C., Franklin, R.J.M., Crang, A.J., Mayer, M., Blakemore, W.F. and Noble, M. (1993) Repair of demyelinated lesions

by transplantation of purified O-2A progenitor cells. Nature 362, 453-455.

14. Gumpel, M., Gout, O., Gansmuller, A. and Baumann, N. (1989) Myelination and remyelination in the central nervous system by transplanted oligodendrocytes using the shiverer model. Dev.Neurosci. 11, 132-139.
15. Kiernan, B.W. and ffrench-Constant, C. (1993) Oligodendrocyte precursor (O-2A progenitor cell) migration; a model system for the study of cell migration in the developing central nervous system. Development Supplement, 219-225.
16. Komoly, S., Hudson, L.D., Webster, H.deF. and Bondy, C.A. (1992) Insulin-like growth factor I gene expression is induced in astrocytes during experimental demyelination. Proc.Nat.Acad.Sci. 89, 1894-1898.
17. Levison, S.W., Chuang, C., Abramson, B.J. and Goldman, J.E. (1993) The migrational patterns and developmental fates of glial precursors in the rat subventricular zone are temporally regulated. Development 119, 611-622.
18. Pfeiffer, S.E., Warrington, A.E. and Bansal, R.(1993) The oligodendrocyte and its many cellular processes. Trends Cell Biol. 3, 191-197.
19. Raff, M.C. (1989) Glial cell diversification in the rat optic nerve. Science 243, 1450-1455.
20. Raff, M.C., Miller, R.H. and Noble, M. (1983) A glial progenitor cell that develops in vitro into an astrocyte or an oligodendrocyte depending on culture medium. Nature 303, 390-396.
21. Rosenbluth, J., Hasegawa, M., Shirasaki, N., Rosen, C.L. and Liu, Z. (1990) Myelin formation following transplantation of normal fetal glia into myelin-deficient rat spinal cord. J.Neurocytol. 19, 718-730.
22. Skoff, R.P. (1990) Gliogenesis in rat optic nerve: astrocytes are generated in a single wave before oligodendrocytes. Dev.Biol. 139, 149-168.
23. Small, R.K., Riddle, P. and Noble, M. (1987) Evidence for migration of oligodendrocyte-type-2 astrocyte progenitor cells into the developing rat optic nerve. Nature 328, 155-157.
24. Takamiya, Y., Kohsaka, S., Toya, S., Otani, M., Mikoshiba, K. and Tsukada, Y. (1986) Possible association of platelet-derived growth factor (PDGF) with the appearance of reactive astrocytes following brain injury in situ. Brain res. 383, 305-309.
25. Vignais, L., Nait Oumesmar, B., Mellouk, F., Gout, O., Labourdette, G., Baron-Van Evercooren, A. and Gumpel, M. (1993) Transplantation of oligodendrocyte precursors in the adult demyelinated spinal cord: migration and remyelination. Int.J.Dev.Neurosci. 11, 603-612.
26. Warrington, A.E., Barbarese, E. and Pfeiffer, S.E. (1993) Differential myelinogenic capacity of specific developmental stages of the oligodendrocyte lineage upon transplantation into hypomyelinating hosts. J.Neurosci.Res. 34, 1-13.

Subject Index

MIX
Papier aus verantwortungsvollen Quellen
Paper from responsible sources
FSC® C105338
www.fsc.org

If you have any concerns about our products,
you can contact us on
ProductSafety@springernature.com

In case Publisher is established outside the EU,
the EU authorized representative is:
Springer Nature Customer Service Center GmbH
Europaplatz 3, 69115 Heidelberg, Germany

Printed by Libri Plureos GmbH
in Hamburg, Germany